CHITIN AND CHITOSAN

CHITIN AND CHITOSAN

Discoveries and Applications for Sustainability

GRÉGORIO CRINI

Environmental Polymer Scientist, Chrono-environnment Laboratory, Université Bourgogne Franche-Comté, France

ELSEVIER

ACADEMIC PRESS

An imprint of Elsevier

Academic Press is an imprint of Elsevier
125 London Wall, London EC2Y 5AS, United Kingdom
525 B Street, Suite 1650, San Diego, CA 92101, United States
50 Hampshire Street, 5th Floor, Cambridge, MA 02139, United States
The Boulevard, Langford Lane, Kidlington, Oxford OX5 1GB, United Kingdom

Notices

Knowledge and best practice in this field are constantly changing. As new research and experience broaden our understanding, changes in research methods, professional practices, or medical treatment may become necessary.

Practitioners and researchers must always rely on their own experience and knowledge in evaluating and using any information, methods, compounds, or experiments described herein. In using such information or methods they should be mindful of their own safety and the safety of others, including parties for whom they have a professional responsibility.

To the fullest extent of the law, neither the Publisher nor the authors, contributors, or editors, assume any liability for any injury and/or damage to persons or property as a matter of products liability, negligence or otherwise, or from any use or operation of any methods, products, instructions, or ideas contained in the material herein.

ISBN: 978-0-323-96119-6

For information on all Academic Press publications
visit our website at https://www.elsevier.com/books-and-journals

Publisher: Nikki Levy
Acquisitions Editor: Anna Valutkevic
Editorial Project Manager: Jose Paolo Valeroso
Production Project Manager: Sreejith Viswanathan
Cover Designer: Miles Hitchen

Typeset by STRAIVE, India

Je vais faire connaître la nature de la substance qui forme le corps ou la base charnue insoluble du champignon, et que je désignerai sous le nom de fongine.

"I going to make known the nature of the substance which forms the body or the insoluble fleshy base of the fungus, and which I will designate under the name of fungin."

Professor Henri Braconnot, November 15, 1810
Professor of Natural History
Director of the Botanical Garden of Nancy, France
Member of the French Académie des Sciences

Dedication

To my wife Nadia.

Contents

About the editor

Grégorio Crini is an environmental polymer scientist at the *Université Bourgogne Franche-Comté, laboratoire Chrono-environnement*, Besançon, France.

He received his PhD degree in organic and macromolecular chemistry from the *Université de Lille* in 1995 under the supervision of Professor Michel Morcellet. He then spent 2 years as a postdoctoral fellow at the G. Ronzoni Institute for Chemical and Biochemical Research in Milan, Italy, working with Research Director Giangiacomo Torri and Professor Benito Casu on the characterization of polysaccharide-based hydrogels by NMR techniques. In 1997, he joined the *Université de Franche-Comté* where he set up a research group working on the use of oligosaccharides (dextrins, cyclodextrins) and polysaccharides (starch, chitin, chitosan, and cellulose) in water treatment. In 2000, he obtained his authorization to supervise research activities. His current interests focus on the design of new functional macromolecular networks and the environmental aspects of oligosaccharide, polysaccharide, and natural fiber (hemp and flax) chemistry for applied research.

Grégorio Crini has published more than 215 papers, a patent, and 16 books (including 1 on chitin and chitosan, the first published in French in 2009). He is a highly cited researcher (h-index of 42, more than 15,000 citations). He has also conducted consulting projects for many companies.

Acknowledgments

It is an honor and pleasure to thank my mentors, colleagues, and friends Professor Michel Morcellet (Université de Lille, France) and Research Director Giangiacomo Torri (Istituto di Chimica e Biochemica G. Ronzoni, Milan, Italy) for inspiring my interest in biopolymers in the mid-1990s, in particular chitin and chitosan polysaccharides. This interest started in 1994 with the first experiments on the adsorption of dyes from the Italian textile industry by chitosan beads, performed during an internship at the G. Ronzoni Institute, during the last year of my PhD under the supervision of Giangiacomo Torri. It was during these years that I also met Professors Benito Casu (1927–2016) and Kjell Morten Vårum (1953–2016), two outstanding scientific authorities in the field of polysaccharides, whose experience proved to be of great benefit to me. I have been fortunate to meet, train, and work with these four colleagues. For encouraging the completion of this book, for much of the information collected, and for the unenviable task of checking the bibliography, I owe a debt of gratitude to Dr. Nadia Morin-Crini, Ms. Sylvie Bastello-Duflot, and Ms. Marlène Gruet (Université Bourgogne Franche-Comté, France). I also sincerely thank Dr. Marc Fourmentin (Université du Littoral Côte d'Opale, France) and Dr. Peter Winterton (Université de Toulouse III, France) for technical assistance, critical reading of early drafts of this book, and final proofreading.

CHAPTER ONE

Historical résumé

Braconnot's discovery is a world treasure.

Professor Saburo Minami, 2012
Professor of Veterinary Surgery at Tottori University, Japan
President of the Japanese Chitin and Chitosan Society 2011–2013

This year marks the 210th anniversary of the discovery of chitin, the second-most important biopolymer on Earth after cellulose and the most abundant substance among renewable polysaccharides in the marine environment.

Chitin is the main structural polymer found in the fungal cell wall, while cellulose is the significant structural polymer in the primary cell walls of the plants. Chitin is also present on the exoskeletons of arthropods and insects. The main source of chitin on an industrial scale today is waste from the fishing industry, in particular the shells of crustaceans (shrimps, crabs, and lobsters). From a classification point of view, cellulose and chitin are polysaccharides consisting of long macromolecular chains of simple sugars (monosaccharides) linked by glycosidic bonds. Cellulose and chitin are, indeed, both β-D-(1 → 4)-linked polymers comprised, respectively, of glucose (Glc) and *N*-acetylglucosamine, that is, 2-acetamido-2-deoxy-D-glucose (GlCNAc). Because of their similarity, research on chitin (β-1,4-GlcNAc) and cellulose (β-1,4-Glc) has often been intertwined, leading to significant advances in their constitution, structure, chemistry, hydrolysis, and biosynthesis (Goosen, 1997; Kurita, 1998, 2006; Ravi Kumar, 2000; Khor, 2001; Ravi Kumar et al., 2004; Rinaudo, 2006; Crini and Badot, 2008; Crini et al., 2009). These two homogenous biopolymers are used industrially in many different applications, usually in modified form.

Chitin and its main derivative, chitosan, have invaded our daily lives. These biopolymers are found in many products and in various industrial applications. Owing to these substances, we can drink clean water, enjoy wine or beer or a dish of noodles, swallow dietary or anticholesterol supplements, brush our teeth, protect them from cavities or improve dental alignment, wash and use alcohol-free disinfectants, nourish and protect our

Chitin and Chitosan
https://doi.org/10.1016/B978-0-323-96119-6.00002-5

hair, apply antiaging cream, enjoy chewing gum, protect ourselves from diseases, prepare materials for wound treatment, sutures, contact lenses, or dental implants, use sprays to protect our pets, treat polluted wastewater, swim in quality swimming pools, improve seed germination and plant growth, protect against pests, extend the shelf life of fruits and vegetables, use nonimpact antifouling paints, wear textiles with antibacterial properties, and use quality photographic paper or edible and intelligent food packaging instead of plastic. Why all these applications? The answers are simple and multiple. Chitin and chitosan have remarkable intrinsic physical, chemical, and biological properties. Chitin is also a substance of high industrial value for the bio-economy because of its low cost and renewable characteristics, in line with the three pillars of sustainable development, namely, economic (myriad applications in particular for chitosan, strong industrial interests, and creation of start-ups promoting new sources for chitin production), social (vector of local job development and health), and ecological (chitin is mainly obtained from wastes, that is, the shells of crustaceans, that are now recovered). These nontoxic, biocompatible, and biodegradable substances are part of our future economic and technological models because they reduce dependence on fossil fuels. Chitin and especially chitosan are, indeed, raw materials that can be substituted for oil in all its energy and nonenergy applications. Finally, public demand for renewable and biodegradable products is growing with the awareness of environmental and health protection. Both substances are part of this demand.

The history of chitin began in France in 1811 with the work of the chemist Braconnot on fungi. By treating fungi with alkali, Braconnot obtained an insoluble residue containing nitrogen which he called "*fongine*" (fungine) and highlighted its "unique" characteristic of resistance to alkali (Braconnot, 1811a,b). Braconnot has just discovered chitin, about 30 years before the discovery of cellulose by Payen in 1838. Using a detailed analytical study, Braconnot proved that *fongine* was a nitrogenous substance and pointed out that its nitrogen content was less than that of protein. However, chitin was probably discovered in 1799 by Hatchett, an English chemist, who, during his experiments on the shells of marine animals, reported the presence in the cuticle of "a material particularly resistant to usual chemical" (Hatchett, 1799). Nevertheless, there is no indication that he knew what he had prepared as a substance, neither in his memoirs nor in the bibliography of the time. Moreover, Hatchett did not investigate further. Thus, the discovery of chitin is usually attributed to Braconnot, who, on the contrary, carried out a detailed chemical analysis on *sa fongine* (his fungine), reporting for the first time the formation of acetic acid when treated with hot acid and concluding that it was *un nouveau matériau* (a new material). *Fongine* was then regarded as "a more or less pure form of cellulose." Today, Braconnot's name is primarily associated with the discovery of chitin, the oldest known polysaccharide (Labrude, 1997; Labrude and Becq, 2003).

Later, in 1823, when found in the elytra of insects by the French chemist Odier, a same substance was named *chitine* (chitin) which was also resistant to the action of alkali, a property that also intrigued Odier. However, Odier and others failed to recognize Braconnot's elucidation of the particular properties of *fongine*, and consequently the term chitin was soon adopted by researchers. Later, this term was applied to both invertebrates and fungi because it signified a mantle, coat, or tunic. For Odier, *chitine est une substance particulière, fort curieuse* (chitin is a particular substance, extremely curious) (Odier, 1823). However, he originally believed that his *chitine* did not contain any nitrogen, unlike Braconnot, and thought that insect cuticles were made of the same substance that forms the cell walls of the plant, cellulose. The following year, Children, a British chemist at the Museum of London, translated Odier's paper (Children, 1824). Children first confirmed that Odier's substance was present in the cuticle of insects and the structure of fungi. In an appendix, Children also published his own observations on the extraction of "pure" chitin from insect origin by decalcification with acids and deproteinization with alkalis. He had repeated much of the work using the "common cantharides of the shops as his material." A percentage of chitin in the elytra of May beetles of 29 has been reported, similar to that of Odier. Children was the first to reveal the nitrogenous nature of chitin. However, at the time, many researchers considered *chitine* (chitin) to be another form of cellulose that did not contain nitrogen. A controversy over the differences between chitin, produced by arthropods, and cellulose, produced by plants, then began. It continued for some time, although French chemist Lassaigne in 1843, working on the exoskeleton of the silkworm butterfly *Bombix morii* (Lassaigne, 1843a), proved that *chitine n'est pas une cellulose mais un nouveau produit contenant de l'azote* (chitin is not a cellulose, but a new product containing nitrogen). The same year, this finding was highlighted by another French chemist, Payen, who had also reported the identification of cellulose in 1838. Payen was the first to question the difference between cellulose and chitin (Payen, 1843), although he initially believed that the two substances were identical. However, the work of Lassaigne and Payen was not unanimously accepted by the scientific community, and the controversy remained a subject of debate until the 1930s, without ever reaching the "feverish heights of the 1890s" according to Tracey (1957).

From 1850, there was copious literature on chitin, particularly on its presence in a wide variety of different species (Frémy, 1847; Peligot, 1858; Städeler, 1859; Baur, 1860; Ludwig, 1861; Bütschli, 1874; Haliburton, 1885; Looss, 1885; Krukenberg, 1886; Ambronn, 1890; Chatin, 1892; Krawkow, 1892; Winterstein, 1893; Gilson, 1894a,b; Fränkel, 1898; van Wisselingh, 1898a,b; Zander, 1897). The first comprehensive collections on the distribution of chitin in fungi and invertebrates were made by van Wisselingh (1898b). They were updated by Wester (1910) for fungi and by von Wettstein (1921) for invertebrates. However, from 1850 to 1930, chitin entered a period of confusion and controversy. During this period, several researchers working on chitin extraction, chemistry, and degradation products, in collaboration or not, have made contradictory arguments in the interpretation of the same experimental works. Some researchers such as Frémy (1855), Peligot (1858), Bütschli (1874), Haliburton (1885), de Bary (1887), Ambronn (1890), Winterstein (1893), and Berlese (1929) considered chitin to be similar to cellulose, others to be a distinct substance (Tiemann, 1886; Schmiedeberg, 1891; Gilson, 1894a; von Fürth and Russo, 1906; Offer, 1907; Ilkewitsch, 1908), and others to be similar to starch (Berthelot, 1859) or glycogen (Zander, 1897). In 1885, Haliburton, studying the chemistry of chitin, claimed that "chitin is a curious substance, a white amorphous body, insoluble in water, in weak acids, and in boiling concentrated alkalis, and soluble in strong acids" (Haliburton, 1885).

Rouget, a French physiologist at the *Faculté de Médecine* in Montpellier, reported in 1859 that treatment of *chitine* with a concentrated solution of caustic potash under reflux gave *une nouvelle chitine modifiée* (a new modified chitin), and that this treatment made it soluble in organic acids (Rouget, 1859). This is the beginning of the history of chitosan, the main derivative of chitin we know today. The name of chitosan was, however, introduced in 1894 by Hoppe-Seyler for *ein säurelösliches Derivat von Chitin* (an acid-soluble derivative of chitin), prepared from the treatment of the shell of crabs, scorpions, and spiders (Hoppe-Seyler, 1894). It seems that Hoppe-Seyler was not aware of Rouget's work (Crini et al., 2009). Hoppe-Seyler was a famous German physiologist and chemist at the *Kaiser Wilhelms-Universität* in Strasbourg, the German Imperial University created after the Franco-Prussian War. In addition to the work of Hoppe-Seyler, chitosan was also prepared from fungal material by both Winterstein, Swiss chemist and physiologist at the *Eidgenössische Technische Hochschule Zürich* (Winterstein, 1894a,b), and Gilson, pharmacognosist at the Belgian *Université de Gand* (Gilson, 1894a,b). Each of the three researchers declared priority on their discovery of the "first soluble derivative of chitin." It is also very probable that both Odier (1823) and Children (1824) isolated chitosan rather than chitin (Muzzarelli, 1977; Roberts, 1992). Nevertheless, the discovery of *chitine modifiée* is attributed to Rouget and the term "chitosan" that was quickly adopted by the scientific community to Hoppe-Seyler.

In 1876, the young German student Ledderhose studied *Chitin und seine Spaltungsprodukt* (chitin and its fission products) at the German Imperial University in Strasbourg (later *Kaiser Wilhelms-Universität Strassburg*), the capital city of the *Reichsland Alsace-Moselle* (Mathews, 1898). Under the supervision of Hoppe-Seyler, Ledderhose hydrolyzed arthropod chitin, which he extracted according to Odier's protocol, and discovered a nitrogen-containing sugar, named *glykosamin* (glucosamine) (Ledderhose, 1876). Glucosamine was the first and easiest chitin derivative to be obtained, differing from glucose by the presence of an amine group instead of a hydroxyl group. This new substance was quickly recognized by the scientific community as an interesting value-added product despite its instability (Ledderhose, 1878). Later, Ledderhose also found that the hydrolysis of chitin not only yielded *glykosamin* but also acetic acid (Ledderhose, 1880a,b). Nevertheless, he did not recognize that these two products were produced in equimolar amounts. However, at the time, the Finnish chemist Sundwick strongly disagreed with Ledderhose's results (Sundwik, 1881). Ten years later, the work of the Germans Tiemann and Schmiedeberg, the Swiss Winterstein, the Belgian Gilson, and the Japanese Araki confirmed Ledderhose's experiments and conclusions on glucosamine (Tiemann, 1886; Tiemann and Landolt, 1886; Fischer and Tiemann, 1894; Schmiedeberg, 1891; Winterstein, 1893, 1894a,b; Gilson, 1894a, 1895a,b; Araki, 1895). At the same time, Fischer, a German organic chemist and Nobel Prize laureate in 1902, suggested for the first time that glucosamine was a dextrorotatory amino sugar to which he recognized the constitution of a α-amino-glucose (Fischer and Tiemann, 1894), which would be demonstrated a few years later not only by Fischer's work but also by those of Lobry de Bruyn and Irvine. Indeed, at the end of the 1890s, Lobry de Bruyn, a Dutch organic chemist and pharmacist at the University of Amsterdam, definitively established the composition of the product obtained by hydrolysis of chitin to be D-glucosamine chloride (chitosamine) and its constitution (Lobry de Bruyn and Franchimont, 1898; Lobry de Bruyn and van Ekenstein, 1899a,b).

Later,Fischer's German research group in Munich (Fischer and Leuchs, 1902, 1903) and Irvine's Scottish group in St Andrews (Irvine et al., 1911; Irvine and Hynd, 1912, 1914), who synthesized glucosamine, finally demonstrated its structure. However, some authors did not consider glucosamine as the basic unit of chitin, but rather mannosamine. This debate continued until the late 1930s. Indeed, numerous attempts to solve this problem of configuration have failed because of the Walden inversion that occurred under certain conditions of deamination to the hexose. It was only at the end of 1930s that both Karrer and Mayer (1937) and Haworth et al. (1939), studying the isolation of glucosamine from chitin under conditions which precluded the Walden inversion, clearly elucidated its structure. The relation of carbon on position 2 in glucosamine to D-glucosamine (and not to D-mannose) was finally demonstrated by unequivocal synthesis. Glucosamine or 2-amino-2-deoxy-D-glucose was then recognized as an amino-monosaccharide derived from the hydrolysis of chitin and its basic unit in its chemical structure.

From the 1880s, many researchers began to consider that chitin was a specific substance well-identified and widely distributed in both the animal kingdom and in the fungi. Remarkable contributions including those of Gilson, Winterstein, and van Wisselingh were published. The first to evaluate the chemical similarity of fungal and animal chitins was Gilson in the mid-1890s (Gilson, 1894a,b, 1895a). There was no detectable chemical difference between products from the two sources. Whatever it occurred, chitin gave the same chemical reactions and had the same physical properties, for example, specific gravity, refractive index, and specific rotation and about the same elemental analysis (Gilson, 1894a,b, 1895a). However, the chemical formula was different depending on the authors (Bounoure, 1919). There was also some confusion in the experimental results and conclusions, which was later explained by the fact that much of this work was based on old histochemical techniques (Hale, 1957; Runham, 1961a,b). At the end of the 1890s, a concept began to emerge that chitin-producing animals had a secretion pattern that was chemically similar to that of cellulose-producing plants. Indeed, the discovery of chitin in the cell membrane of certain plants by Gilson in 1895 brought chitin closer to cellulose. This was the beginning of numerous studies on the chemistry of chitin of various origins which gave rise to famous controversies between researchers, in particular that between Gilson, Winterstein, Hoppe-Seyler, and van Wisselingh. However, the development of an experimental protocol to characterize the presence of chitin in a material has allowed significant progress. Chitin was known to be an extremely stable substance, and this property was used as early as 1898 in the development of microchemical tests for its detection. The colourimetric test, based upon the conversion of chitin into a soluble substance, was developed by van Wisselingh (1898a,b). The substance remaining during the conversion, namely, *chitine modifiée* (chitosan) gave chitosan sulfate crystals that were easily recognizable after treatment with sulfuric acid. Chitosan sulfate also exhibited characteristic coloring reactions with certain chemicals that differentiated it from any other material remaining after hydrolysis. The van Wisselingh's test was then used as a simple and efficient tool to identify the presence or absence of chitin, for example, in plant materials.

At the beginning of the last century, chitin continued to be the subject of numerous fundamental studies on different topics including its distribution, isolation and preparation, detection and determination, molecular structure and composition, physical and chemical properties such as its insolubility, chemistry (modification), biochemistry, and fiber production (Table 1.1). All these studies were useful in the physical, spectroscopic, and microscopic characterization of chitin from the 1930s. In 1912, Brach gave the first "probable empirical formula" $(C_{32}H_{54}N_4O_{21})_n$ for chitin (Brach, 1912). Later, Gonell showed that this formula corresponded to 18 acetylglucosamine residues and proved that chitin was microcrystalline using X-ray data (Gonell, 1926). Studies on the presence of chitin in fungi continued to receive attention (Wester, 1910), with a "new" opinion that "chitin and cellulose are not chemically very different," although this idea had already been put forward a few years earlier by several researchers, including Gilson, in the late 1890s. Many authors have also pointed out that glucose and glucosamine were similar since both substances had the same reducing properties and an identical reaction on phenylhydrazine (Bounoure, 1919; von Wettstein, 1921). This also argued the similarity between chitin and cellulose. Other results published on the chemical modification of chitin also contributed to this concept. For example, in 1907, von Fürth and Scholl (1907), preparing nitrochitins analogous to nitrocelluloses, demonstrated the chemical analogies between chitin and cellulose. Later, studies using van Wisselingh's test showed that chitin was widely distributed among species of *Phycomycetes* (Heyn, 1936a; Nabel, 1939), *Basidiomycetes* (Proskuriakow, 1926; Khouvine, 1932), *Ascomycetes* (Frey, 1950; Blank, 1954), and *Fungi imperfecti* (Khouvine, 1932). These studies were important because they indicated that, in the fungi, chitin played a role similar to cellulose in other plants and that nonchitinous fungi contained cellulose in place of chitin. In addition, von Wettstein (1921) suggested that

Table 1.1 Selected studies on chitin published at the beginning of the last century.

Topic	References
Distribution	Zaitschek (1904), Mirande (1905), Rosenheim (1905), Rippel (1908), Scholl (1908a,b), von Lippmann (1912), Biedermann (1914), Bounoure (1919), von Wettstein (1921), Lacroix (1923), Dous and Ziegenspeck (1926), Proskuriakow (1926), Takata (1929), Anderson and Brown (1930), Rammelberg (1931), Norman and Peterson (1932), Abderhalden and Heyns (1933), Ruser (1933), Diehl and van Iterson Jr (1935), Diehl (1936), and Nabel (1939)
Isolation and preparation	Benecke (1905), Scholl (1908a,b), Bounoure (1911, 1912a,b, 1913), von Wettstein (1921), Dous and Ziegenspeck (1926), and Proskuriakow (1926)
Detection and determination	Wester (1909, 1910), Komori (1926), Proskuriakow (1926), Kühnelt (1928a,b), and Hopkins (1929)
Physical evidence and properties	Holmgren (1902a,b, 1910), Sollas (1907), Irvine (1909), Brach (1912), Schmiedeberg (1920), Eidmann (1922), Herzog (1924), Herzog and Gonell (1924), Becking and Chamberlin (1925), Dous and Ziegenspeck (1926), Gonell (1926), Knecht and Hibbert (1926), Fränkel and Jellinek (1927), Müller (1928), Merker (1929a,b), Yonge (1932, 1936), Diehl and van Iterson Jr (1935), and Alexandrov (1934, 1935)
Molecular structure	Neuberg (1902), Fraenkel and Kelly (1903), Benecke (1905), Biedermann (1907), Brach and von Fürth (1912), Hase (1916)
Chemistry	Steudel (1902), Stolte (1907), von Fürth and Scholl (1907), Alsberg and Hedblom (1909), Vouk (1915), Morgulis (1916, 1917), and Armbrecht (1919)
Biochemistry	Benecke (1905), Wester (1909), Dous and Ziegenspeck (1926), Karrer and Hofmann (1929), Needham (1929), Karrer (1930), Brug (1932), and Norman and Peterson (1932)
Fiber production	von Weimarn (1926a,b, 1927)

the presence of chitin or cellulose in cell walls could be used as a basis for establishing phylogenetic relationships among groups of fungi, especially among the *Phycomycetes*. However, this concept has been criticized by Nabel (1939) who also used microchemical methods, but later supported by Frey (1950) who used X-ray methods of detection.

Using van Wisselingh's colourimetric test, numerous works also demonstrated that, in nature, chitin existed not only in the fungi, the cuticle of insects, and the exoskeleton of crustaceans but also in the cartilage of mollusks and the cell walls of microorganisms (Wester, 1910; Bounoure, 1919; von Wettstein, 1921). All these results were not original, but they demonstrated the studies published half a century before, which would lead to the end of the various controversies on chitin. A consensus then emerged on three points: the first concerned the fact that chitin could come from several sources but was the same substance; the second indicated the structural and chemical similarity between chitin and cellulose; and the last point considered that chitin-producing animals had a secretion mode very close chemically to that of cellulose-producing plants. Becking and Chamberlin (1925) proved that the chemical and physical properties for chitin from fungi, arthropods, annelids, and mollusks were the same. Nevertheless, other researchers considered that there were exceptions, for example, for certain fungi (Dous and Ziegenspeck, 1926), for the marine worm *Eunice* (Gonell, 1926), and arthropods such as *Limulus polyphemus* (Fränkel and Jellinek, 1927). Later, lichens, numerous species of algae and filamentous yeasts, have been found to contain chitin (Schmidt, 1936; Nabel, 1939). However, Schmidt (1936) noted that the amount of chitin present was generally very small, pointing to the difficulty of obtaining high-purity products. From the mid-1930s onward, many studies were, indeed, interested in chitin extraction and purification protocols. Meyer (1942), Richards and Anderson (1942), and Darmon and Rudall (1950) showed that the extraction and purification steps were also fundamental in the interpretation of data obtained from microscopic and X-ray diffraction studies of the chitin of animal or fungal origin. Later, Roelofsen and Hoette (1951) and Houwink and Kreger (1953) also emphasized the fact that for the determination of chitin in yeasts, it was necessary to remove nonchitinous material and purify the samples.

Among the physical and chemical characteristics of chitin, its insolubility or solubility (depending on the researchers) has always intrigued researchers. At the time, it was well-established that chitin was insoluble in water, alcohol, ether, (dilute) acids, and alkalis. In concentrated mineral acids, chitin was hydrolyzed with the formation of glucosamine salts (chitose). At high temperature and in the presence of concentrated solutions of potassium hydroxide, the reaction yielded chitosan and acetic acid (Hoppe-Seyler, 1894). It was this reaction that was used in van Wisselingh's colourimetric test (van Wisselingh, 1898a,b). Chitin was also oxidized and dissolved at room temperature by a solution of sodium hydrochloride, in agreement with the results previously published by Looss (1885). However, Wester (1909), in disagreement with Looss' observations, had proved that "pure" chitin was insoluble in this solution. Nevertheless, this has not been confirmed by other works. Ito (1924) showed the solubility of chitin in cold sodium hypochlorite solution. Later, Campbell (1929) also noted that "pure" chitin disappeared rapidly at room temperature in the same solution containing chlorine, lending credence to the observations of Looss and Ito. Schulze and Kunike (1923) reported that chitin was dissolved in concentrated acids with or without decomposition, while von Weimarn (1926a) observed that chitin formed a colloidal solution in lithium thiocyanate salt solution. Schmidt et al. (1928a,b) reported the ability of liquid ammonia to dissolve chitin. However, studies on the purity and solubility/insolubility of chitin continued to be a source of debate and controversy among researchers for over 50 years (Schulze and Kunike, 1923; Ito, 1924; von Weimarn, 1926a,b; Campbell, 1929). Even today, the solubility of chitin continues to interest the scientific community (Jaworska et al., 2012; Gong et al., 2016; Roy et al., 2017a; Hahn et al., 2020).

In the mid-1920s, research on chitin was also directed toward the study of its degradation by microorganisms. An enzyme promoting the hydrolysis of chitin has been found in 1929 by Karrer in snail (*Helix pomatia*) digestive fluid and in various molds (Karrer and Hofmann, 1929; Karrer and von François, 1929). This work demonstrated that enzymes, then called chitinases, had the ability to degrade (later, the verb "depolymerize" was used) chitin and its components by cleaving β-linkages with the final formation of *N*-acetyl-D-glucosamine. Many types of enzymes (e.g., fungal chitinases, bacterial chitinases, enzymes associated with emulsion) were then prepared and used to degrade chitin, chitosan, and other substances (hyaluronic acid) containing glucosamine units (Zechmeister and Tóth, 1934; Yonge, 1938; Whistler and Smart, 1953; Hackman, 1954; Foster and Stacey, 1958). In the 1930s, with the development of modern techniques such as X-ray diffraction and microscopy, there was also a need for reinvestigation in the structure and functions of "pure" polysaccharides including chitin and cellulose (Khouvine, 1932; Meyer and Pankow, 1935; Clark and Smith, 1936). Their crystal structures have been the subject of X-ray crystallographic studies for over 50 years. This led to similar structures and morphologies, and both polymers were naturally crystalline. Indeed, between 1930 and 1950, the physical structures of the two polysaccharides were highlighted, and these structures were very useful for their biosynthesis from 1960. Meyer published outstanding contributions on chitin and cellulose, demonstrating that there was "a kind of physicochemical parallelism between these two polysaccharides" (Meyer and Mark, 1928; Meyer and Pankow, 1935; Meyer and Wehrli, 1937; Meyer, 1942, 1950). It was recognized that chitin and cellulose were both fiber-forming polymers, and as fibers they served as the load-bearing components of major groups of natural composite tissues (chitin in the skeletal materials of many lower animals and cellulose in many plant-cell walls). The use of X-ray diffraction analysis in the study of chitin elucidated not only the molecular interactions that contributed to the high tensile strength of chitin fibers but also confirmed the type of glycosidic attachments. Furthermore, X-ray diffraction techniques served to show the close similarities in the macromolecular disposition of chitin and the identically linked cellulose. The concept that cellulose was poly-β-(1 → 4)-D-glucose and chitin was its 2-acetamido derivative began to gain consensus (Meyer and Wehrli, 1937; Meyer, 1942; Mark, 1943). In addition, despite the difficulties in isolating chitin from different sources, its origin (animal or fungal) was not in doubt, based on X-ray diffraction studies, their hydrolysis and enzymic reactions, physical properties (viscosity data, optical measurements), and infrared data. Indeed, as early as 1950, the structure of chitin was also identified by infrared spectroscopy coupled with enzymatic reactions (Darmon and Rudall, 1950), and these spectroscopic techniques developed rapidly (Orr, 1954; Brock, 1957; Spedding, 1964). These data also confirmed that this unbranched polysaccharide had in its structure 2-acetamido-2-deoxy-D-glucopyranose (or simply *N*-acetyl-D-glucosamine, GlcNAc) residues linked in the β-(1 → 4) positions. It was also evident that, in nature, chitin was not found alone but formed very complex systems, and this was demonstrated not only from chemical data but also from spectroscopic and microscopic characterizations. Numerous publications reported that chitin was associated in situ with other substances, notably proteins, by hydrogen bonds and covalent linkages. Crustacean shells contained large proportions of calcium carbonate and proteins, whereas insect exoskeletons largely comprised chitin-protein complexes. In fungi, chitin did no exist alone in the cell wall but was in close association with other substances. A few years later, X-ray diffraction patterns recorded from shrimp shells and squid pens also revealed differences that highlighted the concept of chitin polymorphism, with, for example, the presence of α-chitin and β-chitin (Blackwell, 1969; Blackwell et al., 1965, 1967; Walton and Blackwell, 1973; Rudall and Kenchington, 1973). All these techniques, that is, microscopy, X-ray diffraction, infrared, and later NMR were complementary in the studies of chitin and chitosan and their derivatives, and numerous results were then published. Indeed, most of the fundamental information available on these biopolymers, still valid today, was obtained since the 1950s (Richards, 1951, 1958; Whistler and Smart, 1953; Kent and Whitehouse, 1955; Tracey, 1955, 1957; Foster and Stacey, 1958). It is from this time that there was a consensus in the scientific community on many points concerning chitin and chitosan, such as their extraction, nomenclature, constitution, structure, characterization, properties, chemistry, and biochemistry.

From an application point of view, chitin and chitosan attracted considerable attention between 1930 and 1950 with the first industrial patents (Rigby, 1936a,b; McQueen and Merill, 1936; Brubaker, 1937; Heckert, 1937; Lubs, 1937). For instance, two important patents were obtained by Rigby (1936a,b), one for producing chitosan from chitin and the other for making films and fibers from chitosan. Nevertheless, it would take another two decades for industrial applications to develop. Indeed, the early 1970s marked a new stage in the development of chitin and chitosan with the first applications in the food, cosmetics, and pharmaceutical industries, in sludge dewatering and wastewater treatment (flocculation), and in biotechnologies (Hirano, 1989, 1996; Roberts, 1992; Hirano et al., 1994). This development was also made possible by the production of these two products on an industrial scale (Hirano, 1989). There is no doubt that Japan has been the country that has allowed the most significant advances in terms of production, consumption, and applications of chitin and chitosan (Hirano, 1989, 1996; Hirano et al., 1994; Badawy and Rabea, 2017; Bonecco et al., 2017; Crini, 2019). Numerous patents have been filed since the mid-1970s, and an abundant scientific literature across different disciplines has built up (Kurita, 1998, 2006; Khor, 2001; Rinaudo, 2006; Crini et al., 2009; Kim, 2011, 2014). In 1977, the first international conference on chitin and chitosan was held in Boston, and this symposium was a great success (Muzzarelli and Pariser, 1978). Since this event, many national and international conferences have been held regularly, showing the importance of the scientific community working on these biopolymers throughout the world. Over the past 40 years, substantial progress has been made in applied research in chitin and chitosan technology, with applications in virtually all industry sectors (Philibert et al., 2017; Crini, 2019; Crini and Lichtfouse, 2019a,b). Nevertheless, it is important to note that compared to the applications of chitosan, those of chitin are more restricted because its insolubility is a major problem in the development of (bio)materials and processes using chitin.

Today, from Asia to North America and Europe, the chitin and chitosan market is growing rapidly due to strong demand from the water treatment, food (nutrition) and beverages, cosmetics, pharmaceutical and medical, packaging (bioplastics), and agrochemical industries (Crini and Lichtfouse, 2019a,b). Two hundred and ten years after its discovery, chitin continues to interest the scientific community and industry. Studies on the use of chitosan as a protective agent against COVID-19, for example, show that these natural substances have not yet revealed all their secrets. New discoveries are expected in the coming years not only in the biomedical field but also in other industrial sectors (nutraceuticals, cosmeceuticals, cosmetotextiles, biopesticides, nanotechnology, etc.).

This book is a detailed history of chitin and chitosan, reviewing the major historical landmarks in their discovery, characterization, and development reported during the period 1799–2019. I have roughly divided it into five periods: discovery from 1799 to 1894 (this chapter), a period of confusion and controversy from 1894 to 1930 (Chapter 2), exploration in 1930–50 (Chapter 3), a period of doubt from 1950 to 1970 (Chapter 4), and finally the period of application from 1970 to the present day (Chapters 5 and 6). Each period is illustrated by considering examples of studies that appeared in the literature, especially those of several great scientists who have left their mark on the history of these polysaccharides. For this purpose, data from over 815 original publications were compiled and discussed. Obviously, this historical review does not pretend to be exhaustive but aims to show the progressive evolution of scientific knowledge on chitin and chitosan in their historical context from selected works that illustrate this evolution. For readers interested in quickly acquiring information about chitin and chitosan and identifying articles of interest, a final section (Chapter 7) of the book summarizes the fundamental properties, current applications, recent advances, and challenges of chitin and chitosan.

CHAPTER TWO

Discovery: 1799–1894

The contribution of French scientists to our understanding of chitin and chitosan has been considerable from its first isolation by the French chemist Henri Braconnot in 1811.

Professor Benito Casu, 1996, Professor of Chemistry and Biochemistry, G. Ronzoni Institute for Chemical and Biochemical Research, Milan, Italy, President of the International Carbohydrate Organization 1996–1998

The first period, from 1799 to 1894, covers the discovery of chitin. The names that marked this period most were Charles Hatchett, Henri Braconnot, Auguste Odier, John George Children, Jean-Louis Lassaigne, Anselme Payen, Charles Rouget, Georg Ledderhose, Ernst Sundwick, Oswald Schmiedeberg, Ferdinand Tiemann, and Felix Hoppe-Seyler.

2.1 Charles Hatchett

Charles Hatchett (1765–1847; Fig. 2.1) was an English chemist and a self-formed mineralogist and analytical chemist. His father John Hatchett (1729–1806) was a coachbuilder at Long Acre, London, and later became a magistrate in Hammersmith.

The young Hatchett spent a large part of his youth traveling (Russia, Poland) and visiting museums, mines, and geological sites collecting many minerals. He became very interested in chemistry, biology, and mineralogy as a self-taught scientist. For example, Hatchett studied the constituents of marine shells (oysters, mother-of-pearl), fish bones (salmon, skate, and mackerel), fossils, and shark tooth enamel. He also studied the alloying of gold with a wide variety of metals and nonmetals and published results on a new Thibet calomel and artificial agents for leather tanning. However, his name is primarily associated with the discovery of niobium in 1801. While analyzing a mineral sample from the British Museum in London, Hatchett discovered a new element, which he named columbium (now called niobium). For his analytical work on a range of minerals, natural and animal substances, Hatchett was elected a Fellow of the Royal Society

Chitin and Chitosan
https://doi.org/10.1016/B978-0-323-96119-6.00003-7

Fig. 2.1 Engraving of Charles Hatchett. *(Source: Griffith, W.P., Morris P.J.T., 2003. Charles Hatchett FRS (1765-1847), chemist and discover of niobium. Notes Rec. R. Soc. Lond. J. Hist. Sci. 57, 299–316.)*

in London (1857) and of the Linnean Society (1795). Much information about Hatchett's life and career can be found in the references Johnson (1803), Good et al. (1813), Smedley et al. (1845), Griffith and Morris (2003), and Wisniak (2015).

In June 1799, Hatchett decalcified shells of crabs, lobsters, prawns, and crayfish with mineral acids and published his first experiments and observations in the journal Philosophical Transactions, edited by the Royal Society of London. In a 20-page memory, Hatchett described a material particularly resistant to usual chemicals, writing: "it appears that immersion of the shell in acetous or in dilute nitric acid afforded carbonate and phosphate of lime, the former, however, in the largest quantity" (Hatchett, 1799). This is the first mention of calcified chitin in invertebrates (Smedley et al., 1845). In February 1800, in the Journal of Natural Philosophy, Chemistry, and the Arts (Fig. 2.2), Hatchett wrote (in the terms of the time): "Pieces of this fubftance, taken from various parts of thofe animals, was at different times immerfed in acetous and in diluted nitric acid; thofe which had been placed in the diluted

JOURNAL

OF

NATURAL PHILOSOPHY, CHEMISTRY,

AND

THE ARTS.

FEBRUARY 1800.

Experiments and Obfervations on Shell and Bone. By CHARLES HATCHETT, *Efq.* F. R. S.*

SOME experiments, which I lately made at the requeft of Mr. Home, and which he has done me the honor to mention in his ingenious paper on the teeth of granivorous quadrupeds, induced me to turn my attention more particularly to the chemical examination of fhell and bone, efpecially as the former appeared to have been hitherto much neglected.

The time fince thefe experiments were began, has not been fufficient to enable me to enter into all the minutiæ of the chemical analyfis of thefe fubftances; but as fome remarkable facts were afcertained, I have now ventured to bring them forward, with the addition of fome obfervations, although as yet the whole is little more than a very imperfect outline.

The firft of thefe experiments were made on the fhells of marine animals; and to avoid repetition and prolixity, I fhall, in a great meafure, once for all, defcribe the menftrua, the precipitants, and the mode of operation.

When fhells were examined they were immerfed in acetous acid, or nitric acid diluted, according to circumftances, with 4, 5, 6, or more parts of diftilled water; and the folution was always made without heat.

Fig. 2.2 Extract of the article by Charles Hatchett on his experiments and observations on shell and bone in February 1800. *(Source: Hatchett, C., 1800. VI. Experiments and observations on shell and bone. J. Nat. Philos. Chem. Arts. III, 500–506 (Nicholson W, Ed.).)*

nitric acid produced a moderate effervefcence, and in a fhort time were found to be foft and elaftic, of a yellowifh white color, and like a cartilage, which retained the original figure" (Hatchett, 1800a,b). However, Hatchett did not push his scientific investigations any further, preferring to become a noted collector of books, paintings, musical instruments, and musical manuscripts (Wisniak, 2015). This change brought criticism by some of his English colleagues such as Thomas Thomson (1773–1852, mineralogist and chemist, President of the Philosophical Society of Glasgow), Sir Benjamin Collins Brodie (1783–1862, physiologist and surgeon), and Joseph Banks (1743–1820, naturalist and botanist, President of the Royal Society), who lamented the loss of "a gifted researcher." Indeed, after his father's death in 1806, Hatchett inherited a considerable fortune. He begun to put more and more time into the business and little by little decreased his scientific activities (Griffith and Morris, 2003; Wisniak, 2015; Crini, 2019).

2.2 Henri Braconnot

Henri Braconnot (1780–1855; Fig. 2.3) was a French chemist, pharmacist and botanist, professor of natural history, *directeur du Jardin Botanique* de Nancy (director of the Botanical Garden of Nancy), and a member of the French *Académie des Sciences* (Academy of Sciences). The Botanical Garden of Nancy, located in eastern France, was under the patronage of Empress Joséphine and had an international reputation for the number of plant species from around the world that it had. Henri Braconnot was an eminent scientific and is considered the precursor of the science of carbohydrate polymers. In 1973, Rudall and Kenchington wrote: "Modern protein chemistry is regarded as having started from his work" (Rudall and Kenchington, 1973). However, Braconnot's name is mainly associated with the discovery of

Fig. 2.3 Professor Henri Braconnot, 1780–1855. *(Source: Simonin, F., 1856. Notice biographique sur M. Henri Braconnot. In: Compte Rendus des Travaux de la Société de Médecine de Nancy. vols. 1834–1855. pp. 51–79.)*

chitin in 1811 (Labrude, 1997; Labrude and Becq, 2003), the first polysaccharide described 30 years before cellulose and extracted from edible fungi.

Henri Braconnot was born on 29 May, 1780 in Commercy (Meuse), a French commune in the Grand-Est region where his father was *un conseiller municipal au parlement local* (a councilor in the local parliament). In 1787, after his father's death, young Braconnot began his education with private teachers. In 1793, he was placed as an apprentice in a pharmacy in Nancy where he learned pharmacy, botany, and chemistry. Two years later, Braconnot left Nancy for his military service in a hospital in Strasbourg. In 1801, he lived in Paris and studied at the *L'École de Médecine* (School of Medicine) where he followed the lessons of Antoine-François de Fourcroy (1755–1809), an eminent chemist and neurochemist, and of Geoffroy Saint-Hilaire (1772–1844), a famous zoologist and naturalist. A year later, Braconnot returned to Nancy, and until his death, he lived in this city where he was appointed in 1807 director of the Botanical Garden and member of the city's Scientific Academy with teaching duties in natural history. In 1820, *le roi Louis XVIII* appointed him a member of the *Académie Royale de Médecine* (Royal Academy of Medicine). Braconnot was also elected in 1823 as a member of the *Académie des Sciences de Paris* (Paris Academy of Sciences) and promoted to *Chevalier de la Légion d'Honneur* (Knight of the Legion of Honour) in 1828. Braconnot's scientific career covers the period 1806 to 1854, in which he published over 110 *mémoires*. Henri Braconnot died on 13 January, 1855, leaving all his possessions to the city of Nancy. Much information about the life and career of Henri Braconnot is available in the biographies of Simonin (1856, 1870, 2013), Nicklès (1856a,b), Donzelot (1953), Prévost and D'Amat (1956), Labrude (1997), Labrude and Becq (2003), Wisniak (2007), Nwe et al. (2011a), and Muzzarelli et al. (2012).

Braconnot worked all his life in plant chemistry, botany being, indeed, his passion. In his research, he has always shown interest in coupling botany, chemistry, and medicine to address societal issues. It should be remembered that in the early 1800s, France was engaged in the Napoleonic wars and that the soldiers and the people were suffering from famine and contagious diseases. During this difficult period, edible fungi were considered *une manne gratuite et un don de la providence, attendu chaque saison par le peuple avec impatience* (a free manna and a gift of providence, eagerly awaited by the people every season). For Braconnot, the substances extracted from fungi had a great nutritional value. There was, therefore, a scientific and social interest in coupling different disciplines to meet the challenges of the time. Braconnot's work was essentially devoted to the extractive principles of natural vegetables and various aspects of the chemistry and physiology of natural substances, including carbohydrates and alkaloids. Among his many discoveries are gallic acid, ellagic acid, pyrogallic acid (later used in photography), legumin, populin, nitrocellulose, and the first plastic substance created by a chemist. Braconnot also studied pectin; "gelatin sugars" (named later glycocolle); the composition of fats, bile, and gastric juice; the acids produced by the putrefaction of cheese; and the fermentation of milk. Although Braconnot devoted most of his career and energy to botany, fungi, and plant chemistry, he was also interested in other disciplines in which he played an important role, such as color chemistry, toxicology, geology, mineralogy, hydrology, and water treatment. Braconnot was also an eminent pharmacist. However, he was never received as an apothecary, but the *Société Française de Pharmacie* (French Pharmacy Society) has always considered that he belonged to it (Simonin, 1856, 1870; Nicklès, 1856a,b; Klobb, 1907).

In 1811, Braconnot discovered an alkaline-insoluble fraction from *Agaricus volvaceus* by treatment with dilute warm alkali (Braconnot, 1811a, b,c). He published his results in *Recueil de Mémoires concernant la chimie et les arts qui en dépendent et spécialement la pharmacie* (Fig. 2.4). Braconnot was particularly surprised by the fungi's resistance to alkali (Braconnot, 1811c) and regarded this as *une caractéristique unique et fondamentale de cette substance azotée d'une nature particulière* (a unique and fundamental characteristic of this nitrogenous substance of a particular nature). This is the reason why he gave it the name of *fongine* (fungine) (Fig. 2.5). This was the first description of a new substance extracted from fungi, *fongine*, that would later be called *chitine* by Odier (1823). Braconnot analyzed the nitrogen content in the liquid obtained by distillation of the fraction and found that the liquid contained acetate of ammonia contaminated with oil. He suggested for the first time that *fongine* was a nitrogenous substance, but its nitrogen content was less than that of proteins. This fraction also produced acetic acid by degradation with concentrated sulfuric acid (Braconnot, 1811b). However, some colleagues, such as Vauquelin, remained unconvinced by Braconnot's conclusions, considering *fongine* a protein (Vauquelin, 1813). Louis-Nicolas Vauquelin (1763–1829) was a famous French pharmacist, chemist and mineralogist, and assistant to Antoine-François de Fourcroy. Vauquelin, considered as the father of the French analytical mineral chemistry, discovered the elements beryllium and chromium and the substances pectin, amino acid asparagine, and quinic acid.

In a later paper (Braconnot, 1813), Braconnot wrote: *cette substance, de quelques champignons qu'elle provienne, séparée de tout corps étranger par l'eau*

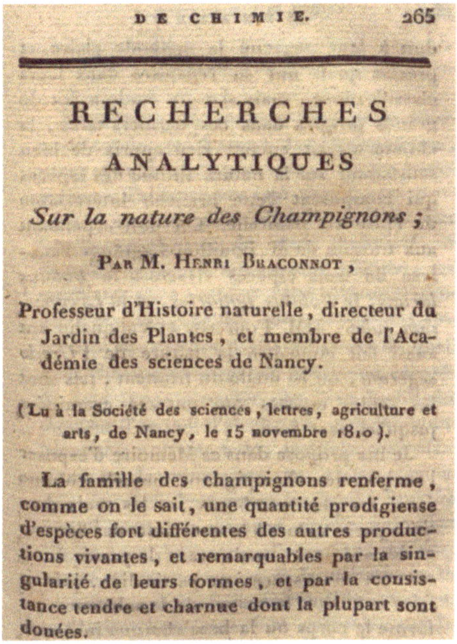
DE CHIMIE. 265

RECHERCHES

ANALYTIQUES

Sur la nature des Champignons ;

PAR M. HENRI BRACONNOT,

Professeur d'Histoire naturelle, directeur du Jardin des Plantes, et membre de l'Académie des sciences de Nancy.

(Lu à la Société des sciences, lettres, agriculture et arts, de Nancy, le 15 novembre 1810).

La famille des champignons renferme, comme on le sait, une quantité prodigieuse d'espèces fort différentes des autres productions vivantes, et remarquables par la singularité de leurs formes, et par la consistance tendre et charnue dont la plupart sont douées.

Fig. 2.4 First page of a *mémoire* of Professor Braconnot published in the journal *Annales de Chimie* or *Recueil de Mémoires concernant la chimie et les arts qui en dépendent et spécialement la pharmacie* in 1811, where he detailed his experiments. *(Source: Braconnot, H., 1811. Sur la nature des champignons. In: Klostermann, J. (ed.), Recueil de Mémoires Concernant la Chimie et les Arts Qui en Dépendent et Spécialement la Pharmacie. 31 Juillet 1811. In: Annales de Chimie, vol. 79. Librairie des Ecoles Impériales Polytechnique et des Ponts et Chaussées Paris, pp. 265–304.)*

bouillante aiguisée d'un peu d'alcali, est, plus ou moins blanche, molasses, fade et insipide (this substance, of whatever fungi it comes from, separated from all foreign bodies by boiling water sharpened with a little alkali, is more or less white, soft, bland, and insipid). Indeed, Braconnot observed that his *fongine* had many different consistencies, more or less soft, leathery, cartilaginous, or cork-like, indicating that there were differences in the proportion of its components. He repeatedly stated that his *fongine* differed from woody materials, containing more nitrogen than wood but much less than wheat protein or in animal materials. His interest then widened and later, he made important contributions to the study of oils, fats, proteins, and materials such as hemp. It is important to note that in this work (Braconnot, 1813), Braconnot did not use extraction with boiling alkali, and some proteins would presumably remain in the water- and alcohol-extracted material. Later, in 1957, Tracey reported that "the material certainly contained

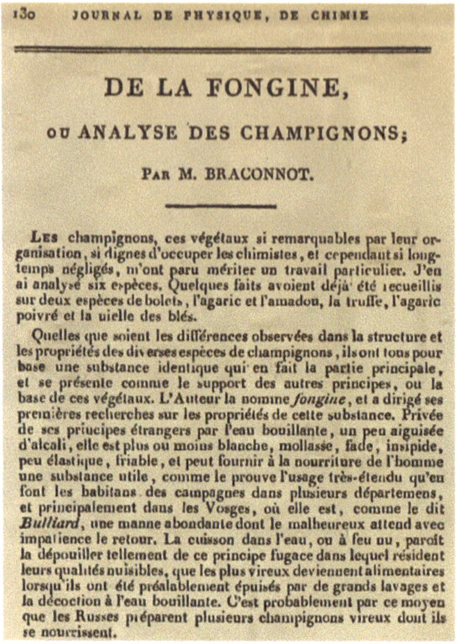

130 JOURNAL DE PHYSIQUE, DE CHIMIE

DE LA FONGINE,

OU ANALYSE DES CHAMPIGNONS;

PAR M. BRACONNOT.

Les champignons, ces végétaux si remarquables par leur organisation, si dignes d'occuper les chimistes, et cependant si long-temps négligés, m'ont paru mériter un travail particulier. J'en ai analysé six espèces. Quelques faits avoient déjà été recueillis sur deux espèces de bolets, l'agaric et l'amadou, la truffe, l'agaric poivré et la nielle des blés.

Quelles que soient les différences observées dans la structure et les propriétés des diverses espèces de champignons, ils ont tous pour base une substance identique qui en fait la partie principale, et se présente comme le support des autres principes, ou la base de ces végétaux. L'Auteur la nomme *fongine*, et a dirigé ses premières recherches sur les propriétés de cette substance. Privée de ses principes étrangers par l'eau bouillante, un peu aiguisée d'alcali, elle est plus ou moins blanche, mollasse, fade, insipide, peu élastique, friable, et peut fournir à la nourriture de l'homme une substance utile, comme le prouve l'usage très-étendu qu'en font les habitans des campagnes dans plusieurs départemens, et principalement dans les Vosges, où elle est, comme le dit *Bulliard*, une manne abondante dont le malheureux attend avec impatience le retour. La cuisson dans l'eau, ou à feu nu, paroît la dépouiller tellement de ce principe fugace dans lequel résident leurs qualités nuisibles, que les plus vireux deviennent alimentaires lorsqu'ils ont été préalablement épuisés par de grands lavages et la décoction à l'eau bouillante. C'est probablement par ce moyen que les Russes préparent plusieurs champignons vireux dont ils se nourrissent.

Fig. 2.5 First page of a *mémoire* of Professor Braconnot published in the *Journal de Physique, de Chimie, d'Histoire Naturelle et des Arts* where he proposed the name of *fongine*. *(Source: Braconnot, H., 1811. De la fongine, ou analyse des champignons. J. Phys. Chim. Hist. Nat. Arts. LXXIII, 130–135.)*

a considerable proportion of chitin," but it was long regarded as a more or less pure form of cellulose (Tracey, 1957). In 1973, Rudall and Kenchington also pointed out that *fongine*/chitin was not pure containing lipids and alkali-resistant polysaccharides (Rudall and Kenchington, 1973).

2.3 Auguste Odier

Auguste Odier (1802–1870) was a little-known French scientist, both naturalist and chemist. Nevertheless, his name has gone down in history because Odier was the first to propose the term chitin. Indeed, in August 17, 1821, Odier presented a *mémoire* to the *Société d'Histoire Naturelle de Paris* (Paris Natural History Society) entitled *une nouvelle substance dans la partie cornée des insectes* (a new substance found in the elytra of insects) (Fig. 2.6). This substance, such as Braconnot's, was particularly resistant to alkalis. Odier gave it the name of *chitine* (chitin). Odier's results, considered *remarquable* (remarkable) in 1822 by the French *Société Française d'Histoire Naturelle*

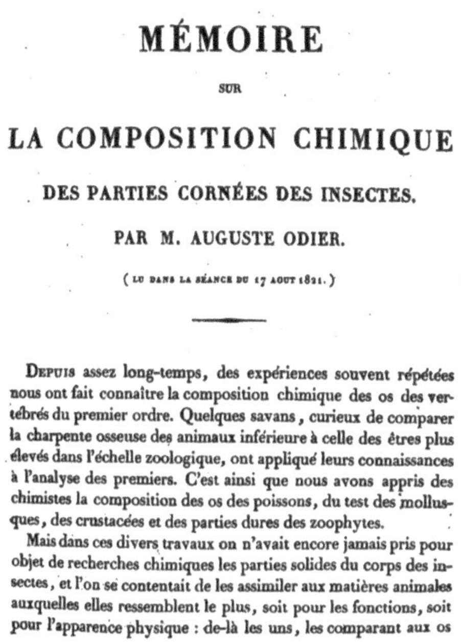

MÉMOIRE

SUR

LA COMPOSITION CHIMIQUE

DES PARTIES CORNÉES DES INSECTES.

PAR M. AUGUSTE ODIER.

(LU DANS LA SÉANCE DU 17 AOUT 1821.)

Depuis assez long-temps, des expériences souvent répétées nous ont fait connaître la composition chimique des os des vertébrés du premier ordre. Quelques savans, curieux de comparer la charpente osseuse des animaux inférieure à celle des êtres plus élevés dans l'échelle zoologique, ont appliqué leurs connaissances à l'analyse des premiers. C'est ainsi que nous avons appris des chimistes la composition des os des poissons, du test des mollusques, des crustacées et des parties dures des zoophytes.

Mais dans ces divers travaux on n'avait encore jamais pris pour objet de recherches chimiques les parties solides du corps des insectes, et l'on se contentait de les assimiler aux matières animales auxquelles elles ressemblent le plus, soit pour les fonctions, soit pour l'apparence physique : de-là les uns, les comparant aux os

Fig. 2.6 First page of the *mémoire* presented by Professor Auguste Odier in 1821, where he described a new substance found in the elytra of insects. *(Source: Odier, A., 1823. Mémoire sur la composition chimique des parties cornées des insectes. In: Mémoires de la Société d'Histoire Naturelle de Paris, vol. 1. Baudouin Frères Libraires-Éditeurs, Paris, pp. 29-42.)*

(French Society of Natural History) (Fig. 2.7), were not published until a year later (Odier, 1823; Brongniart, 1823; Straus-Durckeim, 1828; Latreille, 1831; Bounoure, 1919; Théodoridès, 1962; Roberts, 1992).

Auguste Alfred Odier was born in Hamburg on January 27, 1802. His father Antoine of Swiss origin was an industrialist and banker. Information about his childhood and education is scarce (Théodoridès, 1962). From the correspondence of his friend Audouin, it seems that between 1817 and 1818, Odier attended the chemistry classes of Louis-Jacques Thénard (1777–1857), professor at the *Collège de France* and at the *Faculté des Sciences de Paris* (Faculty of Science of Paris). Jean-Victor Audouin (1797–1841) was a famous naturalist and professor of entomology at the *Muséum de Paris*. In the 1820s, Odier and Audouin worked together on the dissection of arthropods under the guidance of Georges Cuvier (1769–1832), a famous anatomist of the time. Odier then worked in a chemical laboratory belonging to a painted canvas factory owned by his father. In 1826, he prepared for an

RÉSUMÉ DES TRAVAUX

DE LA SOCIÉTÉ D'HISTOIRE NATURELLE DE PARIS,

PENDANT L'ANNÉE 1821;

PAR M. AD. BRONGNIART,

SECRÉTAIRE.

(LU DANS LA SÉANCE DU 12 AVRIL 1822.)

La chimie animale, qui n'a encore fait connaître que peu de choses sur la nature de quelques parties propres seulement à certaines classes d'animaux, est redevable à M. Odier d'observations précieuses sur la nature du test corné des insectes. M. Odier a montré la différence qui existe entre la matière qui compose ces parties, et qu'il a nommée *Chitine*, et les élémens organiques observés dans les animaux vertébrés, tels que la corne, l'épiderme, la gélatine, l'albumine, etc. Il a fait voir ainsi qu'on ne devait pas regarder ce test comme une simple modification dans l'épaisseur et la structure de l'épiderme, mais comme un organe d'une nature particulière, et qui parait propre à tous les animaux articulés. On sent combien il serait curieux d'étendre de semblables recherches aux substances qui composent les autres animaux des classes inférieures, et combien de semblables résultats deviendraient intéressans pour la physiologie générale.

Fig. 2.7 Excerpt from the comment on Professor Odier's work presented by the Secretary of the *Société d'Histoire Naturelle de Paris* in April 12, 1822. *(Source: Brongniart, A.D., 1823. Résumé des Travaux de la Société D'histoire Naturelle de Paris. In: Mémoires de la Société d'Histoire Naturelle de Paris, vol. 1. Baudouin Frères Libraires-Éditeurs, Paris, pp. 15–27.)*

architectural competition at the *École des Beaux-Arts de Paris* (Paris School of Fine Arts), which he passed in 1830. However, his scientific career was very short because, in 1833, Odier abandoned science and architecture to become *un conseiller Référendaire à la Cour des Comptes* (a referendary counselor at the Court of Auditors), a position he held until his retirement. Auguste Odier died in his Parisian home on February 1, 1870.

Odier wrote only two scientific works published in the memoirs of the *Société d'Histoire Naturelle de Paris* (both in Volume I of 1823), a society of which he was a member since 1821 (date of the creation of this society) like his friend Audouin (Théodoridès, 1962). The first memoir concerns an ectoparasitic worm of the gills of the crayfish (*Branchiobdella astaci*). Odier communicated this work to the *Société Philomatique de Paris* (Philomatic Society of Paris) in November 1819, at the age of 17, which was published in 1823 in the *Mémoires de la Société d'Histoire Naturelles de Paris* (Memoirs of the Society of Natural History of Paris). Odier's second publication, entitled *Mémoire sur la composition chimique des parties cornées des insects* (Memoir on the chemical composition of the horny parts of insects), describes his work on the discovery of chitin between 1821 and 1823 (Fig. 2.6).

In 1821, at the age of 19, Odier studied the alkali-insoluble fraction of the cockchafer (May beetle) by repeated treatments with hot solutions of caustic potash. He showed the presence of six substances in the elytra organs: albumin, an extractive material soluble in water, a brown animal substance soluble in potash and insoluble in alcohol, a colored oil soluble in alcohol, three salts, and especially a particular material forming a quarter of the weight of the elytra. Odier isolated this last fraction which he named *chitine*, in reference to the Greek term kiton/chiton meaning envelope. He then studied its chemistry in detail between 1821 and 1823. Odier distinguished this new substance from those constituting the horn, the hair, or the epidermis and compared it to the substance constituting the woody tissue of plants. After several treatments, Odier observed that *chitine* was stable in the external form and soluble in sulfuric acid *avec l'assistance de chaleur* (with the assistance of heat) but did not become yellow by the action of nitric acid (Odier, 1823). However, *chitine* was dissolved when digested in it with heat. Later, de Saint-Vincent (1825), Chevallier and Richard (1827), and Drapiez (1839) reported similar conclusions on *fongine/chitine*. At the same time, a similar substance called *entoméiline* or *entomaderme* was also observed by the young student Lassaigne, with similar chemical characteristics to those of Odier, but Lassaigne's results had not yet been published (Straus-Durckeim, 1828; Latreille, 1831).

Odier noted that *chitine* constituted only a relatively small part of the insect cuticle: the elytra of *Melolontha* contained about 29% (Odier, 1823). There was a certain amount of ash and some oils. Odier considered that the major part of the nonchitinous substance was made of proteins, a conclusion that other researchers also shared (de Saint-Vincent, 1825; Chevallier and Richard, 1827; Straus-Durckeim, 1828; Latreille, 1831; Drapiez, 1839). Later, Krawkow (1892) reported that wherever chitinous structures occurred, proteins were also present. However, it was only in the 1940s that information on the nature of proteins and their associated complexes was available (Fraenkel and Rudall, 1940, 1947; Trim, 1941). Odier also identified *chitine* in another beetle, *Oryctes nasicornis*, and also in demineralized crab carapaces (Odier, 1823). He then suggested that *chitine* was the basic material of the exoskeletons of all insects, and possibly, the arachnids. Odier announced that he would continue his research on other invertebrates (mollusks, zoophytes) and fungi (Odier, 1823), but this research was not successful (Straus-Durckeim, 1828; Latreille, 1831;

Bounoure, 1919). A few years later, Odier's hypotheses concerning the presence of chitin and its wide distribution on the cell walls of fungi, the exoskeletons of arthropods such as crustaceans and insects, the radulas of mollusks, and the beaks of cephalopods and other marine invertebrates would prove to be correct. The only erroneous results concerned the presence of nitrogen in *chitine*. Indeed, Odier tested for nitrogen in this substance and came to the conclusion that it did not contain nitrogen (Odier, 1823). In 1855, Edmond Frémy (1814–1894), another French chemist, reported a similar conclusion for *fongine/chitine*, named by him *metacellulose* (Frémy, 1855). Frémy attributed to chitin and cellulose the same chemical composition because "their reaction with iodine in an acid medium gave the same brown-red coloring."

In his second mémoire (Fig. 2.6), Odier concluded that.

(1) *La chitine est une substance particulière, fort curieuse* (chitin is a peculiar, very curious substance);

(2) *Elle ne contient point d'azote; par ce dernier caractère, elle se rapproche des matières végétales* (chitin contains no nitrogen, which makes it similar to plant material);

(3) *Il est fort remarquable de retrouver dans la charpente des insectes la même substance qui forme celle des végétaux* (it is very remarkable to find in the structure of insects the same substance which forms that of plants).

Odier thought that the frameworks of insects and of plants, indeed, comprised the same substance, cellulose. It is a frequent misapprehension that this statement indicates that he recognized the similarity between *chitine* and *fongine* of Braconnot, but a further statement that "lignin is the only proximate vegetable principle which can be compared to it" clearly shows that he did not (Théodoridès, 1962). Furthermore, there is no reference to Braconnot's paper in that of Odier, and there is no evidence that the latter was aware of the earlier work (Straus-Durckeim, 1828; Latreille, 1831; Bounoure, 1919; Théodoridès, 1962). A century later, Professor George A.F. Roberts (1939–2018), Emeritus Professor of Textile Science (Nottingham Trent University), confirmed these statements in his famous book on chitin (Roberts, 1992).

2.4 John George Children

John George Children (1777–1852) was a British chemist, physicist, mineralogist, astronomer, and zoologist at the British Museum of London. Early in his career, Children (Fig. 2.8) was also interested in mechanics,

Fig. 2.8 Professor John George Children 1777–1852. *(Source: Bettany, G.T., 1887. Children John George. In: Stephen, L. (Ed.), Dictionary of National Biography. Smith, Elder & Co., London, pp. 249–250.)*

electricity, and galvanism. He contributed, for example, to the realization of the largest galvanic battery ever built, and for this work, he received the medal of the Royal Society of London in 1828. Later, Children was a librarian in the antiquities department of the British Museum. During his career, Children translated several scientific works of literature into English. He was a famous scientist at a time when science was still largely the preserve of amateur gentleman and was well-known in London scientific circles. His name is commemorated in a mineral, childrenite.

Children was born on May 18, 1777, in Ferox Hall, Tunbridge, Kent, to Susanna Jordan and George Children (1742–1818). His mother died 6 days after giving birth to him. His father, a banker, devoted himself to his education and then became his benefactor by helping him build his laboratory and by providing with funds for his research (Atkins, 1853; Bettany, 1887). Young Children was described by his daughter as "a man of exceptional intelligence, very spoiled as a child" (Atkins, 1853). Educated at Tunbridge School, Eton, Children entered Queen's College, Cambridge, in 1794 as a scholarship student and studied to become a reverend. However, after becoming engaged in 1798, he left the university without earning a degree but intended to pursue a career in the church. That same year, John George Children married Hester Anna Holwell, and a year later, they had a daughter, Anna. The following year, his wife died unexpectedly and he became a widower like his father. Children then traveled to Europe (Portugal, Spain, and Wales) and North America to mourn his loss. He became interested in minerals and insects, collected during a stone trip in Wales and Cornwall, and began to publish his work. Children also published translations, notes, and works on various domains such as calcination of metals and silver extraction. However, this period of travel was interrupted by health problems. In 1805, Children returned in England to build a laboratory in his father's home at Ferox Hall, where he conducted experiments in mineral chemistry, using his father's fortune to finance them. Two years later, Children was elected, at the age of 30, a fellow of the Royal Society of London and the Royal Society of Edinburgh, of which he later became secretary (between 1826 and 1827 and again from 1830 to 1835). In 1824, he became co-editor of the Zoological Journal then created and later of the Annals of Philosophy. He was also editor for the Quarterly Journal of Science. Children was also actively involved in the creation of the Royal Entomological Society in 1833, of which he was president in 1834–1835 (Atkins, 1853; Bettany, 1887; Gunther, 1978). In 1837, Children became head of the zoology department of the British Museum. Throughout his life, he also wrote many poems. John George Children died on 1 January, 1852 at the home of his daughter, Anna Atkins (1799–1871), a renowned botanist and photographer who wrote a 313-page memoir about her father's life. Atkins wrote: "Like his father, John was a good-natured and generous man. He was of a most lovable disposition, spared by frequent illnesses and misfortunes, without arrogance or vanity, most careful in ascertaining facts, and equally zealous in friendship and in science" (Atkins, 1853). Additional information about the life and career of John George Children is available in the reference Kaminsky (2018).

Upon its publication in 1823, Children carefully read Odier's paper and published an English translation for the Zoological Journal a year later (Fig. 2.9). In concluding this translation (Children, 1824), Children felt that "Odier's conclusion that *chitine* was free of nitrogen because the products of its dry distillation had no effect on test papers was open to debate." It is important to note that Children was not aware of Braconnot's much earlier work on *fongine* (Smedley et al., 1845). Children then described his experiments and his own conclusions on "purified" chitin. First, he repeated the same experiments as Odier, extracting an alkaline insoluble fraction from May bug elytra (Children, 1824). Children observed that "during the action of the alkali, a slight disengagement of ammonia was perceptible." He analyzed by elemental analysis the residue left after repeated extractions with

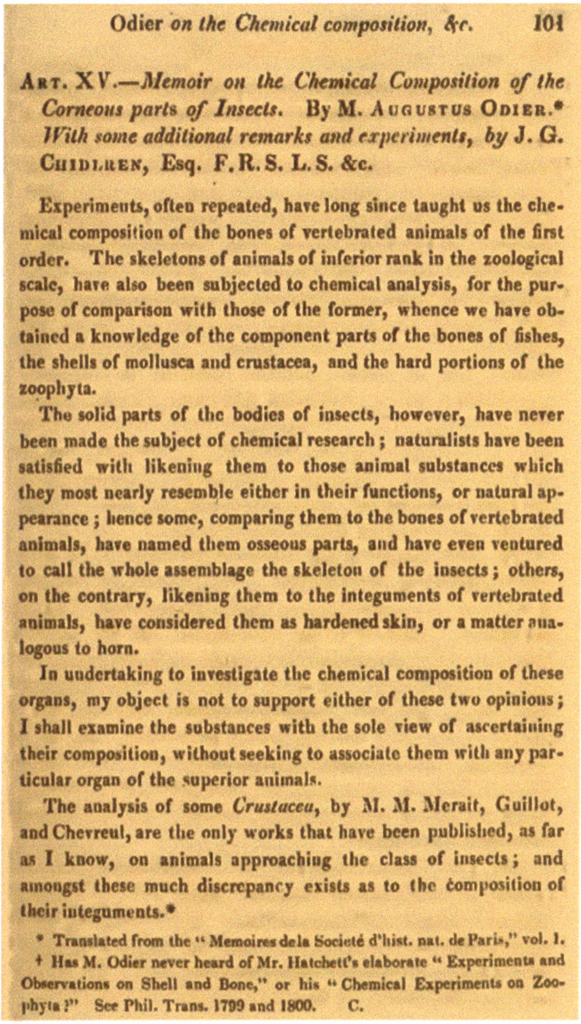

Odier *on the Chemical composition, &c.* 101

Art. XV.—*Memoir on the Chemical Composition of the Corneous parts of Insects.* By M. Augustus Odier.* *With some additional remarks and experiments, by* J. G. Children, Esq. F.R.S. L.S. &c.

Experiments, often repeated, have long since taught us the chemical composition of the bones of vertebrated animals of the first order. The skeletons of animals of inferior rank in the zoological scale, have also been subjected to chemical analysis, for the purpose of comparison with those of the former, whence we have obtained a knowledge of the component parts of the bones of fishes, the shells of mollusca and crustacea, and the hard portions of the zoophyta.

The solid parts of the bodies of insects, however, have never been made the subject of chemical research; naturalists have been satisfied with likening them to those animal substances which they most nearly resemble either in their functions, or natural appearance; hence some, comparing them to the bones of vertebrated animals, have named them osseous parts, and have even ventured to call the whole assemblage the skeleton of the insects; others, on the contrary, likening them to the integuments of vertebrated animals, have considered them as hardened skin, or a matter analogous to horn.

In undertaking to investigate the chemical composition of these organs, my object is not to support either of these two opinions; I shall examine the substances with the sole view of ascertaining their composition, without seeking to associate them with any particular organ of the superior animals.

The analysis of some *Crustacea*, by M. M. Merait, Guillot, and Chevreul, are the only works that have been published, as far as I know, on animals approaching the class of insects; and amongst these much discrepancy exists as to the composition of their integuments.*

* Translated from the "Memoires dela Societé d'hist. nat. de Paris," vol. 1.

† Has M. Odier never heard of Mr. Hatchett's elaborate "Experiments and Observations on Shell and Bone," or his "Chemical Experiments on Zoophyta?" See Phil. Trans. 1799 and 1800. C.

Fig. 2.9 First page of the article of John George Children where he translated Odier's paper and described his own experiments. *(Source: Children, J.G., 1824. Memoir on the chemical composition of the corneous parts of insects; by Augustus Odier. Translated from the original French, with some additional remarks and experiments. Bell T, Children JG, Sowerby JDC and Sowerby GB, editors. London. March, 1824, n I, article XV. Zool. J. 1, 101–115.)*

strong potassium hydroxide solution and found substantial quantities of nitrogen (11.05% and 9.54% in two analyses), giving the empirical formula $C_{11}H_{17}O_7N_2$. Children also found in the cantharides "a small portion of silica and magnesia and a slight trace of manganese." He then published his observations as an appendix in the same journal where he translated Odier's article (Children, 1824). In this appendix, Children suggested that "Odier's test could have failed if the volatile acetic acid was evolved simultaneously with the ammonia." The resulting neutralization would yield negative tests for ammonia, and this had led Odier to conclude that "nitrogen was absent." Later, Smedley et al. (1845) reported that "this suggestion appears to have been made in ignorance of Braconnot's work in which the evolution of acetic acid was reported." However, since the samples burnt by Odier had only been extracted with boiling water and not with hot caustic potash solutions, they would have contained a considerable amount of protein which should have given a positive test for nitrogen on burning. Hence, Children's explanation cannot be correct, and the reason for Odier's negative result is unresolved (Roberts, 1992). It is also important to point out that to obtain "purified" chitin, Children used multiple extractions with hot concentrated caustic potash solutions. From the description of the process, it is probable that Children, and also Odier, isolated chitosan rather than chitin (Roberts, 1992). Indeed, the formula ($C_{11}H_{17}O_7N_2$) proposed by Children was closer to that for the disaccharide repeat unit of chitosan ($C_{12}H_{22}O_8N_2$) than that of chitin ($C_{16}H_{26}O_{10}N_2$).

Finally, in the editorial comments, Children wrote:

(1) "Consequently, M. Odier's conclusion that it (*chitine*) rather belongs to the vegetable than the animal kingdom is erroneous";

(2) "Not everyone agreed with the name *chitine*";

(3) "Every man has a right to name his own child, but we think M. Odier might have made a happier election. Elytrine would have been more significant and at least as euphonius."

Although the nitrogenous nature of *chitine* was first revealed by the experiments of Children in 1824, this discovery is usually attributed to both Lassaigne (1843a) and Payen (1843). However, it was Lassaigne who was the first to really demonstrate that chitin of insect origin contained nitrogen and that it was not a cellulose as Payen believed a few years earlier (Berthelot, 1859; Bounoure, 1911). Both Straus-Durckeim (1828) and Latreille (1831) previously indicated that the presence of nitrogen was highlighted by the young student Lassaigne in *entomaderme* (*chitine*), but he had not yet published his results. At the same time as Lassaigne and Payen, in 1845, the Baltic German chemist Carl Schmidt (1822–1894) studied the extraction of chitin from insects such as *Melontha vulgaris*. Schmidt showed the calcination of insect chitin by potassium resulted in the formation of potassium cyanide, suggesting the presence of nitrogen among its constituent elements. This presence was demonstrated by a detailed elemental analysis. Schmidt claimed to be the first to have demonstrated this discovery (Schmidt, 1845).

2.5 Jean-Louis Lassaigne

The interrogation on the difference between cellulose and chitin was initiated by Lassaigne and presented in *Compte Rendu des Séances de l'Académie des Sciences* in May 15, 1843. Jean-Louis Lassaigne (1800–1859) was a French chemist, professor at the *École Royale Vétérinaire d'Alfort* (Royal Veterinary School of Alfort) in Maisons-Alfort, near Paris.

Jean-Louis Lassaigne was born in Paris on September 22, 1800. His father worked as a mechanic at the *Muséum d'Histoire Naturelle* (Museum of Natural History). Young student Lassaigne studied chemistry under the guidance of Louis-Nicolas Vauquelin (1763–1829), a renowned chemist and pharmacist at the *Université de Paris*. At the age of 17, Lassaigne already co-authored his first papers on green alga with Jean-Baptiste Alphonse Chevallier (1793–1879), pharmacist and assistant to Vauquelin at the *Muséum d'Histoire Naturelle*. In 1821, his brilliant young career as a research chemist was rewarded by the medal of the *Société de Médecine de la Seine* (Seine Medicine Society). Four years later, the French *Académie des Sciences* awarded him an honorable mention for the book *Recherches physiologiques et chimiques pour servir à l'histoire de la digestion*, published in collaboration with François Leuret (1797–1851), a renowned anatomist and psychiatrist. The same year, Lassaigne was appointed professor of chemistry at the *École Spéciale de Commerce de Paris* (Paris Special School of Commerce), and in 1828, professor of chemistry and physics at the *École Royale Vétérinaire d'Alfort*, working in this school until 1854. He was an active member of the *Société de Chimie Médicale* (Society of Medical Chemistry) and the *Académie de Médecine* (Academy of Medicine) and one of the founding members of the *Société Centrale de Médecine Véterinaire* (Central Society of Veterinary Medicine) in 1844. Jean-Louis Lassaigne passed away in Paris on March 18, 1859.

Lassaigne is known for a large number of works not only in pure chemistry (dyes, enamel elaboration for pottery, and chromium salts) but also in vegetable principles (alkaloids, malic acid), agriculture and animal chemistry, physiology, forensic chemistry, hygiene, and toxicology (studies on phosphorus), and in industry (acid production). Like many researchers of the time, Lassaigne was also interested in cellulose (Laroque, 1860; Wisniak, 2014). A controversy arose between him and Payen who claimed that "the cell wall of fungi was a cellulose, and therefore chitin was a cellulose." Lassaigne did not agree with this statement at all. In 1842, Lassaigne developed an analytical procedure, which made it possible "to identify the presence of nitrogen in

CHIMIE. — *Mémoire sur un procédé simple pour constater la présence de l'azote dans des quantités minimes de matière organique ; par* M. LASSAIGNE.

(Commissaires, MM. Thenard, Chevreul, Payen.)

« La présence de l'azote dans les matières organiques est en général indiquée par diverses réactions qui isolent ce corps à l'état gazeux, en brûlant les éléments combustibles auxquels il était combiné, ou qui le transforment en ammoniaque (*azoture d'hydrogène*), dont les propriétés sont alors faciles à constater.

» Ces moyens, que les chimistes emploient chaque jour, toujours suffisants quand il s'agit d'évaluer la proportion d'azote dans une quantité pondérable de matière organique, deviennent presque impossibles, et même impraticables, quand on n'a à sa disposition que des quantités minimes ou presque impondérables de la substance à examiner.

» Le procédé que nous avons mis en pratique, après l'avoir soumis à de nombreux essais, est si sensible, qu'il permet de reconnaître la présence de l'azote dans des quantités de matières azotées aussi petites que celles que les meilleures balances peuvent à peine apprécier. Le principe chimique de ce procédé, déjà acquis à la science, n'avait pas encore, que nous sachions, été appliqué à l'objet que nous traitons ici. Ce moyen repose sur la facilité avec laquelle se forme le cyanure potassique lorsqu'on calcine au rouge obscur, et à l'abri de l'air, du potassium en excès avec une matière organique même très-peu azotée. Le produit de cette calcination, étant délayé dans quelques gouttes d'eau distillée froide, donne une liqueur alcaline qui, mêlée à un

Fig. 2.10 First page of a *mémoire* of Professor Jean-Louis Lassaigne published, where he described the method that permitted to identify nitrogen. *(Source: Lassaigne, J.L., 1843. Mémoire sur un procédé simple pour constater la présence d'azote dans des quantités minimes de matière organique. C. R. Séances Hebd. Acad. Sci. 16, 387-391.)*

quantities of substance that could hardly be weighed with an analytical balance." He described this method in a *mémoire* he presented to the *Académie des Sciences* (Academy of Sciences) on January 2, 1843 (Fig. 2.10). The procedure was based on the fact that potassium cyanide was easily formed when the substance being investigated was calcined to dark red with excess of potassium. The product of the calcination, when diluted with a few drops of distilled water, produced an alkaline liquor, which in the presence of a soluble ferrous-ferric salt, yielded a green blue or yellow precipitate (Lassaigne, 1843a). Much later, Tucker, taking up Lassaigne's work, showed that the precipitate turned blue when treated with a few drops of HCl, which was considered a characteristic reaction (Tucker, 1945). In this first *mémoire*, Lassaigne, studying the preparation and isolation of *chitine* obtained after treatment of the exoskeleton of the silkworm butterfly, also suggested the presence of nitrogen in the substance of Odier, using "his nitrogen identification test."

A few months later, on May 15, 1843, Lassaigne presented a second *mémoire* entitled *Sur le tissu tégumentaire des insects de différents ordres* (on the integumentary tissue of insects of different orders), in which he detailed

CHIMIE. — *Sur le tissu tégumentaire des insectes de différents ordres.* Note de M. LASSAIGNE.

« En examinant sous un point de vue particulier la peau de la chenille du *Bombyx mori* (ver à soie), nous avons été conduit, par une suite d'expériences, à isoler complétement ce tissu et à reconnaître les différences marquées qu'il présentait avec le même tissu dans les animaux supérieurs de l'échelle zoologique.

» Les résultats que nous avons obtenus, d'après le petit nombre d'essais qu'il nous a été possible d'entreprendre, nous ont permis de rapprocher la peau de ces insectes de la substance dure et coriace qui forme les élytres et une partie du corps de certains coléoptères. On sait que c'est à cette substance particulière, étudiée il y a vingt ans par M. Odier, qu'on a donné le nom de *chitine.*

Fig. 2.11 First page of a *mémoire* of Professor Jean-Louis Lassaigne, where he described his experiments on *Bombyx mori. (Source: Lassaigne, J.L., 1843. Sur le tissu tégumentaire des insectes des différents ordres. C. R. Séances Hebd. Acad. Sci. 16, 1087–1089.)*

his conclusions on *chitine*, first named *entomeiline* and then *entomaderme* by him (Fig. 2.11). Lassaigne purified *Bombyx mori* exuviae and the coleoptera elytra and then treated the residues with potassium hydroxide at warm temperature, thus obtaining potassium cyanide that confirmed the presence of nitrogen in *chitine* (Lassaigne, 1843b,c). He proved that *entomaderme* was not a cellulose but a new product containing nitrogen with fiber quality, in disagreement with Payen. Lassaigne wrote *nous sommes arrivés à cette conclusion, tout opposée à celle de ce chimiste* (we arrived at this conclusion, quite opposite to that of this chemist), that is, Payen. Finally, a few weeks later, on August 7, 1843, Payen corroborated Lassaigne's observation during a presentation to the *Académie des Sciences* (Payen, 1843). Curiously, a controversy arrose between the two researchers about the priority of the results; the bibliography of the time attributed the discovery of the presence of nitrogen in chitin to either of the two researchers. Nevertheless, it should be remembered that Payen was convinced that chitin was another type of cellulose that did not contain nitrogen. Straus-Durckeim (1828) and Latreille (1831), and later Berthelot (1859), mentioned that the nitrogenous nature of *chitine/entomaderme* was highlighted in the work of the *jeune et habile chimiste Lassaigne* (a young and skillful chemist). In 1957, Tracey reported that Lassaigne had priority over Payen who confused the issue by his intuitive belief that nitrogen-free tissues were typical of plants and nitrogen-containing tissues of animals (Tracey, 1957).

2.6 Anselme Payen

Anselme Payen (1795–1871) was a French chemist and agriculturist, professor at the *École Centrale des Arts et Manufactures de Paris* (Central School of Arts and Manufactures of Paris) (Fig. 2.12). He was first an industrial and then a multifaceted scientific authority. Indeed, Anselme Payen was a brilliant businessman, extremely knowledgeable, and aware of the issues of the time. As an eminent scientist, Payen has been an active member of the *Académie des Sciences* and the *Académie de Médecine*, publishing more than 200 papers.

To avoid recruitment into the revolutionary army, his father, Jean-Baptiste Pierre Payen (1759–1820), an important industrial with many factories, insisted that the young Anselme enter the *École Polytechnique* of Paris. Anselme then studied chemistry under Professor Vauquelin and quickly developed a passion for this discipline (Phillips, 1940). In Vauquelin's laboratory, Anselme Payen met Pierre-Jean Robiquet (1780–1840) and Alphonse Chevalier (1793–1879), two future famous pharmacists and members of the French *Académie Royale de Médecine*. For many years, these three

Fig. 2.12 Lithograph of Professor Anselme Payen in 1843 by Nicolas-Eustache Maurin (1799–1850), a famous French engraver. *(Source: Public Domain).*

researchers and friends worked together and published numerous works. In 1819, his father, who was seriously ill, insisted that Anselme stop his studies and get involved in the family business, notably by running the borax refinery and managing an industrial pharmaceutical research laboratory created by his father. A year later, when his father died, Anselme Payen, then 21 years old, had to assume full responsibility for the management of the factories. In 1825, while continuing his industrial activities, Payen accepted a teaching position at the *Société Philotechnique de Paris* (Philotechnical Society of Paris). Four years later, he began teaching chemistry at *École Centrale des Arts et Manufactures* as an assistant and later, he was appointed professor at the same school (1835). His management of the sugar beet factory led Payen to take an interest in agricultural chemistry. In 1833, he was elected a member of the *Société Royal et Centrale d'Agriculture* (Royal and Central Society of Agriculture) and became its permanent secretary, a position he held for 26 years. In 1839, Payen accepted a position as professor in chemistry applied to the arts at the *Conservatoire Royal des Arts et Métiers* (Royal Conservatory of Arts and Crafts). He held both teaching positions until his death. Payen was appointed *Chevalier de la legion d'Honneur* (1847), *Officier* (1847), and later *Commandeur* (1863). The *Académie Française de Médecine* elected him an associate member in 1868. A street in the 15th Arrondissement de Paris bears his name. A detailed biography of Anselme Payen can be found in the reviews by Phillips (1940) and Wisniak (2004).

As an industrialist, Payen developed the French process for producing borax from boric acid and sodium carbonate, the utilization of carbon animal as a decolorizing matter for beet sugar, an improved lead chamber process for manufacturing sulfuric acid, the total utilization of animal waste, and the elaboration of bituminous products (for this work, Payen was awarded a silver medal at the Seventh Industrial Exhibition held at the Louvre in 1827). As academic, Payen devoted a large part of his career to the study of plant physiology, and his many achievements in this area include the discovery of diastase, the enzyme for decomposing starch, cellulose and lignin, and the crucial role of nitrogen in vegetable development. Payen is, indeed, known for the "discovery"/identification of cellulose in 1838. The recognition of this polysaccharide as a chemical entity (chemical formula: $C_6H_{10}O_5$) was made by him (Bell, 1949). After treating different woods with nitric acid, Payen obtained a fibrous substance common to all, which he also found in other plants. He published his first results in *Comptes Rendus Hebdomadaires des Séances de l'Académie des Sciences*. A year later, in 1839, Payen highlighted that this chemical entity was the fundamental, uniform constituent of all plant cell walls (Payen, 1839, 1840). Thus, the name cellulose was rightly applied. The Cellulose, Paper and Textile Division of the American Chemical Society gives every year the Anselme Payen Award to the best work in the field.

In the early 1840s, Payen was convinced that Braconnot's *fongine* was a cellulose containing no nitrogen by stating *ce sera donc encore une anomalie éliminée de la science* (so it will be an anomaly eliminated from the science) (Payen, 1840, 1842). On August 7, 1843, during a session at the *Académie des Sciences* (Fig. 2.13), a few weeks after a seminar given by Lassaigne, Payen declared that *mes résultats expérimentaux démontrent des differences significatives entre la cellulose et la chitine, notamment dans la composition élémentaire en azote* (my experimental results demonstrates significant differences between cellulose and chitin, especially in the elemental composition of nitrogen). Payen

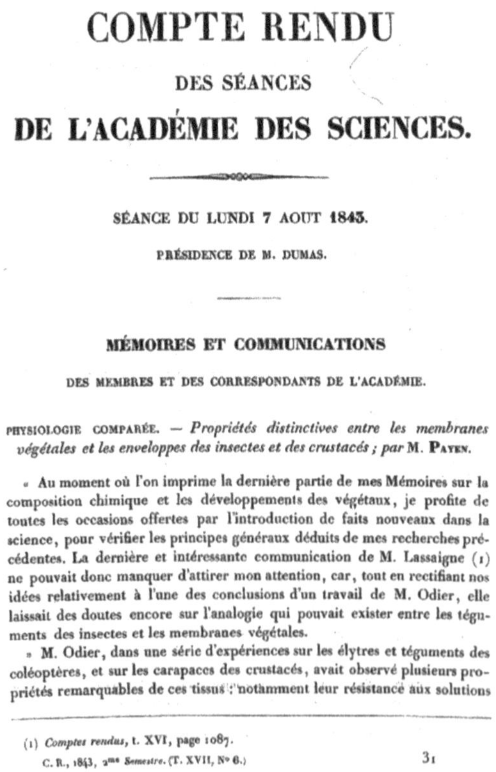
COMPTE RENDU

DES SÉANCES

DE L'ACADÉMIE DES SCIENCES.

SÉANCE DU LUNDI 7 AOUT 1843.

PRÉSIDENCE DE M. DUMAS.

MÉMOIRES ET COMMUNICATIONS

DES MEMBRES ET DES CORRESPONDANTS DE L'ACADÉMIE.

PHYSIOLOGIE COMPARÉE. — *Propriétés distinctives entre les membranes végétales et les enveloppes des insectes et des crustacés; par* M. PAYEN.

« Au moment où l'on imprime la dernière partie de mes Mémoires sur la composition chimique et les développements des végétaux, je profite de toutes les occasions offertes par l'introduction de faits nouveaux dans la science, pour vérifier les principes généraux déduits de mes recherches précédentes. La dernière et intéressante communication de M. Lassaigne (1) ne pouvait donc manquer d'attirer mon attention, car, tout en rectifiant nos idées relativement à l'une des conclusions d'un travail de M. Odier, elle laissait des doutes encore sur l'analogie qui pouvait exister entre les téguments des insectes et les membranes végétales.

» M. Odier, dans une série d'expériences sur les élytres et téguments des coléoptères, et sur les carapaces des crustacés, avait observé plusieurs propriétés remarquables de ces tissus ; notamment leur résistance aux solutions

(1) *Comptes rendus*, t. XVI, page 1087.

C. R., 1843, 2me *Semestre*. (T. XVII, No 6.) 31

Fig. 2.13 First page of the memory of Professor Anselme Payen where he described his conclusions on the chemical differences between cellulose and chitin. *(Source: Payen, A., 1843. Propriétés distinctives entre les membranes végétales et les enveloppes des insectes et des crustacés. C. R. Seances Acad. Sci. 17, 227-231.)*

clearly demonstrated that ammonia was released from purified arthropod exoskeletons upon heating with sodium hydroxide and that nitrogen was ~9%. He also described chemical differences between cellulose occurring in plants and chitin occurring in arthropod exoskeletons (later, at the end of the 1890s, Gilson did not agree with this). This is the reason why Payen is credited with initiating the interrogation of the difference between cellulose and chitin. However, it is Lassaigne that the paternity of the conclusions on this question must be attributed to (Berthelot, 1859; Bounoure, 1911; Tracey, 1957).

2.7 Charles Rouget

The history of chitosan, the main derivative of chitin, dates back to 1859 with the work of the French physician, physiologist, and anatomist Rouget (Fig. 2.14). Charles Rouget (1824–1904) was professor of

Fig. 2.14 Professor Charles Rouget, 1824–1904. *(Source: Gréhant, N., 1904. Charles Rouget, notice nécrologique. Nouv. Arch. Mus. Hist. Nat. 6, I–V.)*

physiology first at the *Faculté de Médecine* in Montpellier and then at the *Muséum d'Histoire Naturelle* in Paris (Gréhant, 1904a,b). Rouget was the first to describe that when *chitine* was boiled in concentrated potassium hydroxide, the resulting substance, called *chitine modifiée* (modified chitin), was soluble in acidic solutions, unlike chitin, which was insoluble in water. This *chitine modifiée* was colored violet by dilute solutions of iodine and acid, whereas *chitine* was stained brown. Rouget had just discovered for the first time the "deacetylation of chitin," opening new possibilities for its use. However, modified chitin did not interest the scientific community, and it was not until 1894 that it was studied again when Hoppe-Seyler named it "chitosan." As already noted, it is highly likely that Odier and Children also obtained chitosan rather than chitin. Nevertheless, Rouget is credited with the discovery of chitosan.

Charles Marie Benjamin Rouget was born on August 19, 1824 in Gisors (Eure, Normandie), son of a surgeon-major of the armies of the First Empire. The young Charles studied at the *Collège Sainte-Barbe*, the oldest college in Paris. He then began studying medicine in Paris where he was admitted to the anatomy competition and did his internship and clinical training at *L'Hôtel-Dieu de Paris* (the Hôtel-Dieu Hospital in Paris). After a doctorate in medicine, Rouget was admitted to the *Concours d'aggrégation* (Competitive examination) in anatomy and physiology. Unable to find a suitable position in Paris, Rouget applied for and obtained the chair of physiology at the *Faculté de Médecine* in Montpellier, in the south of France, in 1860. A few years later, he returned to Paris where he was appointed professor of general physiology at the *Muséum d'Histoire Naturelle*, a position he held from 1879 to 1893. Indeed, following the death of Professor Bernard, a new chair of physiology was created for Rouget at the museum. Claude Bernard (1813–1878) was a prominent physiologist, physician, and epistemologist, known as "one of the greatest of all men of science." Rouget then took over the direction of the laboratory at the *École des Hautes Études* and was elected *Correspondant National* (national correspondent) for the division of anatomy and physiology of the *Académie de Médecine*. When he retired, he moved to the south of France. Charles Rouget died on April 9, 1904, in Saint-Jean-sur-Mer.

Rouget has studied marine animals, birds, reptiles, etc. He has characterized their muscle fibers under microscope, determined their composition, and introduced notions on the contractility of tissues, the physiology of blood capillaries, and on skin perspiration. All this research was used to illustrate his courses which he provided at the universities of Montpellier and

Paris. Rouget is best known as the first to correlate physiology and microscopic anatomical structure. He discovered the branched contractile cells on the outer wall of capillaries in amphibians, structures that would later be known as *cellules de Rouget* (Rouget cells). The circular fibers of the ciliary muscle of the eye are also called the *Muscles de Rouget* (Rouget's muscles), in his honor. In 1887, Rouget received the Physiology Prize awarded by the *Académie des Sciences* for *son oeuvre scientifique dans le domaine et ses qualités de professeur* (his scientific work in the field and his qualities as a teacher). History will also remember that Rouget was the first to discover chitosan (*chitine modifiée*, modified chitin), the main chitin derivative.

In the mid-1850s, many researchers were interested in identifying the presence of chitin in arthropods and attributed the formation of chitin to a process of secretion. Among them, the work of Frémy was recognized. In 1855, Frémy attributed to chitin and cellulose the same chemical composition and found this relationship *remarquable et intéressante* (remarkable and interesting), between two products elaborated, respectively, by groups of beings as different as arthropods and plants (Frémy, 1855). This assertion was shared by other authors (Peligot, 1858; Bütschli, 1874; Ambronn, 1890) who had demonstrated the presence of cellulose in the skin of silkworms and in certain mollusks. Until the 1910s, new experimental facts corroborated this assertion (Berlese, 1929; Bounoure, 1911, 1912a,b). Inspired by Frémy's work, Rouget became interested in the chemistry of *chitine* in order to determine its chemical nature and constitution. At the same time, other researchers, such as Berthelot and Städeler, were working on the same objective. Rouget, Berthelot, and Städeler were familiar not only with the work of Frémy but also with those of Lassaigne and Payen. These three researchers were surprised by the chemical affinities of *chitine* not only with cellulose but also with other substances of the polysaccharide group, such as starch (Berthelot, 1859; Städeler, 1859; Rouget, 1859). In 1859, Rouget found that boiling *chitine* in a concentrated potassium hydroxide solution under reflux rendered it soluble in dilute solutions of organic acids (Rouget, 1859). This new product, called *chitine modifiée*, also yielded a different color on treatment with an acidified iodine solution than did the original *chitine* (Fig. 2.15). Indeed, mixtures of iodine and zinc chloride yielded blue or violaceous colors with *chitine*. Using this method, Rouget proved that the endocuticula of arthropods contained chitin (similar to glycogen) and not cellulose. It was also easier to detect chitin in the endocuticula of insects than in the exocuticula. Later, Zander (1897) studied the chemical modification of *chitine* and detailed similar observations.

MÉMOIRES PRÉSENTÉS.

CHIMIE HYSTOLOGIQUE. — *Des substances amylacées dans les tissus des animaux, spécialement des Articulés (chitine)*; *par* **M. Ch. Rouget.** (Extrait par l'auteur.)

(Commissaires, MM. Pelouze, Milne Edwards, Rayer.)

« Ayant constaté et établi (1) que la substance amylacée signalée dans l'amnios ou le placenta des Mammifères n'est pas le produit d'organes particuliers, qu'elle n'est pas renfermée dans des *cellules glycogènes* spéciales, mais dans les cellules épithéliales mêmes de ces membranes, plus ou moins modifiées, j'ai été conduit à chercher cette substance dans d'autres épithéliums; et je l'ai trouvée, en effet, dans les cellules épidermiques de la peau, du voile du palais, de la langue, dans l'épithélium de l'estomac, dans toutes les cellules cylindriques du revêtement épithélial des villosités de l'intestin grêle et de la surface du gros intestin. J'ai aussi montré le premier que chez certaines espèces, chez les Cobayes en particulier, trois ou quatre jours au plus avant la naissance, tout l'épithélium de l'intestin est rempli de substance amylacée, bien qu'en même temps le foie, depuis longtemps déjà

Fig. 2.15 First page of the memory of Professor Rouget, presented at the *Académie des Sciences* Paris in 1859, where he described his *chitine modifiée. (Source: Rouget, Ch., 1859. Des substances amylacées dans les tissus des animaux, spécialement les articulés (chitine). C. R. Hebd. Seances Acad. Sci. 48, 792-795.)*

For more than 35 years, "modified chitin" has not been extensively studied, unlike another chitin product, namely, glucosamine. Indeed, it was not until the mid-1890s that Hoppe-Seyler again became interested in modified chitin. In 1894, he studied the chemistry of chitin and "discovered" a soluble product that he named chitosan. It seems that Hoppe-Seyler was not aware of Rouget's work. It is also interesting to note that chitosan was not discovered in Nature until 1954, when it was detected in the cell walls of the fungus *Phycomyces blakesleeanus* (Kreger, 1954). Later, in the 1970s, chitosan has been detected in the walls of other fungi and in certain green algae (Muzzarelli, 1977).

2.8 Georg Ledderhose

In the mid-1870s, the work of Ledderhose on the chemistry of chitin was acknowledged to have made an important contribution (Ledderhose, 1876, 1878, 1880a,b). Ledderhose was a young German student in medicine at the German Imperial University in Strasbourg, the capital city of the *Reichsland Alsace-Moselle* (part of the "booty of the victors" in the Franco-Prussian war; Mathews, 1898). Ledderhose (Fig. 2.16) was the first, in

Fig. 2.16 Georg Ledderhose, 1855–1925. *(Source: Public Domain).*

1875, to isolate the first amino sugar, namely, *glykosamin/glycosamin* (glucosamine) and to suggest that the repeating unit of chitin was *glykosamin*. The term glucosamine was, nevertheless, coined by Tiemann in 1884 (Tiemann, 1884). Interestingly, the second amino sugar to be discovered, namely, D-galactosamine (chondrosamine) was reported much later by Levene and La Forge in 1914, as a constituent of cartilage and tendon mucoproteins (Levene and La Forge, 1914, 1915). Georg Ledderhose was also a famous German surgeon who was the first to describe the condition of plantar fibromatosis in 1894, which later became known as Ledderhose disease (Ledderhose, 1894; Al Aboud, 2018).

Georg Ledderhose was born on 15 December, 1855 in Bockenheim, Wiesbaden (Germany). In 1870, his father, Karl Heinrich Ludwig Ledderhose (1821–1899) was appointed *Haut Fonctionnaire Allemand* (a senior German official) in Strasbourg, when Alsace-Lorraine (*Reichsland Elsass-Lothringen*) was annexed by Germany. Later, Karl Ledderhose was *Unterstaatssekretär im Ministerium für Elsass-Lothringen* (Under-Secretary of State at the Ministry of Alsace-Lorraine). Young Georg then accompanied his father to Strasbourg and enrolled at the University of Strasbourg (called *Kaiser Wilhelms-Universität Strasburg* from 1877). Georg Ledderhose made frequent travels back and forth between France and Germany, which later allowed him to teach in both languages. In 1872, the young Ledderhose entered the *Die medizinische Fakultät* (Faculty of Medicine) as a student of Georg Albert Lücke (1829–1894), a renowned German surgeon. One of the subjects Ledderhose was working on was the chemistry of *chitine*, in particular the identification of the products obtained by its hydrolysis with concentrated HCl. In 1875, Ledderhose discovered *glykosamin* (glucosamine), *unter der gütigen leitung des herrn Hoppe-Seyler* (with the kind collaboration of Hoppe-Seyler). Together, they worked on cartilage. Later, this was also the topic of his doctoral thesis presented in 1880 at the faculty of medicine (Fig. 2.17). The President of the jury was Hoppe-Seyler (Ledderhose, 1880a). Ledderhose then worked as a surgeon at the Strasbourg hospital

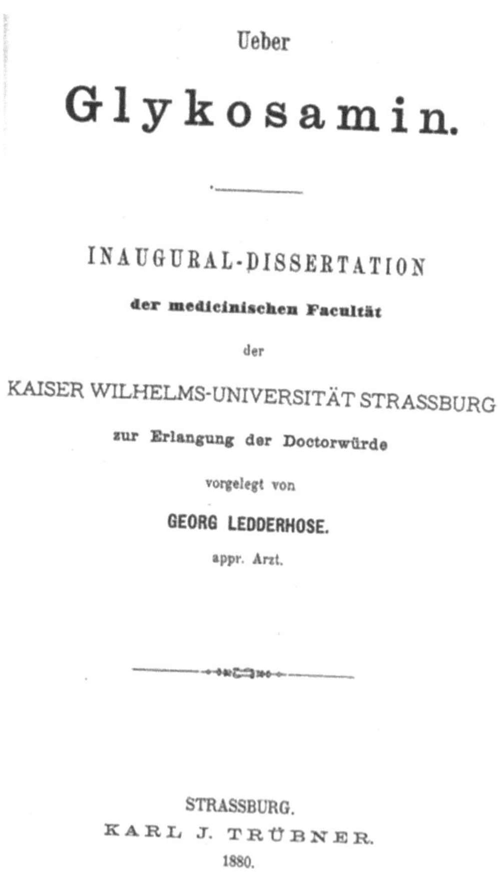
Ueber

Glykosamin.

INAUGURAL-DISSERTATION

der medicinischen Facultät

der

KAISER WILHELMS-UNIVERSITÄT STRASSBURG

zur Erlangung der Doctorwürde

vorgelegt von

GEORG LEDDERHOSE.

appr. Arzt.

STRASSBURG.

KARL J. TRÜBNER.

1880.

Fig. 2.17 First page of the dissertation presented by Georg Ledderhose in 1880. *(Source: Ledderhose, G., 1880. Ueber Glykosamin (Dissertation der medicinischen Facultät der Kaiser Wilhelms). Universität Strassburg zur Erlangung der Doctorwürde, 24 p.)*

and became professor extraordinaire of surgery in 1891. From 1892 to 1899, he was director of the newly founded *Centre de covalescence pour accidentés et de traumatology* (convalescent center for accident victims and traumatology) in Strasbourg. Ledderhose spent his entire career in this city until he was expelled in 1918. Professor Ledderhose then returned in Germany and worked in Munich, where he became Honorary Professor. Georg Ledderhose died on February 1, 1925 in Munich.

In 1875, Ledderhose was working during the summer semester in the laboratory of his uncle, Friedrich Wöhler (1800–1882), German chemist and professor at the University of Göttingen, known for his work in inorganic chemistry and considered a co-founder of organic chemistry. Wöhler was the first to isolate pure aluminum to discover the elements beryllium and yttrium and to synthetize urea. One day, Wöhler had lobsters for lunch and, bringing back the shell to the laboratory, he gave it to his nephew. He told him to "find out what this was" (Foster and Webber, 1961; Brimacombe and Webber, 1964). The next day, the young Ledderhose treated lobsters with hot concentrated hydrochloric with the aim to identify the structure of the products. He found that the claws and the shells dissolved in this solution and that on cooling the solution yielded characteristic crystals. In collaboration with Hoppe-Seyler, Ledderhose identified the crystalline compound as a new nitrogen-containing sugar, which he named *glykosamin/glycosamin*. In 1876, Ledderhose published his first results in the journal *Berichte der Deutschen Chemischen Gesellschaft* (Ledderhose, 1876) where he suggested *glykosamin* as the repeating unit of chitin, demonstrated the presence of nitrogen (6.49%) in its structure, and proposed the formula described in Scheme 2.1, in agreement with that of a hexosamine hydrochloride. The new crystalline product differed from glucose in having an amine group in place of one hydroxyl. Later, Ledderhose, in Hoppe-Seyler's laboratory, found that acetic acid was also a product of hydrolysis of arthropod chitin (Ledderhose, 1878, 1880a). He estimated the quantity of products and arrived at the chemical reaction described in Scheme 2.2, confirming the

$$COH.(CH.OH)_4.CH_2.NH_2 + HCl.$$

Scheme 2.1 *Glykosamin*: structural formula proposed by Ledderhose (1876).

$$\underset{\text{chitin}}{C_{45}H_{26}N_2O_{10}} + 3H_2O = \underset{\textit{glycosamin}}{2C_6H_{13}NO_5} + \underset{\text{acetic acid}}{3C_2H_4O_2}$$

Scheme 2.2 Decomposition of chitin according to Ledderhose (1878).

structural formula $C_6H_{13}NO_5$. Ledderhose also noticed, together with the acetic acid, small quantities of other volatile fatty acids, especially formic and butyric acids. All the results of Ledderhose, summarized in his doctorate presented in 1880 (Fig. 2.17), were in agreement with the previous observations reported by Schmidt (1845) and Städeler (1859). However, Ledderhose did not prove that glucosamine and acetic acid were produced in equimolar amounts. Indeed, the stoichiometry of reaction was only determined in 1912 by Brach and von Fürth (1912).

Two decades after Rouget, the young student Ledderhose observed that *chitine* comprised *glycosamin* (glucosamine) and acetic acid (Scheme 2.2). He was the first to identify these two compounds as structural units of chitin between 1875 and 1876. A few years later, the presence of glucosamine as the repeating unit of chitin was described by the works of Tiemann (Tiemann, 1884; Tiemann and Landolt, 1886), Winterstein (1893, 1894a, b, 1895a,b,c,d), and Gilson (1894a,b,c, 1895a,b). This new substance and its relationship with chitin has, indeed, strongly interested the scientific community. However, it took 70 years before Purchase and Braun (1946) elucidated the final chemical structure of chitin (Brimacombe and Webber, 1964). Following its discovery, glucosamine then deserved particular interest from the fact that it might be regarded as a link between the carbohydrates and α-hydroxy-amino acids. However, considerable controversy existed concerning its structure. The main points at issue were in the structure of its glycosides, absolute configuration, and in the decision between a mannose or a glucose arrangement of the groups (Scheme 2.3). Moreover, some authors insisted that the same compound in arthropod cuticle should be called chitosamine.

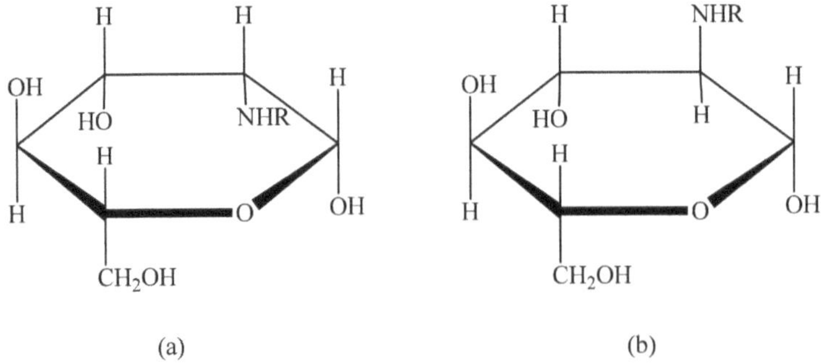

Scheme 2.3 Constitution of glucosamine as (A) a mannose derivative or (B) as a glucose derivative.

2.9 Ernst Sundwick

Not all researchers working on the determination of chitin agreed with the scheme of the decomposition of acid hydrolysis of chitin described in Scheme 2.2 and proposed by Ledderhose. Among them was Sundwick. Ernst Edvard Sundwik (1849–1918) was professor of physiological chemistry and pharmacology at the University of Helsinki.

In 1881, Sundwik published a paper entitled *Zur Constitution des Chitins* (on the constitution of chitin) (Fig. 2.18) where he remained unconvinced by Ledderhose's conclusions, in particular with the structural formula $C_6H_{13}NO_5$ (Sundwik, 1881). The original formula suggested by Ledderhose was based on the assumption that chitin was a glucosidic compound formed by the condensation of amino-glucose and acetic acid. This view was opposed by Sundwik, who recommended various complicated alternatives (Sundwik, 1881), indicating that the analytical data were better satisfied by the formula $C_{30}H_{50}O_{19}N_4$ or its multiples. Sundwik regarded chitin as a complex containing both glucose and amino-glucose residues, but questioned the presence of acetyl groups. He also considered all volatile

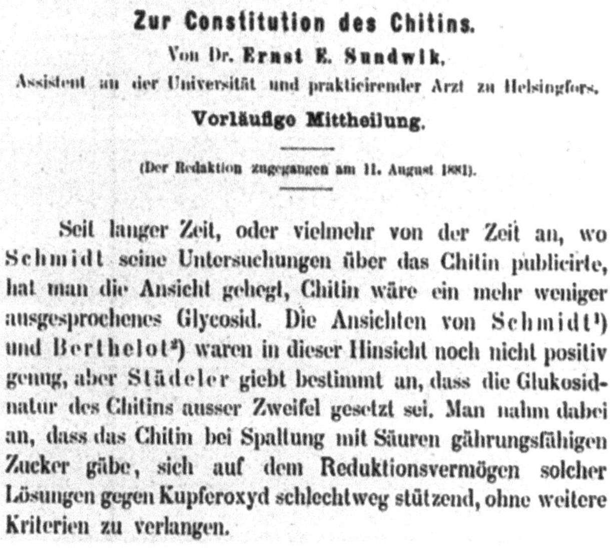

Zur Constitution des Chitins.

Von Dr. **Ernst E. Sundwik**,

Assistent an der Universität und prakticirender Arzt zu Helsingfors.

Vorläufige Mittheilung.

(Der Redaktion zugegangen am 11. August 1881).

Seit langer Zeit, oder vielmehr von der Zeit an, wo Schmidt seine Untersuchungen über das Chitin publicirte, hat man die Ansicht gehegt, Chitin wäre ein mehr weniger ausgesprochenes Glycosid. Die Ansichten von Schmidt[1]) und Berthelot[2]) waren in dieser Hinsicht noch nicht positiv genug, aber Städeler giebt bestimmt an, dass die Glukosidnatur des Chitins ausser Zweifel gesetzt sei. Man nahm dabei an, dass das Chitin bei Spaltung mit Säuren gährungsfähigen Zucker gäbe, sich auf dem Reduktionsvermögen solcher Lösungen gegen Kupferoxyd schlechtweg stützend, ohne weitere Kriterien zu verlangen.

Fig. 2.18 First page of the article of Professor Ernst Sundwik where he remained unconvinced by Ledderhose's conclusions. *(Source: Sundwik, E.E., 1881. Zur constitution des chitins. Z. Physiol. Chem. 5, 384–394.)*

acids, that is, acetic, butyric, and formic acids, formed during the hydrolysis as decomposition products. However, Sundwik's conclusions were contradicted by those published by Tiemann (1884) and especially Schmiedeberg (1891). The latter regarded chitin as an acetic combination of glucosamine, the decomposition of which took place according to Scheme 2.1. This was also the view held by Araki in 1895 after his discovery of a still high molecular weight decomposition product, chitosan, which was considered to be the first stage in the hydrolysis of chitin (Araki, 1895).

2.10 Ferdinand Tiemann

Tiemann published several important works on *Glucosamin, eine wichtige Einheit des Chitin* (Glucosamine as an important unit of chitin) (Tiemann, 1884, 1886; Tiemann and Landolt, 1886; Fischer and Tiemann, 1894). The most important of his discoveries was to show that glucosamine was a dextrorotatory amino sugar to which he recognized the constitution of a α-amino-glucose. However, this discovery, although also confirmed by Lobry de Bruyn in 1898 (Lobry de Bruyn and Franchimont, 1898), Fischer in 1902 (Fischer and Leuchs, 1902, 1903), and Irvine in 1911 (Irvine et al., 1911; Irvine and Hynd, 1912, 1914), was only definitively elucidated at the end of the 1930s with the work of Karrer (Karrer and Mayer, 1937), and above all that of Hatworth (Haworth et al., 1939). The term *chitosamin* (chitosamine) for glucosamine was also probably introduced by Tiemann for the first time. He was also the first to establish the relationship between D-glucosamine and D-glucose, which was not demonstrated until the late 1930s (Karrer and Mayer, 1937; Haworth et al., 1939).

Johann Karl Wilhelm Ferdinand Tiemann (1848–1899) was a German chemist, professor at the University of Berlin (Fig. 2.19). In 1866, Tiemann studied pharmacy at the *Technische Universität Carolo-Wilhelmina Braunschweig* (Carolo-Wilhelmina Technical University of Brunswick) where he graduated in 1869. In the same year, Tiemann accepted a position as chemical assistant at the *Universität Berlin* (University of Berlin) under the direction of August Wilhelm von Hofmann (1818–1892), a renowned organic chemist. Tiemann first worked with Karl Reimer (1845–1883), one of his pupils, on the conversion of phenol to salicylaldehyde. In 1874, with Wilhelm Haarmann (1847–1931; German chemist), he created a company, after he discovered the synthesis of vanillin from coniferyl alcohol. In 1882,

Fig. 2.19 Professor Ferdinand Tiemann, 1848–1899. *(Source: Witt, O.N., 1901. Ferdinand Tiemann. Ein Lebensbild. Ber. Dtsch. Chem. Ges. 4402–4455.)*

Tiemann became professor in chemistry at the University of Berlin. A complete biography can be consulted in the paper by Witt (1901).

As already mentioned, Ledderhose in 1876 reported that after the hydrolysis of chitin with concentrated HCl, a nitrogen-containing sugar could be isolated as a hydrochloride, which he named *glykosamin/glycosamin*. This substance of formula $C_6H_{13}NO_5$ had the particularity to have an amine group in position 6 of the sugar. In 1884, Tiemann published his first results on this sugar, which he called *glucosamin* (glucosamine) for the first time (Fig. 2.20). The chemical formula described in Scheme 2.4 was also suggested (Tiemann, 1884). In this study, *glucosamin* ($C_6H_{11}O_5(NH_2)$ by theory, $C_6H_{13}NO_5$, HCl by experiments) was a α-amino glucose. However, Tiemann attributed the name to Ledderhose (*Mit dem namen salzsaüre glucosamin von G. Ledderhose,* with the name acid salt of glucosamine from G. Ledderhose), stating that the latter had used the name in his original paper (Tiemann, 1884). George A.F. Roberts (Nottingham Trent University) later stated that this attribution was erroneous because Ledderhose referred to chitin monosaccharide as a *glucosamin/glycosamin* and did not attempt to assign a specific configuration to it (Roberts, 1992).

Between 1884 and 1886, Tiemann was interested in the chemistry of *glucosamin* with the aim of establishing its structure and configuration. Tiemann

69. Ferd. Tiemann: Einiges über den Abbau von salzsaurem Glucosamin.

[Aus dem Berl. Univ.-Laborat. No. DXXXXIII; vorgetragen vom Verfasser.]

Mit dem Namen »salzsaures Glucosamin« hat G. Ledderhose[1]) die schön krystallisirte, von ihm zuerst im chemisch reinen Zustande dargestellte Substanz bezeichnet, welche bei dem Kochen von Chitin mit concentrirter Salzsäure entseht. Der Kohlenstoffgehalt des Chitins, des bekannten, von Odier[2]) in den Flügeln der Käfer aufgefundenen Körpers, aus welchem die Hauptmasse der knochen- resp. hornähnlichen Organe der Coleopteren und Crustaceen besteht und der bezüglich der Widerstandsfähigkeit gegen chemische Agentien nur mit der Cellulose verglichen werden kann, erscheint zum weitaus grössten Theil (nach E. Sundwick[3]) bis zu 92 pCt.) in dem daraus gebildeten, salzsauren Glucosamin wieder; ausser diesem werden nur sehr kleine Mengen aliphatischer Säuren, wie Essigsäure und Buttersäure, bei dem Erhitzen von Chitin mit concentrirter Salzsäure gebildet. Da die genannten organischen Säuren bei lang andauernder Einwirkung starker chemischer Agentien aus den verschiedenartigsten Kohlenhydraten entstehen, erscheint es durchaus ungewiss, ob man sie als Spaltungsproducte des Chitins in eine Linie mit dem salzsauren Glucosamin stellen darf, oder ob dieselben nicht vielmehr Producte einer secundären Zersetzung des letzteren sind. Möglich ist es immerhin, dass das Glucosamin zu dem Chitin in einer ähnlichen Beziehung wie Glucose zu Cellulose steht. Die Resultate der Elementaranalysen, welche man bislang mit dem Chitin angestellt hat, können keinenfalls gegen eine derartige Auffassung geltend gemacht werden, da das Chitin eine amorphe, schwer von Beimengungen völlig zu reinigende Substanz ist, welche mithin keinerlei Garantien für chemische Reinheit bietet. Da als Umwandlungsproduct des Chitins bis jetzt nur das salzsaure Glucosamin und von diesem wiederum ausser seinen physikalischen Eigenschaften nur einige Producte einer weit fortgeschrittenen Zersetzung (Essigsäure, Buttersäure, Milchsäure, Ammoniak u. s. w.) mit Sicherheit bekannt sind, so müssen nach meinem Dafürhalten alle Speculationen, welche von verschiedenen Seiten über die Formel, resp. Molekulargrösse des Chitins gemacht worden sind, von einem rein chemischen Standpunkte aus betrachtet, zur Zeit noch als absolut gegenstandslose bezeichnet werden.

[1]) Diese Berichte IX, 1200.
[2]) Berzelius, Jahresb. 4, 247.
[3]) Hoppe-Seyler, Zeitschrift für physiologische Chemie 1881, V, 389.

Fig. 2.20 First page of the article of Professor Tiemann published in 1884 in the journal Berichte where he discussed the name *glucosamin. (Source: Tiemann, F., 1884. Einiges über den abbau von salzsaurem glucosamin. Ber. Dtsch. Chem. Ges. 17, 241–251.)*

$$CH^2.OH — CH.OH — CH.OH — CH.OH — CH.NH^2 — C.HO$$

Scheme 2.4 Chemical formula proposed by Tiemann (1884) for *glucosamin/glykosamin.*

knew the chemistry of phenylhydrazine and their action on sugars discovered in 1884 by his compatriot Emil Fischer (Fischer, 1884). In 1886, Tiemann observed that the reaction of *glucosamin* with phenylhydrazine yielded glucosazone (an osazone) with the elimination of the amino group, suggesting that the sugar was related in configuration to glucose or mannose (Tiemann, 1886; Tiemann and Landolt, 1886). Tiemann found it to be identical to that previously obtained by Fischer from dextrose and laevulose. This proved that the configurations at carbons 3, 4, 5, and 6 were identical to those of D-glucose and that the amine group was attached to carbon 2. Tiemann's conclusions were later confirmed by the synthesis of glucosamine from D-arabinose by Fischer and Leuchs (1903). Their synthesis clearly established its constitution as being that of a 2-amino-hexose but left open the question of the configuration at the carbon-2 position. Indeed, the two possibilities for the structure of glucosamine were 2-amino-2-deoxy-D-glucose or 2-amino-2-deoxy-D-mannose (Scheme 2.3). In the first case, the amino group was below the plane of the pyranose ring (as in Haworth's later formulation), and in the second, it was above. In 1894, in collaboration with Fischer, Tiemann published a paper entitled *Ueber glucosamin* (Fischer and Tiemann, 1894). In this work, 2,5-anhydro hexose was synthetized and named by them chitose. This compound was prepared by the deamination of the naturally occurring *chitosamin* (chitosamine; D-glucosamine, $C_6H_{13}NO_5$). Chitose was characterized by its oxidation derivatives, chitonic acid and isosaccharic acid. It was readily oxidized in two stages to yield an anhydro hexonic acid (chitonic acid) and an anhydro saccharic acid (isosaccharic acid), thus offering a parallel to the oxidation of D-glucose to D-gluconic acid and D-glucosaccharic acid. The problem of determining the configuration of chitose was dependent on a knowledge of the configuration of *chitosamin* from which it was derived (Fischer and Tiemann, 1894). In 1939, it was conclusively proven that the amino sugar of chitin was in fact 2-amino-D-glucose. It then followed that chitose was either 2,5-anhydro-D-glucose or 2,5-anhydro-D-mannose, the first if deamination was not attended by inversion and the second if inversion occurred with the removal of the amino group. Later, the term chitose referred to the anhydro sugar formed by the deamination of glucosamine. Considered an important chitin derivative, it was prepared and studied in detail by Armbrecht (1919) and Schorigin and Makarowa-Semljanskaja (1935a).

2.11 Oswald Schmiedeberg

Ten years later, Schmiedeberg, repeating the experiments of Ledderhose, demonstrated that chitin comprised *glycosamin/glykosamin/chitosamin* and acetic acid (Schmiedeberg, 1891). The formula $C_6H_{13}NO_5$ was confirmed. In agreement with previous conclusions published by Schmidt (1845), Städeler (1859), and Ledderhose (1876, 1878, 1880a,b), Schmiedeberg also proposed the reaction described in

$$C^{18}H^{30}N^{2}O^{12} + 4H^{2}O = 2C^{6}H^{13}NO^{5} + 3CH^{3}COOH$$

chitin — *glycosamin* — acetic acid

Scheme 2.5 Hydrolysis of chitin in order to form *glycosamin* or *chitosamin* according to Schmiedeberg (1891).

Scheme 2.5 (Schmiedeberg, 1891). The terms *glycosamin* and *chitosamin* indicated the same compound. Schmiedeberg also found *glycosamin* among the decomposition products of chondrin, the chief constituent of cartilage. He described a chemical constitution of chitin, similar to that proposed by Tiemann. Schmiedeberg also studied the chemistry of chitin and its behavior in the presence of solvents, and the results have been taken up by several other researchers such as Hoppe-Seyler, Winterstein, and Gilson.

Oswald Schmiedeberg was a Baltic German pharmacologist (Fig. 2.21), considered the "Father of modern pharmacology" (Meyer, 1922; Koch-Weser and Schechter, 1978; Muscholl, 1995, 2001). Schmiedeberg was born on October 11, 1838, in Kurland/Courland, a Baltic province of Russia (now Latvia). He studied medicine at Dorpat University (now Tartu, Estonia). In 1866, Schmiedeberg completed his doctoral thesis in this university under Rudolf Buchheim (1820–1879), a renowned German pharmacologist, at the Institute of Experimental Pharmacology. His dissertation topic was the determination and fate of chloroform in blood (Muscholl, 1995). Schmiedeberg then remained at Dorpat where he was appointed as a lecturer in pharmacology in 1868. When Professor Buchheim

Fig. 2.21 Oswald Schmiedeberg, 1838–1921. *(Source: Public Domain).*

left Dorpat in 1869, Schmiedeberg became his successor. He worked with the renowned German physiologist and physician Karl Friedrich Wilhelm Ludwig (1816–1895) at the University of Leipzig during 1 year. When in 1872 the newly established German Imperial University of Strasbourg needed a professor of Pharmacology, Schmiedeberg was chosen on the recommendation of Professor Ludwig. Schmiedeberg held this position in Strasbourg until 1918, when Alsace-Lorraine was reconquered by France. Although the value of the science of pharmacology has not yet been proven for the medical profession, Schmiedeberg began to build a world-renowned pharmacological institute, where he hosted about 150 pharmacology students from 20 countries. He also published more than 200 publications in his 46-year career. Oswald Schmiedeberg died on July 12, 1921 in Baden-Baden. A detailed biography can be found in the references Koch-Weser and Schechter (1978) and Muscholl (1995).

2.12 Felix Hoppe-Seyler

Felix Hoppe-Seyler (1825–1895) was a German physiologist and chemist (Fig. 2.22). Ernst Felix Immanuel Hoppe, his name at birth, was also a pioneer of biochemistry and molecular biology. Professor Hoppe-Seyler

Fig. 2.22 Professor Felix Hoppe-Seyler, 1825–1895. *(Source: Baumann, E., Kossel, A., 1895. Felix Hoppe-Seyler. Ber. Dtsch. Chem. Ges. 28, 1147–1192.)*

occupied the first chair of physiological chemistry in Germany and directed the first institute devoted to this subject (Baumann and Kossel, 1895a; Gamgee, 1895; Mathews, 1898; Noyer-Weidner and Schaffner, 1995).

Ernst Felix Immanuel Hoppe was born on 26 December, 1825 in Freiburg on the Umstrut (Saxony). He was the tenth child of Pastor Ernst Hoppe and Frederike Nitzsch. He comes from a long line of pastors, teachers, and ministers. His mother died when he was 6 years old and his father 3 years later. Orphaned before the age of 10, Felix Hoppe was adopted by the clergyman Dr. Seyler, husband of his older sister, and in 1864, he added the name of his benefactor to his original name of Hoppe. However, Felix was soon sent to an orphan asylum in Halle. It was in this institute that he learned about chemistry with the "help of a simple, rigid, and elderly pharmacist" (Mathews, 1898). During this period, Felix was also an assiduous student of natural sciences and mathematics. In 1846, Felix Hoppe enrolled as a medical student at the University of Halle, where he began chemical laboratory work. A year later, he worked in the laboratory of Ernst Heinrich Weber (1795–1878) in Leipzig and studied chemistry, medicine, and physiology with Otto Linné Erdmann (1804–1869; German chemist) and with Karl Gotthelf Lehmann (1812–1863; German physiological chemist). Hoppe then completed his medical studies in Berlin and worked on the chemical and histological aspects of cartilage structure (Baumann and Kossel, 1895a,b; Mathews, 1898; Fruton, 1990). In 1850, Hoppe obtained his doctoral dissertation entitled *De cartilaginum structura et chondrino nonnulla* (On the structure of cartilage and on chondroin), which he dedicated to Professors Weber and Lehmann. In 1851, Hoppe was licensed as a practical physician and spent some time in Prague studying obstetrics. The same year, he returned to Berlin and began to practice. However, Hoppe preferred scientific research to medicine and held research posts in Germany and France (Mathews, 1898). Indeed, he was not satisfied with his medical role. Three years later, he applied for a research position at the University of Greifswald and was appointed prosecutor in anatomy. He later became *Privatdozent* (Privat-docent). In 1855, Hoppe accepted a position as director of the chemical laboratory at the newly founded Charité Institute of Pathology in Berlin. In 1861, he was called to the chair of applied chemistry in the medical faculty of Tübingen, where he was shortly made Full Professor. From 1861 and 1872, Professor Hoppe taught physiological chemistry in this faculty. His research laboratory, located in the old kitchen of an old castle on top of a hill, quickly became the center of physiological chemistry in the World (Noyer-Weidner and Schaffner, 1995). In 1872, after the conclusion of the Franco-German war, Hoppe-Seyler also became Full Professor in physiological chemistry and hygiene in the Medical Faculty of the German University (*Kaiser Wilhelms-Universität*) in Strasbourg. Professor Hoppe-Seyler was appointed rector of the University just 1 year later. At that time, as part of the gigantic works undertaken by the Germans in Strasbourg, remarkable university buildings were constructed. Strasbourg became one of the main research centers in Europe, and because of its prestige, the *Kaiser Wilhelms-Universität* attracted many foreign students, including many Japanese (e.g., Torasaburo Araki). Hoppe-Seyler spent the rest of his life in this town, but he continued to work with the institutes in Berlin and Tübingen. Felix Hoppe-Seyler died suddenly of heart disease in his summer residence on Lake Constance on August 10, 1895 (Baumann and Kossel, 1895a; Gamgee, 1895; Mathews, 1898; Thierfelder, 1926; Fruton, 1990; Noyer-Weidner and Schaffner, 1995; Brock, 2013).

Hoppe-Seyler was the founder of the *Zeitschrift für Physiologische Chemie* in June 1877, which today is known as Biological Chemistry (renamed in 1996), devoted exclusively to biochemistry. Hoppe-Seyler argued that biochemistry should be an independent science. However, the term "biochemistry" only became popular in the early twentieth century. The new journal was also an important contribution to the recognition of physiological chemistry as an independent academic discipline (Baumann and Kossel, 1895a,b; Fruton, 1990). In 1881, Hoppe-Seyler summarized all that was then known of biochemistry in the 1000 pages of *Physiologische Chemie* (Hoppe-Seyler, 1881; Brock, 2013). Among his students and collaborators were Friedrich Miescher (1844–1895), Eugen Baumann (1846–1896), Ernst E. Sundwik, Georg Ledderhose, Torasaburo Araki, Albrecht Kossel (1857–1927), and Eugène Gilson (1862–1908). The most extensive biography of Professor Hoppe-Seyler is the obituary notice by Baumann and Kossel, which includes a complete list of his personal publications (Baumann and Kossel, 1895a,b). The following references can also be consulted: Gamgee (1895), Thierfelder (1926), Fruton (1990) and Brock (2013).

In 1894, Hoppe-Seyler treated the shells of crabs, scorpions, and spiders with potassium hydroxide at 180°C and found a "new" product (Hoppe-Seyler, 1894, 1895). Hoppe-Seyler, who did not refer to the *chitine modifiée* of Rouget, gave it the name of chitosan on January 14, 1895 with the chemical formula $C_{14}H_{26}N_2O_{10}$ (Fig. 2.23). Hoppe-Seyler wrote: *La chitine des arthropodes, traitée par la potasse à 180°C, puis lavée à l'eau, devient soluble dans l'acide acétique étendu, et la solution donne avec la potasse un volumineux précipité. Ce corps, nommé chitosane, présente des propriétés basiques*

610. F. Hoppe-Seyler. Ueber Chitin und Cellulose.

(Eingegangen am 4. December; mitgeth. in der Sitzung von Hrn. A. Pinner.)

In diesen Berichten 27, Heft 17, S. 3113—3115 ist von E. Winterstein soeben eine Mittheilung, betitelt »Ueber ein stickstoffhaltiges Spaltungsproduct der Pilzcellulose« veröffentlicht, die mir besonders durch ihren letzten Satz Veranlassung zu folgenden Bemerkungen giebt:

Unter Benutzung einer vor mehreren Jahren beschriebenen Isolirungsmethode, welche von Winterstein auch in der citirten Mittheilung erwähnt ist, habe ich mich seit längerer Zeit mit den der Cellulose verwandten Kohlehydraten von Thieren und Pflanzen beschäftigt und einige Resultate erhalten, deren kurze vorläufige Schilderung wohl jetzt zweckmässig sein wird, während die ausführlichen Mittheilungen erst später gegeben werden können.

Das Tunicin der Tunicaten hat sich auch bei der Behandlung mit Aetzkali bis 180° als übereinstimmend mit gewöhnlicher Cellulose erwiesen, dagegen hat das Chitin der Gliederthiere (untersucht wurden Panzer von Insecten, Krebsen, Scorpionen, Spinnen) eine recht merkwürdige Abweichung gezeigt. Während bei dem Erhitzen mit Aetzkali und ein wenig Wasser im Oelbade bis 180° (als Maximum) die Formen der Chitingewebe so wenig wie die Zellengewebe der Pflanzen bis hinab zu den Algen eine wesentliche Aenderung erkennen lassen, wird das Chitingewebe nach dieser Behandlung und sorgfältigem Auswaschen des Aetzkalis mit kaltem Wasser leicht löslich in verdünnter Essigsäure zur klaren Flüssigkeit gefunden und durch Alkalilauge wird aus dieser Lösung ein reichlicher voluminöser Niederschlag gefällt. Der Stickstoffgehalt des Chitins ist bei dieser Behandlung unverändert geblieben, aber im aufgelösten Aetzkali, mit dem das Chitin erhitzt war, fand sich Essigsäure und zwar so rein, dass nach Uebersättigen mit Schwefelsäure aus dem Destillate sofort das Baryumsalz und aus diesem das Silbersalz von berechneter Zusammensetzung erhalten wurden. Das Umwandlungsproduct des Chitins, welches neben Essigsäure entsteht, sich in Essigsäure, auch in äusserst verdünnter Salzsäure sehr leicht löst, dem ich den vorläufigen Namen Chitosan gegeben habe, zeigt insofern basische Eigenschaften, als es sich mit den Säuren leicht verbindet, beim Verdunsten der wässrigen Lösung der salzsauren Verbindung diese in quadratischen Krystallen liefert, die

Berichte d. D. chem. Gesellschaft. Jahrg. XXVII. 214

Fig. 2.23 First page of the article of Professor Hoppe-Seyler published in 1894 in the journal Berichte where he introduced the name chitosan. *(Source: Hoppe-Seyler, F., 1894. Ueber chitin und cellulose. Ber. Dtsch. Chem. Ges. 27, 3329–3331.)*

et fournit des sels cristallisables, solubles dans l'eau (Arthropod chitin, treated with potassium hydroxide at 180°C and then water-washed, becomes soluble in diluted acetic acid, and the solution yields a bulky precipitate upon addition of potassium hydroxide. This substance, named chitosan, has basic properties and provides crystallizable salts that are soluble in water). Later, von Fürth and Russo (1906) showed that chitosan formed crystallizable salts with acids, behaving as an alkaloid. Löwy (1909) demonstrated that chitosan sulfate was a stable, insoluble, and crystalline chitosan salt. Chitosan sulfate had exactly the same tetragonal form as chitosan hydrochloride (von Fürth and Russo, 1906; Löwy, 1909; Brunswick, 1921; Campbell, 1929). These results were in agreement with those of Hoppe-Seyler's results.

Hoppe-Seyler has noted various observations (Hoppe-Seyler, 1894): (i) chitosan was readily soluble in dilute acetic acid, in agreement with the observation by Rouget (1859), and in hydrochloric acid solution, (ii) it could be precipitated from such solutions by addition of alkali; and (iii) it began decomposing at temperature 184°C and was stated, surprisingly, to have the same nitrogen content as the original chitin. One year later, his student Araki (1895) and later Löwy (1909) also reported similar conclusions. Using Schmiedeberg's conclusions on the constitution of chitin (Schmiedeberg, 1891), Hoppe-Seyler clearly demonstrated the relationship between chitin and chitosan and the dependence of the reactions on temperature and alkali concentration (Fig. 2.24). When chitosan is treated with concentrated hydrochloric acid it, like chitin, yielded glucosamine (Hoppe-Seyler, 1894, 1895). If heated with acetic anhydride, it yielded a substance resembling chitin, which, when heated with potassium hydroxide at 180°C, was resolved into chitosan and acetic acid. The higher the temperature and alkali concentration, the faster and more complete the results can be obtained. Hoppe-Seyler showed that the highest concentration of alkali that could theoretically be used is that at which potassium hydroxide crystallizes at 180°C. He indicated that cellulose, when fused with potassium hydroxide at 180°C remained intact, but at 200°C, it decomposed. Chitosan and propionic anhydride also yielded a chitin-like substance. However, Hoppe-Seyler did not indicate if the chitin contained *N*-acetyl or *O*-acetyl groups. This was first studied and suggested by Araki (1895) and resolved by Fränkel and Kelly (1901a). Following the work and conclusions of Hoppe-Seyler, the characteristics of chitosan (its solubility in dilute acids, its color

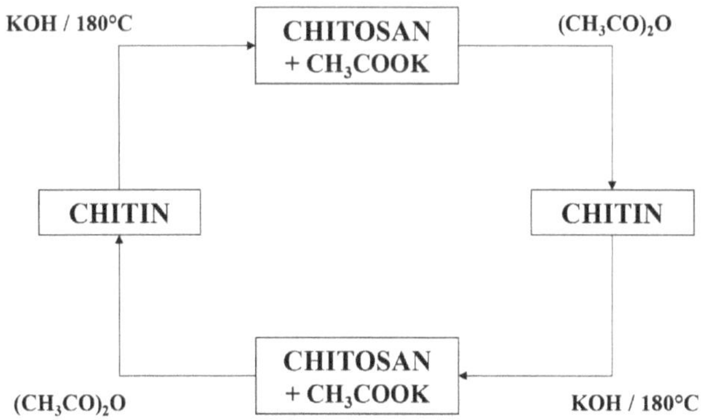

Fig. 2.24 Interconversion of chitin and chitosan proposed by Professor Hoppe-Seyler. *(Source: Hoppe-Seyler, F., 1894. Ueber chitin und cellulose. Ber. Dtsch. Chem. Ges. 27, 3329–3331.)*

reaction to iodine in acidic solution, and its ability to form salts) were used to develop experimental protocols as microchemical tests for chitin.

At the same time, the product described by Hoppe-Seyler as a partially "deacetylated," acid-soluble derivative of chitin was also prepared from fungal material by both Professors Winterstein (Winterstein, 1893, 1894a,b, 1895a,b,c,d), Gilson (Gilson, 1894a,b,c), Araki (Araki, 1895), and Lobry de Bruyn (Lobry de Bruyn and Franchimont, 1898; Lobry de Bruyn and van Ekenstein, 1899a,b). There were heated debates between these researchers, with Hoppe-Seyler claiming priority over discoveries about chitin and chitosan. For example, referring to Winterstein's work, Hoppe-Seyler often stated in his publications that he has been *engagé depuis quelques années dans une étude des hydrates de carbone de type cellulose et d'autres substances provenant de plantes et d'animaux, contrairement à d'autres chercheurs* (engaged for some years in a study of cellulose-type carbohydrates and other substances from plants and animals, unlike other researchers) (Hoppe-Seyler, 1881, 1894).

CHAPTER THREE

A period of confusion and controversy: 1894–1930

À partir des champignons Claviceps purpurea et Agaricus campestris, on obtient un produit insoluble dans la liqueur de Schweizer et ne se colourant en bleu ni par le chlorure de zinc iodé, ni par SO^4H^2 + I. Fondu avec 4 parties de KOH et chauffé à 180-190°C, ce corps, que je nomme mycosine, se modifie. Chose curieuse, il se dissout dans HCl dilué et se précipite lorsqu'on ajoute à cette solution de l'acide chlorhydrique concentré.

From the fungi Claviceps purpurea *and* Agaricus campestris, *a product is obtained which is insoluble in Schweizer's liquor and is not coloured blue either by iodinated zinc chloride or by SO^4H^2 + I. Melted with 4 parts of KOH and heated to 180-190°C, this substance, which I call* mycosine, *changes. Curiously, it dissolves in dilute HCl and precipitates when concentrated hydrochloric acid is added to this solution.*

Professor Eugène Gilson, 1894, Professor of Chemistry and Physiology, University of Gand, Belgium

From 1894 to 1930, chitin entered a period of confusion and controversy. The names that most marked this period were Ernst Winterstein, Eugène Gilson, Cornelis van Wisselingh, Trasaburo Araki, Cornelis A. Lobry de Bruyn, Sigmund Fränkel, Emil Fischer, James C. Irvine, Paul Karrer, and Albert Hofmann.

In 1838, Anselme Payen isolated a "new" substance from plant matter and coined the name cellulose. However, at that time, the structure of cellulose had not yet been well-established, nor had those of starch and dextrins (Haworth, 1946; Bell, 1949). The structure of cellulose was only determined by Hermann Staudinger (1881–1965) in 1920. In addition, it was not clearly established that starch was a macromolecule comprising several units of glucopyranose, although, in the four decades from 1880s to 1920s, both Emil Fischer (1852–1919), James C. Irvine (1877–1952) and Walter N. Haworth (1883–1950) have contributed significantly to our current understanding of the structure of polysaccharides and oligosaccharides, including glucose and maltose (Seetharaman and Bertoft, 2012). Similarly, the structure of chitin was controversial.

https://doi.org/10.1016/B978-0-323-96119-6.00003-0

In the period 1894–1930, research on chitin was not only focused on its structure but also and above all on the study of its occurrence in living organisms, its determination, and its chemistry. These works were even marred by frequently contradictory results and hot debate between the numerous and different laboratories, mainly due to confusion arising from the terminology of the different polysaccharides, that is, cellulose, chitin, starch, and dextrins and compounds such as *glycosamin*, *glykosamin*, glucosamine, *chitosamin*, chitosamine, *mycetosamine*, chitose, chitobiose, and chitosan studied during the 19th Century; the lack of a systematic nomenclature; and also the lack of certainty concerning their structure (Haworth, 1946; Bell, 1949). The confusion was certainly due to Payen's erroneous statement in 1840 that the cell wall of fungi was cellulose having entered the texts and "become dogma" as commented by Tracey (1957), although several observations had previously indicated that chitin was insoluble in Schweitzer's reagent, while cellulose was soluble. This was apparently not an obstacle to belief in fungal cellulose. The recognition of fungal chitin was then delayed. There was also some confusion between cellulose and chitin owing to Odier's conclusion that the structural material in the exoskeletons of insects and in plants was similar (Tracey, 1957). Some researchers considered cellulose and chitin as two similar substances. Others also believed that chitin was a protein chemically related to vertebrate cartilage and mucin.

3.1 Nomenclature

Chitin was first named *fongine* by Braconnot (1811a) and then *chitine* by Odier (1823). However, other names have been proposed but have not gained acceptance and/or led to extensive debate. These include elytrine by Children (1824), *entomeiline* and later *entomaderme* by Lassaigne (1843a), *metacellulose* by Frémy (1855), *entomeiline* by Packard (1886, 1898), *pilzcellulose* or fungus-cellulose (fungo-cellulose, fungal cellulose) by de Bary (1887) and Winterstein (1893), pupine by Griffiths (1892a,b), and later mycetin by Ilkewitsch (1908) and Dous and Ziegenspeck (1926). The American entomologists frequently used the term chitinous or chitinized in morphological or taxonomic descriptions (Imms, 1924; Ferris and Chamberlin, 1928; Campbell, 1929).

The term *chitine* (spelled chitin in German and English) first proposed by Odier was introduced in 1823 in the Natural History Dictionary and in 1827 in the Medical Natural History Dictionary (Fig. 3.1). This term is derived from Greek word chiton/kiton ("χιτών"), meaning a covering, tunic, or

DICTIONNAIRE

DES DROGUES

SIMPLES ET COMPOSÉES,

OU

DICTIONNAIRE

D'HISTOIRE NATURELLE MÉDICALE,

DE PHARMACOLOGIE

ET DE CHIMIE PHARMACEUTIQUE.

CHITINE. M. Auguste Odier (2) a nommé ainsi la substance solide des élytres et autres parties cornées des insectes. Elle forme à peu près le quart de ces enveloppes, que l'on avait considérées comme analogues à la matière cornée des animaux vertébrés. On l'obtient en traitant à chaud les élytres par la potasse, dans laquelle elle est insoluble, ce qui la distingue au premier abord de la corne, des cheveux et de l'épiderme. Elle est soluble dans l'acide sulfurique à chaud; elle ne jaunit point dans l'acide nitrique; elle brûle sans boursoufflement, c'est-à-dire en laissant un charbon qui conserve la forme de l'organe; enfin, elle ne contient point d'azote. Par ce dernier caractère, elle se rapproche des matières végétales, et l'auteur la compare au ligneux. M. Auguste Odier a retrouvé la chitine dans la carapace des Crustacés. (A. C.)

Fig. 3.1 Introduction of the word *"chitine"* in the Medical Natural History Dictionary. *(Source: Chevallier, A., Richard, A., 1827. Dictionnaire des Drogues Simples et Composées. Bécher Jeune, Libraire de l'Académie Royale de Médecine, Bruxelles, vol. 2, p. 64.)*

envelope (Odier, 1823; de Saint-Vincent, 1825; Chevallier and Richard, 1827; Straus-Durckeim, 1828; Drapiez, 1839; Payen, 1843), and later "a coat of mail" (Whistler and Smart, 1953). On May 15, 1843, before the members of the French *Académie des Sciences*, Lassaigne claimed that the term *entomaderme* was more appropriate that the term *chitine* (Fig. 3.2). In 1855, Frémy stated that there were three varieties of cellulose in the tissues of vegetables, one of which, called *metacellulose* (*fongine*), could be identified by its insolubility in Schweitzer's reagent (Frémy, 1855; Bounoure, 1919; Hopkins, 1929). Later, A. de Bary also reported that *his fungo-cellulose* differed from true cellulose in that it was insoluble in Schweitzer's reagent and did not give the color with iodine, which was characteristic of cellulose (de Bary, 1887). In 1891, Franz Ferdinand Schulze (1815–1873) proposed to

COMPTE RENDU

DES SÉANCES

DE L'ACADÉMIE DES SCIENCES.

SÉANCE DU LUNDI 15 MAI 1843.

PRÉSIDENCE DE M. DUMAS.

MÉMOIRES ET COMMUNICATIONS

DES MEMBRES ET DES CORRESPONDANTS DE L'ACADÉMIE.

» **La présence de cette matière organique dans la peau des larves, et l'enveloppe tégumentaire des insectes de différents ordres, nous a engagé, d'après l'étude que nous en avons commencée, à la distinguer par le nom d'*entomaderme*, ce nom nous paraissant plus convenable pour désigner cette matière que celui de *chitine*, donné d'abord par M. Odier.**

Fig. 3.2 Extract of the reports of the French *Académie des Sciences* of May 1843, where Professor Lassaigne stated that the term "*entomaderme* was more appropriate that the term *chitine*."

use the name cellulose only for the following case: "the constituent of the cell wall was not dissolved in dilute mineral acid or dilute alkali" (cited in the references: Bounoure, 1919; Hopkins, 1929). In the mid-1890s, Ernst Winterstein highlighted the "fungal cellulose" (*pilzcellulose*) which permitted to obtain glucosamine (Winterstein, 1893, 1895c,d). However, Winterstein concluded that it was difficult to establish that fungal cellulose and chitin were identical. In 1927, Gwynne-Vaughan and Barnes also made the statement that the cell wall of fungi was usually of cellulose or of a special variety known as fungal cellulose (Gwynne-Vaughan and Barnes, 1927). At the same time as Winterstein, Zander (1897) reported that chitin was similar to glycogen because in both cases, the reaction with iodine in potassium iodide solution colored the substance reddish brown. Later, in 1909, the Italian entomologist Antonio Berlese (1863–1927) wrote *La chitina non rappresenti altro che una cellulose azotata* (Chitin was a cellulose containing nitrogen) (Berlese, 1929).

From the mid-1900s, different structures of chitin began to be proposed (Bounoure, 1919). For Otto von Fürth and Michele Russo, chitin was a polymerized mono-acetylglucosamine (von Fürth and Russo, 1906) and for Th.R. Offer, a polymerized acetyl-di-glucosamine (Offer, 1907), although initially he thought that chitin was a mono-acetyl-diglucosamine (Bounoure, 1919). James C. Irvine, studying the identity of the chitins derived from various invertebrate animal structures (Irvine, 1909; Irvine and Hynd, 1912), considered chitin to contain acetylamino-glucose and amino-glucose residues in the proportion of three to one, in agreement with the formula $(C_{30}H_{50}O_{19}N_4)_n$. For Sergius Morgulis, chitin, whose empirical formula was $C_{64}H_{120}N_8O_{48}$, must be a polymerized acetylglucosamine (Morgulis, 1916), but he doubted that the acetyl group was a constituent of the chitin substance. Morgulis, indeed, considered acetic acid, obtained after acid or alkaline hydrolysis, a secondary decomposition product. Brach (1912) gave $(C_{32}H_{54}N_4O_{21})_x$ as the formula, later also accepted by Karrer and Smirnoff (1922) and Gonell (1926). In 1924, Augustus Daniel Imms (1880–1949), professor of entomology at Cambridge, wrote: "the chitin is related chemically to the cartilage of vertebrates and also to mucin," considering that chitin was a protein (Imms, 1924).

The confusion and controversy of the nomenclature, composition, and structure of chitin have then continued. Nevertheless, the idea crept into the literature as far back the nineteenth century that (fungal) chitin was a polymer of acetylglucosamine residues and also that the amino sugar was believed to have D-glucose configuration (Bounoure, 1919; Richards, 1951).

3.2 Ernst Winterstein

Ernst Winterstein (1865–1949), born on June 17, 1865 in Ernstthal (Germany), was a Swiss chemist and physiologist (Fig. 3.3). Between 1887 and 1890, Winterstein studied chemistry at the *Eidgenössische Technische Hochschule Zürich* (Polytechnic School of Zurich). In 1892, he obtained his doctorate at the University of Zurich. After a position as *Privatdozent* in 1894 and as a chemistry assistant in 1902, Winterstein was appointed in 1912 professor of General Chemistry and Physiology at Polytechnic School until 1935. Ernst Winterstein died on July 4, 1949 in Zürich.

In August 1893, Winterstein, who removed fats and proteins from fungus (*Boletus edulis, Agaricus campestris, Morchella esculenta, Botrytis cinerea,* and *Polyporus officinalis*), found that the residue was insoluble in Schweitzer's reagent (Winterstein, 1893). Winterstein concluded that it was a cellulose differing from that in tissues of higher plants and named it *pilzcellulose* (fungus cellulose). Araki (1895) agreed with this conclusion. Winterstein wrote that *Un produit de décomposition azoté de la cellulose de champignons et de l'acide acétique se retrouvent parmi les produits obtenus à partir de l'hydrolyse acide de la chitine*

Fig. 3.3 Professor Ernst Winterstein, 1865–1949. *(Source: Public Domain.)*

fongique (a nitrogenous decomposition product of fungal cellulose and acetic acid are among the products obtained from the acid hydrolysis of fungal chitin) (Winterstein, 1893). This statement confirmed the observations previously reported by Ledderhose and Schmiedeberg. Winterstein reported similar reactions to those described in Scheme 2.5 (Chapter 2). On the November 9 of the next year, Gilson, in France, reported the presence of *chitine* in fungi (*Claviceps purpurea, Agaricus campestris*) and studied its chemistry and its conversion to *mycosine* (mycosin or chitosan). These results were published 1 year later (Gilson, 1894a). Gilson, a student of Professor Hoppe-Seyler, also noted the presence of *une nouvelle substance azotée* (a new nitrogenous substance), whose formula $C_6H_{13}NO_5$ corresponded to glucosamine. Six days later, Winterstein presented another work (also published in 1894) paper dealing with "fungus cellulose," the nitrogen-containing material obtained from the same fungi by fusion with caustic potash solution at 180°C (Winterstein, 1894a). Winterstein also confirmed identification of glucosamine (formula $C_6H_{13}AzO_5{\cdot}HCl$; $[\alpha]_D=73.7$) in the products when heated with hydrochloric acid, the same glucosamine previously described by Ledderhose (Winterstein, 1894b,c; Fig. 3.4) and by Gilson. However, Winterstein was not convinced by Gilson's conclusions on glucosamine

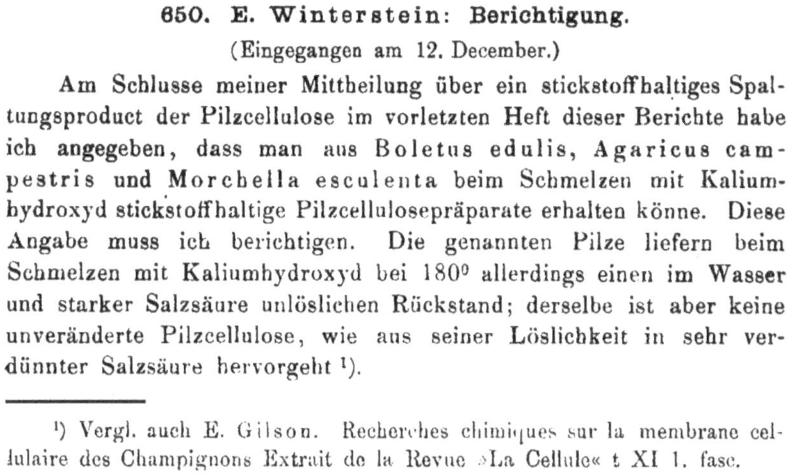

650. E. Winterstein: Berichtigung.

(Eingegangen am 12. December.)

Am Schlusse meiner Mittheilung über ein stickstoffhaltiges Spaltungsproduct der Pilzcellulose im vorletzten Heft dieser Berichte habe ich angegeben, dass man aus Boletus edulis, Agaricus campestris und Morchella esculenta beim Schmelzen mit Kaliumhydroxyd stickstoffhaltige Pilzcellulosepräparate erhalten könne. Diese Angabe muss ich berichtigen. Die genannten Pilze liefern beim Schmelzen mit Kaliumhydroxyd bei 180⁰ allerdings einen im Wasser und starker Salzsäure unlöslichen Rückstand; derselbe ist aber keine unveränderte Pilzcellulose, wie aus seiner Löslichkeit in sehr verdünnter Salzsäure hervorgeht [1]).

[1]) Vergl. auch E. Gilson. Recherches chimiques sur la membrane cellulaire des Champignons Extrait de la Revue »La Cellule« t XI 1. fasc.

Fig. 3.4 Extract of the article by Professor Winterstein on his experiments and observations on the hydrolysis of fungal cellulose where he claimed priority. *(Source: Winterstein, E., 1894. Berichtigung. Ber. Dtsch. Bot. Ges. 27, 3508–3509.)*

obtained from "fungus cellulose" nor indeed on those concerning *mycosine*, the soluble derivative of chitin. Then, a controversy arose between the two researchers and also with Hoppe-Seyler.

Between 1893 and 1894, Winterstein prepared an acid-soluble derivative of chitin (Winterstein, 1893, 1894a,c) and determined its formula to be $C_{14}H_{26}N_2O_{10}$, but published this result a year later (Winterstein, 1895c,d). This formula was the same as that published by Hoppe-Seyler on January 14, 1895, which claimed priority (Hoppe-Seyler, 1895). The next year, Winterstein demonstrated that *mycosine*/mycosin (term proposed by Gilson) was decomposed in acid solution into D-glucose, other hexoses, and then into acetic acid and an undetermined nitrogenous organic substance (Winterstein, 1895c,d). When heated with concentrated hydrochloric acid, it yielded a crystallizable fission product, which proved identical with the hydrochloride of chitosamine, at that time erroneously termed glucosamine. The same behavior was exhibited by *chitine*, the substance discovered by Odier. As also shown by Ledderhose in 1876, this substance furnished under similar treatment glucosamine, and that too, in the state of hydrochloride.

The question of whether fungal cellulose was identical to cellulose in higher plants was hotly debated, with evidence based on the solubility behavior and staining reactions summarized by Winterstein (1894b). The use of the name cellulose with reference to fungal chitin as proposed by

Winterstein continued for several years. Despite this error, Winterstein's work led to significant advances in the chemistry of chitosan, which were confirmed a few years later by van Wisselingh (1898a,b) when he used Winterstein's work to propose a microchemical test protocol for the detection of chitin. van Wisselingh also relied on the results of Hoppe-Seyler to develop these tests.

3.3 Eugène Gilson

Since Odier's discovery in 1823 on the presence of chitin in the horny parts of insects, the work has fallen into two categories: that of researchers interested in the histological origin of chitin and that of those interested in its chemical nature and constitution. At the time, histologists were also debating whether chitin was the product of a secretion process by epidermal cells or the result of chemical transformation of the surface layer of the protoplasm (Schmidt, 1845). From a physiological point of view, chitin was regarded as a secretion (a product of de-assimilation from which the cell gets rid). It was known that chitin had been discovered in the cell membrane of some plants and that there were many facts showing chemical analogies between chitin and cellulose. There were also studies on the influence of diet on chitin production in beetles. The production of chitin in carnivorous beetles was greater than in vegetarian beetles. The established concept was that adaptation to the vegetarian diet determined an overproduction of the chitinous material. The insolubility of chitin was an interesting feature and a very simple way to prepare the insect skeleton. It consisted in dissolving with hot hydroxide potassium all the soft tissues other than the chitinous formations which remained intact at the end of the operation. The chitin, washed and dried, was then simply weighed. This protocol allowed in the early 1900s to establish a chitin coefficient according to the diet adopted by the beetles. Gilson, professor in pharmacognosy at the Belgian *Université de Gand*, knowing all these research studies, considered that plants and arthropods were different but that had the same chemical secretion mode. Gilson's research was strongly focused on the extraction of cellulose from different sources. He studied the presence of chitin in the cell membrane of plants and in the cell wall of fungi in order to compare it and above all to demonstrate that this chitin was not a cellulose. Indeed, Gilson believed that fungus tissue did not contain cellulose and argued that Winterstein's findings on fungal chitin were inappropriate. For Gilson, *les réactions caractéristiques de la chitine étaient également données par la membrane cellulaire de nombreuses espèces de champignons* (the characteristic reactions of chitin were also given by the cell membrane of many species of fungus).

Eugène Gilson (1862–1908) started his studies of biology in Gand/Ghent (a port city located in the north-west of Belgium) and finished them in Louvain (a city near the capital Brussels) where he obtained the *Grade de Pharmacien* (pharmacist's diploma) in 1886. The young student was interested by medicinal drugs extracted from plants and other natural sources. Winner of a university competition and a travel grant, Gilson then traveled to Europe to study chemistry and physiology, notably at the *Kaiser Wilhelms-Universität* in Strasbourg, where, under the direction of Professor Hoppe-Seyler, he obtained his Ph.D. in 1890. The same year, Gilson accepted a position as professor at the *Université de Gand* (Leboucq, 1913). Professor Gilson was first in charge of teaching pharmacognosy and then, from 1892, theoretical and practical pharmacy. He was appointed Professor Extraordinary in 1896 and promoted to the Ordinary in 1901. His research work focused on plant and animal chemistry for applications in both the pharmaceutical and medical fields. In 1906, Professor Gilson received a prize from the Belgian *Académie de Médecine* for his outstanding career, and 1 year later, he was awarded the rank of *Chevalier de l'Ordre de Léopold* (Knight of the Order of Leopold).

As soon as he was appointed to the University of Gand, Gilson continued the research on chitin and cellulose that he had initiated in France under the direction of Hoppe-Seyler. He tried to demonstrate that in a large number of fungi, chitin had completely replaced cellulose. In 1893, Gilson was unable to obtain crystalline cellulose from *Mucor vulgaris*, *Thamnidium vulgare*, and *Agaricus campestris*, while he succeeded easily with plant tissue (Gilson, 1893). One year later, he studied the presence of chitin in *Agaricus campestris* and *Claviceps purpurea* (Gilson, 1894a,b,c) and noted that its elemental composition was in close agreement with previously reported analyses for chitin of insect origin. However, the results were in disagreement with those published by Payen. The chitin in fungi had the same role as did the cellulose in the phanerogams and many of the cryptogams. Gilson also noted that the residue obtained after treating certain fungi with dilute sulfuric acid and then dilute sodium hydroxide under reflux yielded glucosamine ($C_6H_{13}AzO_5$) on hydrolysis with hydrochloric acid and that, just as in the case of chitin, acetic acid was produced during hydrolysis. As a result of his work, at the end of the 1890s, it was recognized that chitin occurred in both animals and plants and that there was no detectable chemical difference between products from the two sources. In whatever form it occurred, chitin had the same physical properties including refractive index, specific rotation, and elemental analysis. In 1895, Gilson stated that *la substance squelettique de la membrane cellulaire des champignons, traité par l'acide chlorhydrique concentré et par la potasse caustique à 180°C, fournissait les mêmes produits de transformation que la chitine, qu'elle était donc très probablement identique à celle-ci* (the skeletal substance of the cell membrane of fungi, treated with concentrated hydrochloric acid and potash at 180°C, yielded the same transformation products as chitin, and was therefore most probably identical to it) (Gilson, 1895a). This was the first evidence that chitin from fungal and animals was chemically similar. In 3 years of research, Gilson demonstrated that in a large number of fungi, chitin completely replaces cellulose; these two substances never coexisting. He then established a large list of fungi for which he characterized the chitin in the cell walls. This list was completed by van Wisselingh 3 years later (van Wisselingh, 1898a,b). Gilson also indicated that *La nature chimique de la chitine est difficile à déterminer car sa grande stabilité et la difficulté de décomposer sa molécule en groupes plus simples n'ont pas permis de clarifier sa constitution* (the chemical nature of chitin is difficult to determine because its great stability and the difficulty of breaking down its molecule into simpler groups did not allow us to clarify its constitution).

Gilson also became interested in the chemistry of chitin, "discovering" a soluble derivative in acidified aqueous solution. By fusing cell preparations (*Agaricus campestris*, ergot of rye *Secale cornutum*) with caustic potash at 180°C, according to the Hoppe-Seyler method, Gilson obtained a residue in sulfate or chlorhydrate form, not of cellulose but a substance insoluble in Schweitzer's reagent (Gilson, 1894a,b,c) and to which he gave the name of *mycosine* (mycosin) with the formula $C_{14}H_{28}Az_2O_{10}$. In fact, two other structures were also proposed: $C_{14}H_{30}Az_2O_{10}$ and $C_{18}H_{26}Az_2O_{10}$. One year later, Araki, another pupil of Professor Hoppe-Seyler, studied the formation of chitosan/mycosin from chitin and proposed another composition for chitosan, that is, $C_{18}H_{30}N_2O_{10}$ (Araki, 1895). Gilson showed that *mycosine* was soluble in 2% to 3%. hydrochloric acid or in very dilute acetic acid (Gilson, 1895a,b), in agreement with the observations previously published by Hoppe-Seyler. A solution of iodine in potassium iodide, containing a trace of free acid, gave a reddish violet stain. Zinc-iodo-chloride solution varied in action in accordance with the amount of zinc chloride present, 50%. producing a blue to blue-violet coloration. These reactions closely resembled those of cellulose. Winterstein (1895b) and van Wisselingh (1898a,b) reported similar experiments and conclusions. Finally, Gilson confirmed his formula for chitosan, that is, $C_{14}H_{28}Az_2O_{10}$. He was also among the early researchers to point out that chitin may be associated with other carbohydrate materials and substances, called by the German term *Inkrusten*, that is, associated substances, analogous or identical to those found in phanerogams. Later, Schulze (1921) used the same German term for the associated substances present in chitin.

Gilson claimed priority over Winterstein and Hoppe-Seyler for findings on the sources of chitin extraction, the conclusions on glucosamine, and the discovery of *mycosine*/chitosan (Fig. 3.5). A scientific controversy then opposed the three research groups on the authorship of these results and conclusions. This controversy was characterized by the persistent and public division of these three researchers through publications and *mémoires*, in which the work was carried out, and the results were the same, but the arguments advanced were contradictory in the interpretation of these results. However, the name *mycosine* (mycosin) for chitosan, like the alternative names for chitin, has never come into general usage. In addition, the paternity of the discovery of chitosan is attributed to Hoppe-Seyler.

173. E. Gilson: Das Chitin und die Membranen der Pilzzellen.
(Eingegangen am 8. April.)

Von verschiedener Seite werden zur Zeit das Chitin und die chemischen Bestandtheile der Pilzzellenmembranen untersucht.

Da ich selbst, seit März 1893[1]), mehrere Notizen über diese Frage geliefert habe, möchte ich Folgendes darüber bemerken:

1. Ich habe zuerst in den Membranen der Pilze das Vorkommen eines Körpers bewiesen, welcher dieselben charakteristischen Umwandlungsproducte liefert, wie Chitin, und also mit Chitin aller Wahrscheinlichkeit nach identisch ist. Und zwar, nachdem ich im Juli 1894[2]) auf die zahlreichen Analogien hingewiesen hatte, welche zwischen der stickstoffhaltigen Gerüstsubstanz der Pilze und dem Chitin und zwischen Mycosin und Glucosamin existiren, habe ich in einer Notiz, welche am 9. November 1894[3]) der Société chimique de Paris mitgetheilt wurde, angezeigt, dass die Gerüstsubstanz der Pilze, mit concentrirter Salzsäure behandelt, Glucosamin sowie Chitin liefert, und dass dieses, nach Schmelzung mit Aetzkali (180—190°), Mycosin erzeugt, gleich wie die Gerüstsubstanz der Pilze.

Fig. 3.5 Extract of the article by Gilson where he claimed priority over Winterstein and Hoppe-Seyler with regard to the discovery of *mycosine*/chitosan and to the findings on glucosamine. *(Source: Gilson, E., 1894. Recherches chimiques sur la membrane cellulaire des champignons. Bull. Soc. Chim. Fr. 3, 1099–1102.)*

3.4 Cornelis van Wisselingh

In 1898, van Wisselingh, Dutch pharmacist and chemist at the *Rijksuniversiteit Groningen* (University of Groningen), published a concept that "in the fungal cell wall, either chitin or cellulose could prevail." Cornelis van Wisselingh was born on July 30, 1859 in Utrecht, his parents being Johannes Pieter van Wisselingh (1812–1899) and Jacoba Terwogt (1818–1880). Cornelis van Wisselingh was professor at the University of Groningen between 1906 and 1925 and Rector Magnificus between 1916 and 1917. His outstanding scientific contributions in the field of chemistry were highlighted by the Royal Swedish Academy of Sciences in nominating him for the Nobel Prize in Chemistry. Cornelis van Wisselingh died on November 30, 1925 in Apeldoorn.

To demonstrate his concept, van Wisselingh proposed a colorimetric test, based upon conversion of chitin to chitosan, to detect chitin in the cell wall of fungi (van Wisselingh, 1898a,b). Indeed, van Wisselingh, using the findings of Hoppe-Seyler and Winterstein, was the first to prepare chitosan specifically for microchemical testing. He studied the presence of chitin in a wide variety of materials such as in Myxomycetes, Peronosporales, Saprolegniales, Chitridiales, Entomophtharales, Mucolares, and in almost all of the higher fungi. To characterize the samples, van Wisselingh proposed the following procedure: the material was mixed with 50% (or 60%) potassium hydroxide and heated in a sealed tube containing glycerol at 160–180°C for 5–10 min, or until colorless. After that, the alkali was removed by washing with 90% alcohol, dilute alcohol, and water. The partially degraded product, that is, modified chitin or chitosan was then stirred in water and placed on a slide, saturated with an iodine-potassium iodide reagent. The development of a violet color on addition of 1% sulfuric acid was taken as a positive indication of chitin. In fact, with iodine, chitosan gave a brown coloration, turning violet upon the addition of sulfuric acid (van Wisselingh, 1898a,b). The van Wisselingh test was then widely used for over 40 years.

In the mid-1920s, new tests appeared, and the microchemical test and conclusions of van Wisselingh then began to be criticized, including by Kühnelt (1928a), Hopkins (1929), and Campbell (1929), who claimed that "he had failed to improve the test in 26 years." Schulze also had his doubts earlier (Schulze, 1921, 1922a,b; Schulze and Kunike, 1923). However, in 1924, van Wisselingh rightly criticized all the simplified methods of others and defended his own (van Wisselingh, 1924). In fact, all the criticisms were based on the fact that in order to apply van Wisselingh's test, a reasonable degree of purification was necessary before the test could be satisfactorily applied for the detection of chitin. The van Wisselingh test was reasonably specific for chitin, but other products (polysaccharides) were capable of giving greenish stains. It was not suitable for quantitative studies because the same degree of coloration did not appear in successive experiments. In fact, the strongly alkaline conditions induced considerable degradation of chitin, far beyond simple deamination. A few years later, a critical assessment of this topic was published in two comprehensive reviews (Richards, 1947b; Foster, 1949) and a book (Kent and Whitehouse, 1955), while pointing out the remarkable contribution of Professor van Wisselingh.

3.5 Torasaburo Araki

Like Schmiedeberg (1891), Araki, a Japanese student of Hoppe-Seyler, was convinced by Ledderhose's conclusions on glucosamine and more surprisingly by Winterstein's observations on the constitution, structure, and hydrolysis of chitin and chitosan. Araki was the first to demonstrate that the acid hydrolysis of chitin was carried out in two successive steps, chitin giving chitosan and then glucosamine.

Torasaburo Araki was born on November 23, 1866 in Kyoto. The young Torasaburo first studied medicine at Kyoto Imperial University. In 1889, he continued his studies in Germany at the *Physiologisch Chemischen Institut* (Physiological Chemical Institute) of the *Kaiser-Wilhelms Universität* in Strasbourg, working in the laboratory of Hoppe-Seyler. In 1891, Araki defended his doctoral thesis entitled *Über die Bildung von Milchsäure und Glycose im Organismus bei Sauerstoffmangel* (on the formation of lactic acid and glucose in the body in case of oxygen deficiency) (Fruton, 1990). Araki continued to work with Hoppe-Seyler until 1895 at Strasbourg. He then returned to Japan and became first an assistant at Okayama Medical School where he continued his studies on the chemistry of chitin in order to determine its composition and structure. In 1896, Araki then became professor of medical chemistry and abandoned his research on chitin, devoting himself to teaching medicine and to administrative responsibilities, such as director of the Medical Department. Three years later, Professor Araki accepted the position as physician and educationist at Kyoto Imperial University. He then returned to France as visiting professor at *Université de Paris* between 1902 and 1903, during which time he also lectured in Strasbourg. Between

1915 and 1929, Professor Araki was the 7th President of Kyoto Imperial University. In 1929, he was appointed Emeritus Professor of this prestigious university and Director of The Peers' School in Tokyo (an institution of higher learning founded in 1842 by Emperor Ninko), a position he held until 1937. Professor Araki then received the Order of the Rising Sun, a Japanese order established by Emperor Meiji in 1875. In 1937, he was elevated to membership in the Privy Council. Torasaburo Araki died on January 28, 1942.

Araki agreed with Winterstein that the tissues of fungi contained a substance very closely allied to chitin. In 1895, he published his first results on the hydrolysis of chitin. When the dried fungus material is extracted first with alcohol and diluted soda and then with concentrated soda, a colorless, friable product was obtained. This product consists of fungal chitin (containing a high percent of nitrogen) and other complex non-nitrogenous substances. After 3 h in concentrated hydrochloric acid and a concentration step under reduced pressure, *glykosamin/glycosamin* crystals separated. Elemental analysis confirmed the formula $C_6H_{13}NO_5$, which corresponded to glucosamine, the same compound described by Winterstein and Gilson a year earlier. Araki also observed that the release of acetic acid always accompanied the hydrolysis reaction. The products obtained were highly dependent on the conditions used during hydrolysis. Under mild conditions, the degradation was incomplete with the formation of *N*-acetyl-*glycosamin*, proving the linkage of the acetyl group to the amine group. Araki proposed the reactions described in Scheme 3.1 and pointed out three important points: (i) chitin was a fungal cellulose (*pilzcellulose*) different from that of higher plant tissues; (ii) *glykosamin/glycosamin* was the unit of chitin and chitosan, and the ultimate product of these substances during hydrolysis; and (iii) in the chitin structure, the acetyl group was attached to the amine (Araki, 1895). This last point was in agreement with the previous conclusions by Schmidt (1845), Städeler (1859), and Ledderhose (1876). Araki also proposed a chemical formula identical to that proposed by Schmiedeberg a few years earlier, but

$$\underset{\text{chitin}}{C_{18}H_{30}N_2O_{12}} + 2H_2O = \underset{\text{chitosan}}{C_{14}H_{26}N_2O_{10}} + \underset{\text{acetic acid}}{2(C_2H_4O_2)}$$

$$\underset{\text{chitosan}}{C_{14}H_{26}N_2O_{10}} + 2H_2O = \underset{\textit{glycosamin}}{2C_6H_{13}NO_5} + \underset{\text{acetic acid}}{C_2H_4O_2}$$

Scheme 3.1 Two steps during the hydrolysis of chitin to obtain chitosan and *glycosamin*/glucosamine according to Araki (1895).

different from that of Gilson and, more surprisingly, from that of Hoppe-Seyler (Schmiedeberg, 1891). In this article (Araki, 1895), Araki wrote as conclusion: "chitosan is chitin-less acetic acid." Later, Fränkel and Kelly (1901a) and Offer (1907) confirmed the reactions described in Scheme 3.1 (but also with different formulae for the compounds) and arrived at a similar conclusion. Araki showed that chitosan is colored violet by dilute iodine solution, but its acetyl derivative and chitin were not. Finally, Araki pointed out that chitosan was not a single chemical entity, but rather a family of chitin derivatives (Araki, 1895). Later, Löwy (1909), Kotake and Sera (1913), and Proskuriakow (1926) also suggested the same assumption. This was only conclusively demonstrated by the diversity of X-ray diffraction patterns obtained in the mid-1930s by Clark and Smith (1936) and by Meyer and Wehrli (1937).

During the same period in which Araki obtained his results on chitin, three important contributions were made independently by three European researchers, namely, Cornelis A. Lobry de Bruyn, Sigmund Fränkel, and Emil Fischer. Reaction schemes and conclusions similar to those of Araki were published by these authors.

3.6 Cornelis A. Lobry de Bruyn

Chitosamine/glucosamine in agreement with Schmiedeberg's results was also obtained by Lobry de Bruyn using decomposition in the presence of sodium methoxide (Lobry de Bruyn and Franchimont, 1898; Lobry de Bruyn and van Ekenstein, 1899a,b). First, Lobry de Bruyn confirmed the work on Schmiedeberg, Tiemann, Winterstein, and Araki on chitin and glucosamine. From Crustacean chitin, glucosamine is isolated as a highly crystalline hydrochloride in the acid hydrolysis. Then, using his data, Lobry de Bruyn definitively demonstrated in 1898 the composition of the product obtained by the hydrolysis of chitin, namely, D-glucosamine chloride, then called *freie chitosamin* or *chitosamine libre* (free chitosamine) (Fig. 3.6). He wrote: *Lorsqu'on traite la chitine des crustacés ou des insectes ou des insectes par l'acide chlorhydrique puis pas le méthylate de sodium, on obtient un précipité cristallin de chitosamine libre qui fond à 105–110°C* (when chitin from crustaceans or insects is treated with hydrochloric acid and then not with sodium methylate, a crystalline precipitate of free chitosamine is obtained which melts at 105–110°C) (Lobry de Bruyn and Franchimont, 1898). Lobry de Bruyn noted that this compound altered with time by yielding ammonia and sugars including fructose. He also observed that chitosamine hydrochloride heated

Sur la chitosamine ; C. A. LOBRY de BRUYN et **W. Alberda VAN EKENSTEIN** (*R. tr. ch. P. B.*, t. **18**, p. 77). — Lorsqu'on traite la chitine des crustacés ou des insectes par l'acide chlorhydrique, on obtient le chlorhydrate de chitosamine. Celui-ci, traité par le méthylate de sodium donne, après séparation de NaCl, un précipité cristallin de chitosamine libre. La chitosamine fond à 105-110° et répond à la formule $C^6H^{13}AzO^5$; $\alpha_D = +44°$. Cette base s'altère spontanément sous l'influence du temps, à l'état sec ou dissous, en donnant un composé identique à la combinaison am-

Fig. 3.6 Extract of the publication of Professor de Bruyn where he detailed the composition and chemistry of chitosamine. *(Source: Lobry de Bruyn, C.A., van Ekenstein, A.W., 1899. La chitosamine libre. Recl. Trav. Chim. Pays-Bas. 18, 77–85.)*

in the presence of sodium acetate and acetic aldehyde yielded a mixture of two pentacetates, one much more soluble than the other (Lobry de Bruyn and van Ekenstein, 1899a). From his experiments, Lobry de Bruyn concluded that *chitosamine* (glucosamine) was the primary unit of chitin and chitosan (Lobry de Bruyn and van Ekenstein, 1899a,b). The same reactions as in Scheme 3.1 were reported, but the nomenclature was not the same as that reported by Araki. The formula $C_6H_{13}AzO_5$ proposed by Schmiedeberg was also demonstrated. In the same year, Robert Breuer (1869–1936), Austrian physician working at the *Allgemeine Poliklinik Wiener* (Vienna Polyclinic) published an article titled *Ueber das freie Chitosamin* (about the free chitosamine) where, using hydrolysis and decomposition in the presence of diethylamine, also reported an identical formula for glucosamine, also called by him *chitosamin* (Breuer, 1898), and similar conclusions to those of Lobry de Bruyn. Breuer, studying the chemistry of glucosamine, also prepared different derivatives. This research was later taken up by von Fürth and Scholl (1907).

Cornelis Adriaan Lobry van Troostenburg de Bruyn (Fig. 3.7) was born on January 1, 1857, at Leeuwarden, The Netherlands, where his father was a physician in practice. In 1875, Lobry de Bruyn entered the University of Leiden, where he became an assistant to Antoine Paul Nicolas Franchimont (1844–1919, professor in organic chemistry). Eight years later, Lobry de Bruyn obtained a doctorate under the supervision of Professor Franchimont. The subject of the thesis was the interaction of dinitrobenzenes with potassium cyanide in alcoholic solution (Spiller et al., 1905; Speck Jr, 1958). After two postdoctoral internships in Paris with Charles Wurtz (1817–1884; chemist, physician, and member of the *Académie de Médecine*) and Charles

Fig. 3.7 Professor Cornelis Adriaan Lobry van Troostenburg de Bruyn, 1857–1904. *(Source: Spiller, J., Tilden, W.A., McLeod, H., Mills, E.J., Carey Foster, G., 1905. Obituary notices. J. Chem. Soc. Trans. 87, 565–618.)*

Friedel (1832–1899; chemist and mineralogist), Lobry de Bruyn returned to Leiden in 1884 and was appointed chemist to the Government Department of Marine, studying the manufacture and properties of explosives. During this period, like Fischer, Lobry de Bruyn became interested in the chemistry of chitin and phenylhydrazine. For instance, he was the first to prepare hydroxylamine and hydrazine in anhydrous form (Speck Jr, 1958). In 1896, Lobry de Bruyn was appointed professor of Organic Chemistry and Pharmacy at the University of Amsterdam where he studied the hydrolysis of chitin. Professor Lobry de Bruyn was elected an Honorary Foreign Member of the English Chemical Society. He also published several works on the behavior of chitin in water and solvents and on its chemical modification. Many of his investigations were also related to nitrogenous compounds, both organic and inorganic. Furthermore, he was a prolific researcher whose scientific interests extended beyond the field of chemistry, for example in the biology of substances for industrial applications or the physic of matter. Despite offers of professorships from Austrian and German universities and an offer of position as State Chemist from the Transvaal Government, Professor Lobry de Bruyn remained at the University of Amsterdam until his death on July 27, 1904.

3.7 Sigmund Fränkel

Sigmund Fränkel (1868–1939) was a Polish-born physiological chemist and pharmacist who worked in the University of Vienna. Fränkel (or Fraenkel) studied medicine at the Universities of Prague, Freiburg, and Vienna. In 1892, he obtained a doctorate in medicine at the University of Vienna.

Fränkel played an important role in chitin chemistry research (Fränkel, 1898; Fränkel and Kelly, 1901a,b; Fraenkel and Kelly, 1903; Fränkel and Jellinek, 1927). He ascribed to chitin a molecular complexity similar to that of starch or glycogen. Fränkel also showed that chitin could be converted into acetylglucosamine, indicating that in the parent substance, acetyl groups were linked to nitrogen. Later, this was confirmed by Irvine (1909). Scheme 3.2 reports the hydrolysis of chitin and chitosan according to Fränkel and his colleague, Agnes Kelly. In 1901, using a milder acid hydrolysis of purified chitin (cold concentrated H_2SO_4 for 2 days, room temperature), Fränkel and Kelly isolated five fractions by precipitation with alcohol followed by ether; the last and most soluble fraction was shown to be *N*-acetylglucosamine (Fränkel and Kelly, 1901a,b). They wrote: *La chitine se dédouble sous l'influence de HCl à l'ébullition en chitosamine (formule* $C_6H_{13}NO_5$*) et acide acétique, la fusion avec KOH fournit la chitosane et de l'acide acétique* (chitin splits under the influence of boiling HCl into *chitosamine* (formula $C_6H_{13}NO_5$) and acetic acid; fusion with KOH provides chitosan and acetic acid). The two reactions are shown in Scheme 3.2, which shows that the chemical formula of *chitosamine* was the same as that of *glykosamin* (Scheme 3.1). Again, the release of acetic acid accompanied the hydrolysis of chitin, and most researchers agreed with this fact. Fränkel and Kelly demonstrated that complete acid hydrolysis of chitosan resulted in the release of 4 mol of chitosamine and 4 mol of acetic acid. They also pointed out the fact that the hydrolysis conditions could have a degradation effect on the products of hydrolysis. At that time, almost all researchers also agreed on the

$$\underset{\text{chitin}}{C_{18}H_{30}N_2O_{12}} + 4H_2O = \underset{\text{chitosamine}}{2(C_6H_{13}NO_5)} + \underset{\text{acetic acid}}{3(C_2H_4O_2)}$$

$$\underset{\text{chitosan}}{C_{14}H_{26}N_2O_{10}} + 2H_2O = \underset{\text{chitosamine}}{2(C_6H_{13}NO_5)} + \underset{\text{acetic acid}}{C_2H_4O_2}$$

Scheme 3.2 Hydrolysis of chitin and chitosan according to Fränkel and Kelly (1901a,b).

proposed chemical formulas for chitin, chitosan, and glucosamine and the reaction schemes, but there was no consensus on the terminology to be used. Moreover, some considered glucosamine (*chitosamine*) and glucose to be similar in terms of their chemistry (Ambronn, 1890; Breuer, 1898; von Fürth and Russo, 1906; Berlese, 1929), while for others, the two substances were different (Lobry de Bruyn and Franchimont, 1898; Fränkel and Kelly, 1901a,b; Fraenkel and Kelly, 1903; Fischer and Leuchs, 1902, 1903; Ilkewitsch, 1908; Irvine, 1909).

Fränkel and Kelly then focused their research on the conditions of the hydrolysis reaction. They demonstrated that reaction yielded not only *chitosamine* (glucosamine) and acetic acid but also small amounts of mono-acetyl-chitosamine (later mono-acetylglucosamine) (formula $C_8H_{15}NO_6$, Scheme 3.3) and of mono-acetyl-di-chitosamine (formula $C_{14}H_{26}N_2O_{10}$), depending on the conditions used during the hydrolysis. The mono-acetyl-chitosamine was identical with the compound previously obtained by *N*-acetylation of D-glucosamine by Araki (1895) and Breuer (1898). Fraenkel and Kelly (1903) concluded that chitin was an acetylated and aminated polymer. There was one acetyl group for each *chitosamine* residue in chitin. A detailed discussion on the hydrolysis of chitin and on the structure of products can be found in the *mémoire* by Bounoure (1919). The milder hydrolytic conditions proposed by Fränkel permitted isolation of 2-acetamido-2-deoxy-D-glucose, the true structural component of chitin. Because of the tendency to compare chitin with cellulose, it can be readily understood why it was so soon postulated that chitin was a polymer/polysaccharide comprising 2-acetamido-2-deoxy-D-glucose units.

From Fränkel's results, chitin was, indeed, regarded as an acetylated amino polymer of high molecular weight. The configuration of its "molecule," and that of glucosamine, remained to be determined. Again, there was no consensus (von Fürth and Russo, 1906; Offer, 1907; Irvine, 1909; Brach, 1912). Moreover, many authors still believed that chitin and cellulose were related because they had chemical analogies, for example, the same reducing properties (iodine reaction) and an identical reaction on phenylhydrazine. For example, nitrochitins analogous to nitrocelluloses were prepared by

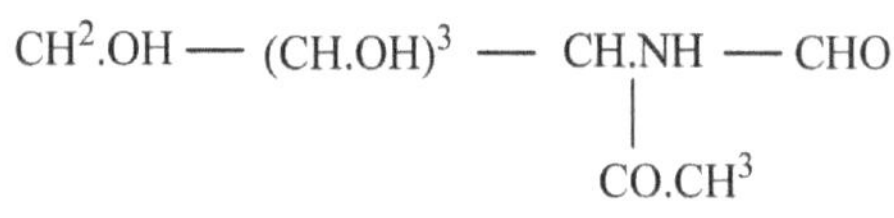

Scheme 3.3 Structure of the acetylglucosamine proposed by Fränkel and Kelly (1901a,b).

von Fürth and Russo (1907), suggesting chemical analogies between cellulose and chitin and also between glucose and glucosamine. However, Fränkel's results were confirmed by the work of Fischer and Irvine groups, which definitively demonstrated the exact configuration of glucosamine as a hexose amine.

3.8 Emil Fischer

Emil Hermann Fischer (1852–1919), Nobel Prize in 1902, was a German organic chemist and a pioneering figure in biochemistry. With Sir James C. Irvine and Sir Walter N. Hatworth, Professor Fischer contributed significantly to our current understanding of the structure and configuration of glucose and maltose, and subsequently, polysaccharides. A detailed biography of Emil Fischer (Fig. 3.8) can be found in the references Freudenberg (1967) and Lichtenthaler (2002).

The chemistry of phenylhydrazine was discovered accidentally by Fischer in 1884 while working in Strasbourg (Fischer, 1884). During his research at the University of Munich in the mid-1880s, Fischer found that phenylhydrazine converted sugars into osazones whose crystals had characteristic forms that could be identified. Later, in 1893, when at the University of Berlin, Fisher first described the glycosidic bond by synthetizing

Fig. 3.8 Professor Emil Hermann Fischer, 1852–1919. *(Source: Lichtenthaler, F.W., 2002. Emil Fischer, his personality, his achievements, and his scientific progeny. Eur. J. Org. Chem. 4095–4122.)*

methyl-glucoside and the ring of structure of glucose having an O-bridge between C-1 and C-4. He suggested the nomenclature of α- and β-isomers of the glycosidic linkage. Fisher also discovered caffeine and other related purines and studied the molecular structures of numerous sugars and proteins. He also proposed the synthesis of glucosamine and wrote of the value of this discovery: "The synthesis of glucosamine showed it to be an intermediate grape sugar and the α-amino acids, hence providing one of the longest sought-for bridges between the carbohydrates and the proteins" (Fischer, 1912). A discussion on this can be found in Bunge (1912).

Fischer was convinced by Tiemann's conclusions. The osazones from both sugars were laevorotatory, so glucosamine was assigned to the D-carbohydrate series. Fischer was also aware of the work of Lobry de Bruyn's and that of Fränkel on the composition of the product obtained by the hydrolysis of chitin and its identity. At the beginning of 1900s, Fischer and its Ph.D. student Hermann Leuchs (1879–1945) proposed the synthesis of glucosamine and established its constitution (Fischer and Leuchs, 1902, 1903). By the combination of D-arabinose and ammonium cyanide, or D-arabinose-imine with hydrogen cyanide, D-glucosaminic acid was obtained, and its lactone reduced to glucosamine. Glucosamine formed a penta-acetyl derivative and also an oxime, semicarbazone, and phenyl hydrazine. However, it could not be converted into glucose, though it yielded glucose phenyl osazone when heated with phenyl hydrazine (Fischer and Leuchs, 1903). Nitrous acid converted it into a new compound regarded as a sugar and also termed chitose ($C_6H_{10}O_5$), in agreement with the formula previously proposed by Fischer and Tiemann (1894). This formed chitonic acid when oxidized. Glucosamine ($C_6H_{13}NO_5$) was then regarded as a derivative of chitose and also termed chitosamine by Fischer, confirming the conclusions previously reported by Fränkel and Kelly (1901a,b). In another contribution, chitose was shown by Fischer and Andreae (1903) to be a hydrated furfurane derivative rather than a true sugar, formed by simultaneous elimination of the amino group and anhydride formation.

3.9 James C. Irvine

Like Fischer, Irvine attempted to resolve the question of the configuration at the carbon-2 position by de-amination of glucosamine to a hexose (Irvine et al., 1911; Irvine and Hynd, 1912, 1914). James Colquhoun Irvine (1877–1952) was a Scottish organic chemist, professor of chemistry at the University of St Andrews (Fig. 3.9). Irvine studied chemistry at the

Fig. 3.9 Professor James Colquhoun Irvine, 1877–1952. *(Source: Hirst, E.L., 1953. James Colquhoun Irvine 1877-1952. Adv. Carbohydr. Chem. 8, XI–XVII.)*

University of Leipzig under the supervision of Friedrich W. Ostwald (1853–1932, Nobel Prize in chemistry in 1909). Returning to St Andrews, Irvine obtained a doctorate and was appointed professor of chemistry in 1909. Professor Irvine was elected a Fellow of the Royal Society of Edinburgh in 1917 and of the Royal Society of London in 1918. Sir James C. Irvine was made a Commander of the Order of the British Empire in 1920 and knighted 5 years later (Hirst, 1953).

At that time, the measurement of optical rotation was commonly used to characterize preparations of a substance to compare different fractions from fractional precipitation or to study products obtained by partial hydrolysis. Irvine first reported the specific rotatory power of chitin in hydrochloric acid solution ($^{20}[\alpha]_D = -14.1°$) and its index of refraction (1.525). He also studied the action of acetyl-bromide on glucosamine hydrochloride with the object to clarify the constitution of the structural unit of chitin (Irvine, 1909). Although the theoretical amount of nitrogen was released during the decomposition of glucosamine by nitrous acid, the product of the transformation was not a simple hexose, but a hydrated furan derivative known as chitose. Chitose was formed under all conditions when nitrous acid acted on glucosamine, which explained the alternative name chitosamine (Irvine, 1909; Irvine et al., 1911). Pursuing his work on chitin derivatization into aceto-halogen derivatives, Irvine with his pupil Alexander Hynd concluded that glucosamine may be derived from either glucose (Irvine and Hynd, 1912) or mannose (Irvine and Hynd, 1914) according to the method of preparation. Glucosamine as an α-amino-derivative had D-glucose configuration. Their conclusions were in agreement with those published by Fischer and Leuchs (1903) and by Hamlin (1911). However, Irvine and Hynd stated that no rigorous proof could be offered in support of this latter assumption and "the displacement of any group by any other group may be accompanied by a Walden inversion." The stereochemical arrangement of the amino-group was thus uncertain.

In 1925, in his comprehensive monograph Hexosamines and Mucoproteins, Levene commented that "much of the chemical structure of the sugar was formulated correctly rather by instinct than by experimental evidence." To avoid confusion, he proposed to use systematically the name of chitosamine for the sugar now known as glucosamine (Levene, 1925). Phoebus Aaron Levene (1869–1940) was an American biochemist who studied the structure and function of nucleic acids (Tipson, 1957). Chitosamine, 2-amino-2-deoxy-D-glucose, was then considered an amino-hexose from chitin and alternatively described as glucosamine or mannosamine. However, there was a period of confusion for the nomenclature related to this designation until the work of Karrer.

3.10 Paul Karrer

In the early 1920s, the Swiss chemist Paul Karrer (1889–1971), Nobel Prize winner in 1937 for his work on vitamins (the prize was shared with British chemist Norman Haworth), published several studies on the chemistry and biochemistry of chitin (Karrer and Smirnoff, 1922; Karrer and Hofmann, 1929; Karrer and von François, 1929; Karrer et al., 1924; Karrer, 1930; Karrer and White, 1930; Karrer and Mayer, 1937).

Paul Karrer (Fig. 3.10) was born on April 21, 1889 in Moscow to Swiss parents. In 1892, his family returned to Switzerland. Karrer studied chemistry at the University of Zurich with Alfred Werner (1866–1919; Nobel Prize in chemistry in 1913) and obtained his doctorate in 1911. After a position of researcher at the Georg-Speyer Haus Foundation in Frankfurt working with Paul Ehrlich (1854–1915; Nobel Prize in medicine in 1908), Dr. Karrer in 1919 accepted the position of professor of Organic Chemistry at the *Chemisches Institut Zürich* and succeeded Werner as director of this institute (Dahn et al., 1969; Eugster, 1972; Roche, 1972; Wettstein, 1972; Beer, 1977). Karrer started studies on the chemistry of sugars and polysaccharides (starches, glycogen, inulin, cellulose, chitin, etc.) and their derivatives

Fig. 3.10 Professor Paul Karrer, 1889–1971. *(Source: Foulk, C.W., 1934. Paul Karrer. Ohio State University, 1932. J. Educ. 11, 1.)*

(amylose, dextrins, and glucosamine). He published more than 1000 papers, most of them in the journal Helvetica Chimica Acta (Crini, 2021).

At the time of Karrer, the work on chitin and chitosamine by Tiemann, Fränkel, Fisher, and Irvine was not unanimously accepted, with some researchers not accepting the name and conclusions on chitosamine, preferring to use the term mycetosamine (Ilkewitsch, 1908; Bounoure, 1919; Dous and Ziegenspeck, 1926). They also preferred the term mycetin instead of chitin. For example, Dous and Ziegenspeck (1926) prepared mycetin from the tissues of *Boletus edulis* and concluded that mycetin was similar to crab chitin, but "it was not exactly the same chitin." Furthermore, mycetin and chitin differed in the proportion of nitrogen and, when hydrolyzed, products such as glucosamine also differed in the amounts of reducing sugar formed. Therefore, different terms had to be used for these two forms of chitin. The work of Dous and Ziegenspeck confirmed that of Ilkewitsch (1908) published a few years later. However, Karrer disagreed with these conclusions, considering that the product extracted from *Boletus edulis* was indeed chitin, as did Gilson, at the end of the last century, and other researchers such as Irvine, Fisher, Fränkel, and Tiemann.

In the early 1920s, Karrer studied the chemistry (degradation, modification) and biochemistry (degradation, biosynthesis, biology/"bioactivity") of chitin (Karrer and Smirnoff, 1922; Karrer et al., 1924). Karrer first observed that chitin was a colorless, nitrogenous carbohydrate comprising an unknown number of residues, linked through nitrogen. He also studied the solubility of chitin, reporting similar conclusions as Schulze and Kunike (1923) and Ito (1924), such as the fact that chitin was dissolved with or without decomposition by concentrated acids. Analyzing the data, Karrer considered that the residues were acetylglucosamine units. He then prepared different derivatives such as glucosamine or chitosamine and *N*-acetylglucosamine. The powerful acid hydrolysis of chitin led to the removal of the acetyl group from the (macro)molecule, giving thereby both glucosamine and acetic acid, confirming previous works on this topic. Karrer demonstrated that complete acid hydrolysis of chitin yielded D-glucosamine and acetic acid in nearly theoretical amounts if it was assumed that polysaccharide comprised only of mono-acetyl-D-glucosamine units (Karrer and Smirnoff, 1922; Karrer et al., 1924), in agreement with the result of Brach and von Fürth (1912). Later, Zechmeister and Tóth (1931) and Bierry et al. (1939) using chemical procedures and enzymatic reactions reported a similar conclusion. Karrer and Mayer (1937), like Tiemann (1886) and Fischer and Leuchs (1902, 1903), confirmed that D-glucosamine behaved in a similar manner to D-glucose in many of its chemical reactions. The relationship between the carbon at position 2 of D-glucosamine and D-glucosamine was then suggested and finally demonstrated 2 years later by Haworth et al. (1939) using an unequivocal synthesis.

In 1929, Karrer and his student Hofmann also reported the formation of crystalline *N*-acetylglucosamine (in 50% yield) by the action of the gut contents of the edible snail *Helix pomatia* on lobster chitin (Karrer and Hofmann, 1929). Indeed, they found the first enzyme (chitinase) promoting the hydrolysis of chitin in snail digestive fluid. Chitosan was also degraded by the enzyme but yielded products other than 2-amino-2-deoxy-D-glucose and 2-acetamido-2-deoxy-D-glucose. The same year, Karrer and another pupil, Götz von François, reported the isolation of 80% of the theoretical amount of acetylglucosamine from an enzymatic digest of a fungal chitin from *Boletus edulis* (Karrer and von François, 1929). Karrer then confirmed that chitin extracted from *Boletus edulis* tissues and that from crab shells was the same. When the products were hydrolyzed, they yielded chitosamine. Karrer's group also showed that re-acetylated chitosan was hydrolyzed by chitinases with the production of acetylglucosamine. Chitosan re-acetylated remained a substrate for *Helix* chitinase, while chitosan derivatives containing formyl, propionyl, butyryl, and benzoyl groups were not substrates (Karrer and von François, 1929; Karrer, 1930; Karrer and White, 1930). Araki (1895) previously reported that chitosan can be reacetylated more or less satisfactorily to yield products that may be similar to chitin except for a greatly shortened chain length, in agreement with the interconversion of chitin and chitosan proposed by Hoppe-Seyler (1894). Later, Meyer and Wehrli (1937) also reported that the reacetylation, which may be accomplished with acetic anhydride plus acetic acid and zinc chloride, may result in simply the reacetylation of the amine groups or in the addition of acetyl groups in place of some of the hydroxyls as previously reported by Karrer (Karrer, 1930; Karrer and White, 1930). In 1930, Karrer published his conclusions on the preparation of chitosan, its purification, and structure (Karrer, 1930). He pointed out that chitosan gave low nitrogen values in a Kjeldahl analysis due to the hydrolysis of some of the amine groups to hydroxyls during the synthesis of chitosan. Later, this was confirmed by Clark and Smith (1936). Like Araki (1895), Karrer pointed out that chitosan was rather a family of chitin derivatives but considered it a mono-acetyl-di-glucosamine. This was demonstrated by the fact that chitinase acting on chitosan gave both *N*-acetylglucosamine and glucosamine (Karrer, 1930). In 1950, Darmon and Rudall confirmed it by infrared studies (Darmon and Rudall, 1950). Finally, Karrer prepared several derivatives of chitin. For instance, he suggested that phosphoric esters might be of interest in the metabolism of chitin (Karrer et al., 1943). He also studied their anticoagulant activity.

3.11 Albert Hofmann

Albert Hofmann (1906–2008) was born on January 11, 1906 in Baden, an industrial Swiss town near Zurich. In 1926, he completed his Matura studies at a private school in Zurich, while working as a commercial apprentice. At the age of 20, Hofmann began his chemistry degree at the *Chemisches Institut* of the University of Zurich, studying the elucidation of the sugar components under the supervision of Karrer. Hofmann presented his doctorate entitled *Über den enzymatischen abbau des chitins und chitosans* (about the enzymatic degradation of chitin and chitosan) in the spring of 1929 at the age of 23 (Hofmann, 1929). He received his doctorate with distinction. The same year, after leaving university, Hofmann took a job with Sandoz Laboratories in Basel, where he stayed for more than 4 decades. His main interest was the chemistry of plants and animals. Hofmann also conducted important research on the chemical structure of chitin. His research yielded commercially important products for pharmaceutical and medical applications. Hofmann later became head of the research department for natural medicines and stayed with Sandoz Pharmaceuticals until his retirement in 1971. He is known for the synthesis of lysergic acid diethylamide (LSD) and the description of its hallucinogenic properties. Hofmann has received several honorary doctorates from European Universities (Stockholm University, Berlin Free University). In 2007, he was elected one of the "10 best living geniuses in the World" by the English magazine The Guardian (Sticher and Hamburger, 2008). Albert Hofmann passed away on April 29, 2008, at the age of 102 (Finney and Siegel, 2008; Sticher and Hamburger, 2008; Hagenbach et al., 2013).

Hofmann's thesis described the first enzyme that very efficiently degraded chitin (Hofmann, 1929), obtained from the crude extracts of a common snail (*Helix pomatia*), and "collected during Hofmann's walk on the Wienberg, near the University campus" (Finney and Siegel, 2008). Hofmann and Karrer proposed the name chitinase for the active principle of the gastrointestinal juice of the vineyard snail used (Karrer and Hofmann, 1929). Its action on partially deacetylated chitin (chitosan) led to the formation of oligosaccharides, resistant to acid hydrolysis, containing *N*-acetyl groups and forming insoluble sulfates. Disaccharides such as chitobiose and trisaccharides such as chitotriose were isolated. Two years later, using a similar procedure, Bergmann et al. (1931a,b,c) reported an acetylated disaccharide, octa-acetyl chitobiose, which is structurally similar to cellobiose octa-acetate. Zechmeister and Tóth (1932) also isolated chitotriose as its crystalline acetate, in agreement with Hofmann's conclusions. The existence of these compounds suggested that in the structure of chitin, there were no branched chains. In 1935, Freudenberg and Eichel have reported that the same chitinase, proposed by Hofmann, destroyed the blood activity of human urinary mucosubstances and liberated *N*-acetyl-aminosugar from them (Freudenberg and Eichel, 1935). Later, there was evidence for the existence also of chitodextrinases in snail digestive fluid (Jeuniaux, 1950). This type of enzyme differed from chitinase in facilitating the hydrolysis of oligosaccharides derived from chitin, for example, chitotriose. Karrer and Hofmann (1929) conclusively observed that chitin was a polymer of *N*-acetylglucosamine and chitosan a polymer containing both glucosamine and *N*-acetylglucosamine. However, it was only in 1946 that Earl R. Purchase and Charles E. Braun clearly elucidated the chemical structure of chitin using hydrolyzing experiments. Purchase and Braun (1946) demonstrated that D-glucosamine could be obtained in 60%–70% yield by heating crab chitin with concentrated hydrochloric acid on the water bath for 2.5 h. Later, in 1977, Muzzarelli reported its distribution, along with that of its derivative chitosan, in the living species (Muzzarelli, 1977).

3.12 Other studies

In 1901, Müller showed that the hexosamine hydrochlorides prepared from chitin and mucin respectively, yielded the same phenylosazone, identical with D-glucosazone (Müller, 1901). Though the evidence for the identity of the mucin hexosamine with chitosamine was conclusive, Müller realized that the configuration at carbon 2 of the compounds, that is, whether D-glucosamine or D-mannosamine, was still ambiguous. In 1902, Neuberg claimed to have correlated glucosamine with the mannose configuration by conversion to an α-aminohexoic acid on reduction with hydriodic acid (Neuberg, 1902; Neuberg and Wolff, 1903). However, later, Karrer and Mayer (1937) were unable to confirm this.

Löwy (1909) proposed a "simple" protocol to prepare chitosan (conditions: fused potash treatment at 170–180°C for 30 min), a polymeric mono-acetyl-di-glucose-amine. He considered that only one acetyl group had been removed for every low sugar residues and pointed out that chitosan was not a single substance. Scheme 3.4 shows the structure of chitosan suggested by Löwy (1909) compared with that of chitin and cellulose. Chitosan was identical with cellulose (and also chitin), except that the

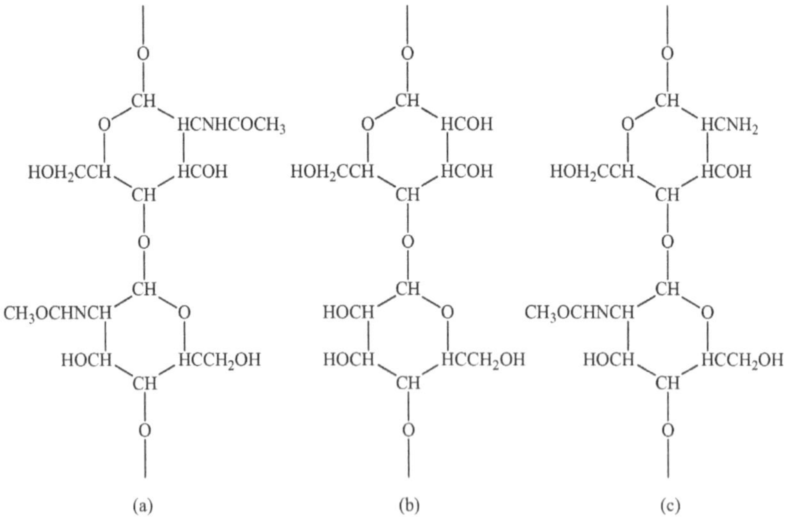

Scheme 3.4 Structures suggested by Löwy in 1909 for (A) chitin, (B) cellulose and (C) chitosan. *(Source: Löwy, E., 1909. Über kristallinisches chitosansulfat. Biochem. Z. 23, 47-60.)*

secondary hydroxyl on the alpha-carbon of the latter was substituted by an amine group. Löwy studied chitin derivatives and chitosan chemistry, in particular the elementary composition and crystalline structure form of chitosan sulfate, which was one of the most stable and insoluble of the chitosan salts. The partially deacetylated nature of chitosan has been further demonstrated by Karrer and White (1930). These authors pointed out the fact that, during chitinase reaction, 2-acetamido-2-deoxy-D-glucose was liberated.

In agreement with the structure suggested by Löwy (1909), Brach (1912) provided an empirical formula for chitin: $(C_{32}H_{54}N_4O_{21})_n$. Later, Gonell (1926), studying chitin by X-ray analysis, proved that this formula corresponded to 18 acetylglucosamine residues. Using the experimental protocol proposed by Fränkel and Kelly (1901a,b), Brach also carried out the acid hydrolysis of chitin to obtain *N*-acetyl-D-glucosamine in order to determine its structure. Brach observed both the presence of *glykosamin* and acetic acid and obtained conclusions in agreement with the results of Schmidt (1845), Ledderhose (1878), Tiemann (1886), Schmiedeberg (1891), Araki (1895), and Fränkel and Kelly (1901a,b). With von Fürth, Brach also determined the stoichiometry of D-glucosamine and acetic acid production. They first demonstrated that the two products were formed in equimolecular proportions and concluded that chitin was a polymerized mono-acetylglucosamine (Brach and von Fürth, 1912). One year later, Kotake and Sera (1913), studying the stoichiometry of D-glucosamine and acetic acid production, also found it to be a 1:1 ratio. Then, it became clear that chitin consisted of acetylated glucosamine, and hence it was different from cellulose. However, others researchers such as Morgulis remained unconvinced by these conclusions. Sergius Morgulis (1885–1971) was an American biochemist and marine biologist of Russian descent, professor of Biochemistry and Physiology in the College Medicine, University of Nebraska, Omaha. Morgulis (1916, 1917) doubted that the acetyl group was a constituent of the chitin molecule. For him, acetic acid from acid or alkaline hydrolysis of chitin was a secondary decomposition product. In the journal Science, Morgulis (1916) assumed that the chitin was "a polymere of eight $C_8H_{15}NO_6$ molecules" and its hydrolysis yielded seven molecules of glucosamine ($C_6H_{13}NO_5$), one molecule of glucose ($C_6H_{12}O_6$), and one of an unidentified nitrogenous fraction. The volatile acid produced in hydrolysis of chitin was a mixture of lower fatty acids, and its production was associated with a decomposition of the glucose molecule. One year later, Morgulis used a detailed hydrolytic study of chitin and demonstrated that "the evidence is against the accepted

AN HYDROLYTIC STUDY OF CHITIN[1]

SERGIUS MORGULIS

WITH THE COLLABORATION OF E. W. FULLER

From the United States Fisheries Biological Station, Woods Hole, and the Department of Physiology of the College of Medicine, Creighton University

Received for publication March 17, 1917

Since its discovery in 1823, chitin has aroused a great deal of interest. This substance has a very wide distribution throughout the invertebrate kingdom, particularly in the skeletons of arthropods, but the mode of its formation and its nature are still shrouded in mystery For the solution of this problem the first and essential thing is a clear understanding of the composition of chitin. Although much attention has been given to this matter by previous investigators neither the constitution of chitin nor indeed its empirical formula may be said to be beyond dispute, and this uncertainty is due to the fact that the subject has not been approached from the proper point of view.

4. The evidence presented in this paper is against the accepted view that chitin is a polymerized acetylglucosamine.

Fig. 3.11 Extract of the article by Professor Morgulis where he claimed that chitin is not a polymerized acetylglucosamine. *(Source: Morgulis, S., 1917. An hydrolytic study of chitin. Am. J. Phys. 43, 328–342.)*

view that chitin is a polymerized acetylglucosamine" (Fig. 3.11). He commented that "the substance isolated by Fränkel and Kelly probably has nothing to do with chitin, and the correspondence between the empirical formula and that for acetylated glucosamine is entirely fortuitous" (Morgulis, 1917).

The work of van Wisselingh published in 1898 on the preparation of chitosan used in microchemical tests to identify chitin was confirmed 20 years later by that of Scholl, who prepared pure chitin from the tissues of *Boletus edulis* (Scholl, 1908a,b). Scholl proposed a simplified experimental protocol for the purification of fungal chitin. He demonstrated that the colorimetric test proposed by van Wisselingh (1898a,b) was efficient and perfectly reproducible despite the difficulties of the purification steps. Scholl's method is detailed and discussed in the review by Tracey (1957). After multiple purification steps, the final product from *Boletus edulis* had a nitrogen content of 5.9%–6.1%, in agreement with van Wisselingh's data. Later, similar results were published by Blank (1954), while Zechmeister and Tóth (1934) and Schmidt (1936) had previously found the nitrogen content to be only 4%. All the authors pointed out serious problems in the purification steps. To avoid degradation, Schmidt (1936) recommended dilution of the

HCl solution used by pouring into dilute NaOH or Na_2CO_3 as the presence of NaCl assisted flocculation of the precipitate. Later, Fraenkel and Rudall (1940, 1947) also reported similar problems of purification for animal chitin due to the association between chitin and protein. The preferred methods involved prolonged treatment with hot alkali. However, at the same time, Scholl's method was challenged by Ilkewitsch (1908) who claimed that the skeletal substance of the cell membrane of fungi was neither chitin nor cellulose, but a nitrogenous substance close to these two substances. Later, Dous and Ziegenspeck (1926) also concluded that the product from *Boletus edulis* was not chitin. Chitin prepared from *Boletus edulis* and crab shells contained 2.09%N and 6.05%N, respectively. The nitrogen values for fungal chitin were much lower than those reported by van Wisselingh and Scholl, but no explanation was given. Ilkewitsch (1908) and Dous and Ziegenspeck (1926) then proposed to call chitin mycetin. When mycetin was hydrolyzed by hydrochloric acid, it yielded mycetosamine (a hexosamine), which was different from the chitosamine obtained from crab chitin. The corresponding sugars were also different, with mycetosamine giving a methyl pentose, called mycetose, while from chitin, the sugar chitose ($C_6H_{10}O_5$) was obtained. The main conclusion was the fact that animal chitin and fungus mycetin were not identical. Scholl totally disagreed with these conclusions. Over a hundred different fungi were then examined for the presence of chitin and cellulose by color tests, and the two were never found to occur together (Kühnelt, 1928a,b; Hopkins, 1929; Campbell, 1929). The tests proposed by van Wisselingh and Scholl were also applied later by Wester (1910) to the cuticle of arthropods. His work has shown that chitin was present in all arthropods, invariably in the exoskeleton and lining the respiratory system and except in some arachnids, in the greater part of the gut. It never occurred in protozoa, echinoderma, worms other than annelids, or in vertebrates (Wester, 1909, 1910). Wester proved conclusively that cellulose was not present in the cuticles of several species of insects, which had been suggested by Krawkow (1892) and Zander (1897).

In 1915, Vouk proposed a "much simpler method" but based on van Wisselingh's protocol for detecting the presence of chitin in a wide variety of materials, including *Agaricus fusipes*, *Amanitopsis plumbea*, *Boletus sanguinus*, and *Russula aeruginosa* (Vouk, 1915). The method consisted in boiling the material for 20–30 min in a saturated aqueous solution of potassium hydroxide in a beaker and washing the product with 90% alcohol. The product was then treated with iodine-potassium iodide reagent and sulfuric acid. As in van Wisselingh's chemical microtest, chitin also gave a violet color. Vouk's method is discussed in the references Hopkins (1929) and Campbell (1929). At the beginning of the 1920s, interest in chitosan was principally due to its property of forming a number of insoluble salts (Brunswick, 1921; Schulze and Kunike, 1923; von Weimarn 1926a). The least soluble salts were the phosphate, sulfate, and chromate. The nitrate and sulfate were not soluble in 5%–20% HCl, while the phosphate and chromate were soluble, in agreement with the previous results published by Löwy (1909). The hydrochloride and acetate were freely soluble in water. Based on these observations, Brunswick then proposed in 1921 another "simple test" to detect chitin, which consisted of first boiling chitosan (without alkali) in 10% sulfuric acid and then allowing the solution to cool slowly. During the cooling, sphaerocrystals of chitosan sulfate were formed which showed a color violet with iodine (Brunswick, 1921). Chitosan sulfate formation from chitin used as a test for the presence of chitin was later studied by Tauber (1934), and Roelofsen and Hoette (1951). These later authors also indicated detailed directions for carrying out the test and in particular the use of a polarizing microscope was suggested (the sphaerocrystals of chitosan sulfate showed a white cross in polarized light). For histological investigations, another test was proposed by Schulze and his student Kunike (Schulze, 1921, 1922a,b, 1926; Schulze and Kunike, 1923; Kunike, 1925, 1926a). It consisted in treatment of chitin with a solution of chlorine dioxide in 50% acetic acid ("diaphänol" solution, also used for delignifying cellulose) in the dark for several days. The chitin became soft and white and gave a blue/violet color with zinc chloride and iodine. The color was provided by pure, or nearly pure, chitin and not by chitosan. The conditions of the test proposed by Schulze and a full discussion can be found in Richards' book (Richards, 1951). Schulze considered his test to be less problematic to use than the test proposed by Brunswick and by van Wisselingh. However, Kühnelt (1928a,b), Hopkins (1929), and Campbell (1929) proved that Schulze's test (zinc chloriodide test) was less specific and not entirely reliable as other substances also behaved similarly, and some tissues known to contain chitin failed to show a positive reaction. Later, other works confirmed that the zinc chloriodide test was not specific for chitin but was given by other polysaccharides and by certain inorganic substances (Wigglesworth, 1948; Whistler and Smart, 1953).

The main criticism of these qualitative microchemical tests used for chitin detection was the presence of chitosan in the starting material, which could lead to misinterpretations. Kreger (1954) and Hackman (1960), indeed, indicated that samples should be treated to remove chitosan by acid extraction, prior to deacetylation with alkali. The authors indicated that although there was no doubt about the reliability of chitosan tests in establishing the presence of chitin when the results were positive, negative results were not conclusive evidence of the absence of mucopolysaccharide. Several authors also indicated the difficulty of establishing clear interpretations, preferring enzymatic or physical (X-ray diffraction, electron microscopy) detection methods (Kreger, 1954; Tracey, 1955; Hackman, 1960; Runham, 1961a,b; Aronson, 1965).

von Weimarn (1926a,b, 1927) reported that aqueous solutions of certain neutral salts that were capable of a high degree of hydration can be used to disperse polymers such as chitin, cellulose, and fibroin. The dispersal efficiency of suitable salts followed the order $LiCNS > Ca(CNS)_2 > CaI_2 > CaBr_2 > CaCl_2$ and $LiCNS > LiI > LiBr > LiCl$. Dispersion of chitin in aqueous calcium chloride was very difficult. Chitin can be recovered by dilution with alcohol (von Weimarn, 1927) or acetone (Clark and Smith, 1936). Schmidt et al. (1928a,b) reported that chitin dissolved in a solution in liquid ammonia "with difficulty" yielded a monosodium product, and in an anhydrous hydrofluoric acid, it yielded a water-soluble compound of unknown nature. Chitin dissolved slowly when treated with hypochlorite solutions at room temperature and was soluble in Schweitzer's solution, differing in both respects from cellulose. Payen previously noted the disintegrating effect of bleaching powder on insect cuticle (Payen, 1843).

At the end of the 1920s, chitin from the cell walls of fungi, from the cuticle of arthropods, and elsewhere in the animal kingdom began to be regarded as the same substance. Nevertheless, the complete identity of chitin from different sources was not universally accepted. The chitin of *Limulus* was said by Fränkel and Jellinek (1927) to contain more carbon and less nitrogen than that of crustaceans, but later this claim was not confirmed by Lafon (1941a,b, 1943a,b,c, 1948). Furthermore, there was a debate on the fundamental constituent of the arthropod cuticle, chitin, or protein (Imms, 1924; Forbes, 1930), and also on the structure of glucosamine, later resumed and discussed by Richards (1947a,b).

3.13 Literature review

The first reviews on the occurrence, distribution, and chemistry of chitin were those by van Wisselingh (1898a,b), Wester (1909, 1910), Biedermann (1914), von Wettstein (1921), Kunike (1925), Levene (1925),

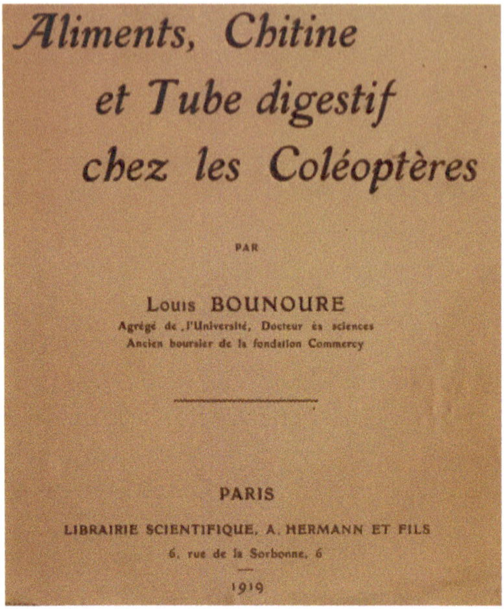

Fig. 3.12 Cover of Bounoure's book published in 1919 where Professor Louis Bounoure described the distribution, production, characterization, chemistry, and biology of chitin. *(Source: Bounoure, L., 1919. Généralités sur la chitine. In: Houssay, F., (Ed.), Aliments, Chitine et Tube Digestif chez les Coléoptères. Collection de Morphologie Dynamique. Librairie Scientifique A. Hermann et Fils, Paris. 294 p.)*

and Campbell (1929). The first comprehensive *mémoire* (294 pages and 242 references) on chitin was edited by Louis Bounoure in 1919 (Fig. 3.12; Bounoure, 1919).

CHAPTER FOUR

Exploration: 1930–1950

The biological analogy between chitin and cellulose led us to the conclusion that the units of glucosamine are united in the same manner as the glucose units in cellulose, that is, in 1,4-β linkage, each chain having a diagonal screw axis.

Professor Kurt H. Meyer, 1936,
Professor of Inorganic and Organic Chemistry,
Director of the Organic Chemical Institute,
University of Geneva, Switzerland

Between 1930 and 1950, X-ray analysis became the most reliable physical method for differentiating chitin from cellulose in fungal cell walls (Farr and Sisson, 1934; Clark, 1934; Mark, 1943; Frey, 1950). X-ray analysis was used not only as a means for their detection but also for the purpose of determining molecular orientations in their structure. Rapidly, the use of X-ray powder diffraction offered simplicity and ease of interpretation for the detection and characterization of crystalline substances (Frey, 1950; Blank, 1954; Kreger, 1954; Rudall, 1955). For example, the technique was used to detect chitin on the walls of filamentous fungi, yeasts, and pathogenic fungi, and the results were compared to those obtained from chitin from crustaceans. Chitin was also thought to be present in some green algae, which intrigued mycologists. Kreger et al. studying the components in the walls of *Phycomyces blakesleeanus* claimed that X-ray analysis was ideal for detecting chitin and chitosan instead of chemical tests, but the analysis required purification steps before the chitin reflections began to appear (Houwink and Kreger, 1953; Kreger, 1954). Orr (1954) also reported a similar conclusion for infrared analysis. It is important to point out that much of the knowledge of the properties and structure of chitin has been derived from studies on "purified" chitin. This was important because the removal of the associated protein increased the degree of orientation of polysaccharide chains (Picken and Lotmar, 1950), and the attraction between them was such that they aggregated to give a continuous sheet of chitin. This was clearly demonstrated by X-ray diffraction. Using this technique, several studies showed that chitins of different biological origin displayed different degrees of crystallinity. In fibrous chitin found in the laminae of beetle elytra

Chitin and Chitosan
https://doi.org/10.1016/B978-0-323-96119-6.00007-4

or in sporangiophores of Phycomycetes, a high level of molecular orientation was found. During the same period, chitin was the object of numerous studies not only on its (macro)molecular structure, chemistry (chitosan production) and biochemistry but also on the fiber production. Indeed, chitin and chitosan attracted considerable attention with the exploration of natural fibers, with George W. Rigby's first patents on chitosan films and fibers in the mid-1930s (Rigby, 1936a,b,c,d,e, 1937). The first potential applications are also beginning to be explored. For example, chitosan was proposed as a laundering-resistant coating, sizing agent, and cement. There was also the first use in papermaking industry (Lubs, 1937; Larson, 1939, 1940), textile (Heckert, 1937; Arnold, 1939), photography (Marasco, 1938, 1939; White, 1944), and as adhesives (Maxwell, 1939). Fundamental research was also directed toward the study of the biology of the arthropod cuticle. From a fundamental point of view, the names that marked this period were Kurt H. Meyer, Max Bergmann, László Zechmeister, Norman Haworth, and Albert G. Richards.

4.1 Kurt H. Meyer

Kurt H. Meyer (1883–1952) was a German chemist of Baltic origin who published outstanding contributions on natural and synthetic polymers and their chemistry. Many of them were devoted to chitin (Meyer and Mark, 1928; Meyer and Pankow, 1935; Meyer and Wehrli, 1937; Meyer, 1942, 1950). Bütschli (1874) previously emphasized the resemblance of chitin to cellulose, and that similarity had become still clearer in the 1930s with the data published by Meyer obtained using X-ray diffraction analysis. For Meyer, there was "a kind of physicochemical parallelism between chitin and cellulose."

Kurt Otto Hans Meyer was born on September 29, 1883 in Dorpat. His father Hans Horst Meyer was a renowned pharmacologist who taught at the University of Dorpat. Young Kurt was educated in Germany and began his chemistry studies in Marburg with Ernst Theodor Zincke (1843–1928; German chemist and a student of August Kekulé, 1829–1896, German organic chemist who was the principal founder of the theory of chemical structure) until 1901 and finished them in Leipzig under Arthur Rudolf Hantzsch (1857–1935; German chemist and a student of Johannes Wislicenus, 1835–1902, known for his work on stereochemistry). Meyer studied both chemistry and medicine. In 1907, he obtained his doctorate entitled "*Untersuchungen über Halochromie*" (research on halochromie) under the supervision of Professor Hantzsch. Meyer then traveled with his father throughout America and Europe. He spent a year in Rutherford's laboratory in Manchester (Ernest Rutherford, 1871–1937, New Zealand-British physicist and chemist, considered the Father of nuclear physics) and in Ramsay's department of chemistry at London (Sir William Ramsay 1852–1916, British inorganic chemist, Nobel Prize in 1904). After his return to Germany, Meyer worked at Bayer's laboratory in Munich. During the First World War, Meyer spent 3 years as an officer in the artillery. In 1917, he was called upon to carry out research on warfare at the *Kaiser Wilhelm-Berlin Institute* under the supervision of Fritz Haber (1868–1934; German chemist, Nobel Prize in Chemistry in 1918). After the war, Meyer returned to Munich as professor in chemistry until 1920. In 1921, Professor Meyer left university life to become director of a research laboratory at the *Badische Anilin- und Sodafabrik* in Ludwigshafen and in 1926 at the *I. G. Farbenindustrie*. However, the impending success of Nazism forced Meyer to leave Germany. In 1932, Professor Meyer accepted a position at the University of Geneva where he succeeded Professor Pictet as head of the Laboratory of Organic and Inorganic Chemistry. He then became director of the Organic Chemical Institute, where he continued research on natural polymers and lectured in German and French. Among his many students was Edmond H. Fischer, an American biochemist who obtained in 1992 the Nobel Prize in Medicine. Kurt Otto Hans died on holidays in Menton on April 14, 1952 (Mark, 1952; van der Wyk, 1952; Hopff, 1953; Jeanloz, 1956).

In 1924, Herzog was the first to show the crystalline nature of chitin using X-ray diffraction (Herzog, 1924). Reginald O. Herzog (1878–1935), born in Vienna, first worked at the *Technische Hochschule* at Karlsruhe, Germany, from 1905 to 1912. He was then professor in physiology chemistry at the German Technical High School in Prague (1912–1919) in organic chemistry at the *Kaiser Wilhelm-Institut für Faserstoffchemie* in Berlin-Dahlem (1919–1933), and in chemical engineering at the University of Istanbul (1934–1935). Herzog was also the first to demonstrate the microcrystalline structure of cellulose using X-ray diffraction. The fibers of cellulose were nearly "ideal" fibers in that the crystallites had a perfectly random distribution about the axis of the fibers. Herzog's work opened up the modern concepts on high macromolecular weight substances. The crystalline nature of chitin was confirmed and amplified 2 years later by Gonell, who arrived at a hexagonal unit cell (Gonell, 1926). Indeed, Gonell worked in the same institute of fiber chemistry and studied the X-ray data of chitin (*Goliathus giganteus*) under the supervision of Herzog. Gonell proposed and discussed a rhombic cell with the dimensions $a = 19.42\,\text{Å}$, $b = 10.40\,\text{Å}$, and $c = 11.58\,\text{Å}$ (the cell contained 10 acetylglucosamine units). However, Gonell eliminated it because such a cell could not contain the 10 sugar units which were indicated by density measurements (Gonell, 1926). He preferred a hexagonal unit cell with 18 glucosamine units (Fig. 4.1). His work has been taken up, examined in detail, and discussed extensively by Meyer.

Meyer and Mark (1928), aware of the biological analogy between chitin and cellulose and of the similarity of their structure, proposed a rhombic unit cell. The analogy between the two polysaccharides was apparent in the main features of the X-ray diagram. The units of glucosamine were united in the same manner as the glucose units in cellulose, that is, in 1,4-β linkage, each chain having a diagonal screw axis. Meyer and Mark concluded that chitin was identical with cellulose, except that the secondary hydroxyl on the alpha-carbon atom of the latter was substituted by an acetamide group. The authors ascribed to chitin the constitution shown in Scheme 4.1 and assumed that chitin had a micellar structure of parallel oriented chains similar to that of cellulose. At the beginning of the 1930s, this structure received important support when Bergman isolated chitobiose, a product of acetolysis, considered the descriptive unit of chitin (Bergmann et al., 1931a,b,c). The structure of chitin has been made clear from this chemical investigation. The assumption that chitin had a micellar structure of parallel oriented chains similar to that of cellulose was also confirmed in the same period by the demonstration of the presence of β-glucosidic linkages by Zechmeister using incomplete acidic hydrolysis (Zechmeister et al., 1932;

Röntgenographische Studien an Chitin.

Von

H. W. Gonell.

Mit 1 Tafel.

(Aus dem Kaiser Wilhelm-Institut für Faserstoffchemie, Berlin-Dahlem.)

(Der Redaktion zugegangen am 4. November 1925.)

Vergleich verschiedener Chitinarten.

Fig. 4.1 First page of the article of Professor Gonell where he describes for the first time the crystalline nature of chitin and cell dimensions. *(Source: Gonell H.W., 1926. Röentenographische studien an chitin. Hoppe-Seylers Z. Physiol. Chem. 152, 18–30.)*

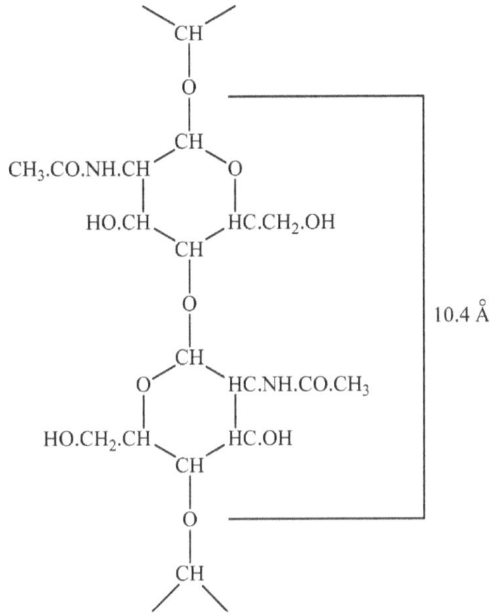

Scheme 4.1 Structural formula of chitin proposed by Meyer and Mark (1928) in agreement with that suggested by Löwy (1909).

Zechmeister and Tóth, 1933). Later, the results of more detailed X-ray examination and of the study of the mode of reaction were also in agreement with this formula. However, in the mid-1930s, Schorigin stated that the unit of the structure was neither glucose nor mannose, but another sugar (named chitose), differing in configuration (Schorigin and Hait, 1934, 1935; Schorigin and Makarowa-Semljanskaja, 1935a,b).

A detailed analysis by Meyer and Pankow (1935) provided the basis for the structure commonly accepted in the 1950s (Meyer, 1950; Whistler and Smart, 1953). This analysis was made on a tendon of the rock lobster (*Palinurus vulgaris*) and led to a rhombic cell comprising four chitobiose units with the dimensions $a = 9.40$ Å, $b = 10.46$ Å (fiber axis), and $c = 19.25$ Å. Indeed, the cell contained eight acetylglucosamine units (7.9 calculated from the density = 1.415). The analogy with cellulose was apparent in the main features of the diagram proposed for the unit cell of chitin (Fig. 4.2): equal numbers of chains were arranged in opposing directions, the rings followed one another in a diagonal screw sequence and all rings lay flat in one plane. These results have been confirmed 1 year later by Clark and Smith (1936) but stated the dimension of the a axis as 9.25 Å. The crystalline regions were of approximately the same size and shape as those of cellulose. Later, Lotmar

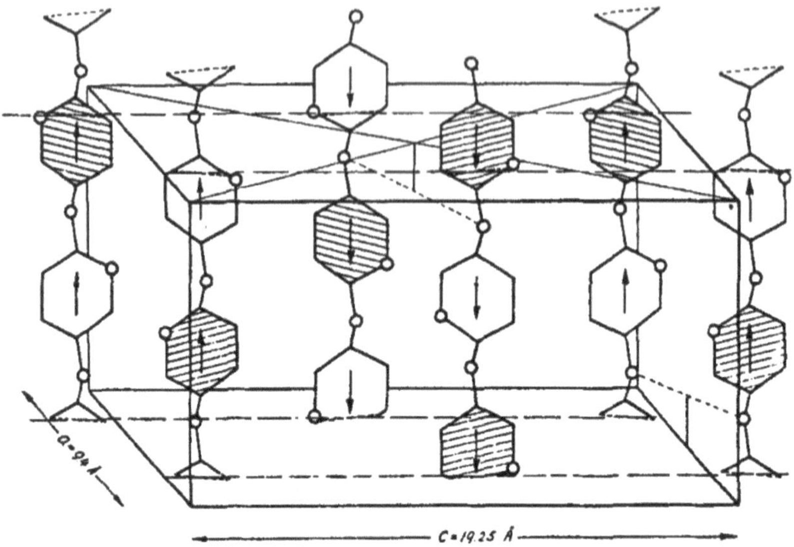

Fig. 4.2 Diagram of the unit cell of chitin proposed by Meyer and Pankow (1935).

and Picken (1950), using chitin from the same source (*P. vulgaris*), confirmed the *a* and *c* values but preferred 10.27 Å for *b*. Similar figures were also obtained using fungal and insect chitin. Heyn (1936a,b) stated dimensions of $a=9.70$, $b=10.4$, and $c=4.6$ Å for fungal chitin, and the diagrams obtained were almost the same as those from control chitin from cockroaches. Similar X-ray patterns were obtained from chitin in plant cell walls (Khouvine, 1932; Heyn, 1936c; van Iterson Jr et al., 1936).

Meyer and Pankow (1935) also indicated that acetyl groups alternated from one side of the chain to the other on passing from one residue to the next (evidence in favor of β-linking) and that while chains along the a axis ran in the same direction; the direction was reversed in their neighbors in the direction of the c axis. This alternation in the direction of contiguous chains was also found in cellulose, and the similarity in the glycosidic links in the two materials was supported by the fact that the heat of activation for acid hydrolysis of both was about 30 kcal/mole (Meyer and Wehrli, 1937; Meyer, 1942). This heat of activation was the same as that of the linkage in cellulose (Meyer and Wehrli, 1937). Later, the crystal structure determined by Meyer and Pankow (1935) was slightly modified by Darmon and Rudall (1950), who also suggested that there were C=O···H—N and C=O···H—O hydrogen bonds and free O—H and N—H groups in the crystal. Pursuing his studies on chitin chemistry, Meyer reported that solutions of fungal chitin in acids had the same viscosity as those of animal chitin (Meyer and Wehrli, 1937). It was noted that the viscosity of solutions of chitin in nitric acid solution was of the same order as that of similar solutions of cellulose, indicating a degree of polymerization of more than a hundred. Fikentscher (1932) previously reported a similar conclusion. Studying chitosan preparation, Meyer and Wehrli (1937) obtained almost complete deacetylation (conditions: 50% sodium hydroxide solution for 35–40 min at 100°C), in disagreement with the result published by Löwy (1909).

To describe polymers of animal origin containing hexosamines, either in pure form or as salts, Meyer also first used the term mucopolysaccharide. This term was intended to denote the relationship of these polysacharrides to protein content. However, little was known about the structure of protein complexes, and the term mucopolysaccharide was first applied only to pure polysaccharide and then to protein-polysaccharide complexes with high protein content. Meyer reviewed and classified these complexes, introducing the term glycoproteins (Meyer, 1938). Meyer's classification was revised by Stacey 5 years later (Stacey, 1943), and the following definition was accepted: a mucopolysaccharide is a polysaccharide-protein (or peptide) complex in which the carbohydrate is the major constituent while a mucoprotein comprises the same substances but with protein as the major constituent. Later, extension of the classification included lipoproteins and lipopolysaccharides (Kent and Whitehouse, 1955). It was also accepted that the mucopolysaccharides performed various functions in the animal body, and for example, chitin, as mucopolysaccharide containing 2-acetamido-2-deoxy-D-glucose, was similar in structure in many invertebrates to cellulose in the plant world. Chitin closely resembled cellulose in many of its properties and in structure, and both consisted of long primary valence chains of glucose residues (Picken, 1940; Meyer, 1942). Both polysaccharides served as structural and defensive materials in nature. By the early 1940s, it was also clear from all the results published that animal and plant chitins were essentially similar, as suggested by Meyer's studies. The constitution of chitin shown in Scheme 4.2 has been recognized. Later, this

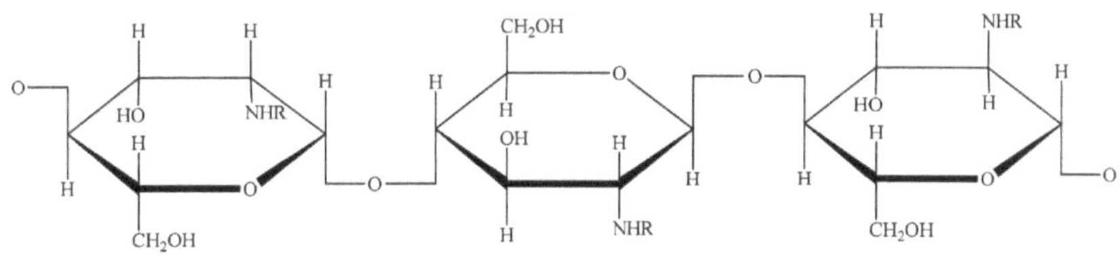

Scheme 4.2 Structural formula of chitin accepted in the 1940s.

structure was also identified by enzymatic analysis and infrared spectroscopy (Darmon and Rudall, 1950; Orr, 1954; Brock, 1957; Pearson et al., 1960; Spedding, 1964).

4.2 Max Bergmann

At the end of the 1920s, there seemed little doubt that nearly all of a purified chitin can be regarded as a polymeric form of *N*-acetyl-2-amino-2-deoxy-D-glucose (Tracey, 1957; Foster and Webber, 1961; Kent, 1964). However, it remained to determine the mode of linkage. On this topic, Bergmann published several papers (Bergmann and Zervas, 1931; Bergmann et al., 1931a,b,c). He was also the first to suggest that glucosamine had the typical pyranose ring structure (Bergmann et al., 1934). In the proposed structure, the six atoms of the ring were nearly coplanar, the oxygen atom being slightly displaced out of the plane of the carbon atoms. This was later confirmed by Cox (Cox et al., 1935; Cox and Jeffrey, 1939).

Max Bergmann (1886–1944; Fig. 4.3), a German biochemist, received his Ph.D. in 1911 and became the assistant to Fischer at the University of Berlin. After Fischer's death in 1919, Bergmann automatically became his scientific executor, assuming responsibility for the completion and publication of unfinished research studies (Helferich, 1969; Katsoyannis, 1973;

Fig. 4.3 Professor Max Bergmann, 1886–1944. *(Source: Helferich B., 1969. Max Bergmann 1886–1944. Chem. Ber. 102, I–XXVI.)*

Lichtenthaler, 2002). In 1920, Bergmann habilitated at the University of Berlin and was appointed to head the newly established *Kaiser-Wilhem Institut für Lederforschung* in Dresden. In 1933, he left Germany for the United States and accepted a position of researcher at Rockefeller University in New York City. Max Bergmann is considered a pioneer of applied sciences and of molecular biology.

Like Meyer, Bergmann first observed that chitin was not found alone and formed a part of very complex systems. Insect exoskeletons comprised chitin-protein complexes, whereas crustacean shells contained large proportions of calcium carbonate in addition to protein. In the fungi, chitin was also found in close association with other substances. The choice of experimental conditions for extraction and purification of chitin was fundamental to avoid degradation of the products. Bergmann concluded that chitin could not be strictly considered a natural chemical entity but as a degradation product of the chitin-protein complex (the term glycoprotein proposed by Meyer was used later). Bergmann then studied the partial degradation of purified chitin by acetolysis, consisting in a treatment with a mixture of sulfuric acid and acetic anhydride. With his student Leonidas Zervas (Katsoyannis, 1973), Bergmann in 1931 was the first to isolate a crystalline disaccharide, as the octa-acetate, by acetolysis of chitin (Bergmann et al., 1931a). Bergmann assigned it the name chitobiose, considered the main descriptive unit of chitin. The reducing disaccharide chitobiose was the first low molecular weight polymer obtained from chitin. The product contained two *N*-acetyl and six *O*-acetyl groups. Further acetylation confirmed the presence of six hydroxyl groups. Its configuration was identical with the cellobiose molecule except for the substitution of acetylamine groups for the hydroxyl group on carbon-2. All results were in agreement that chitin had a similar structure to that of cellulose, and the same formula showed in Scheme 4.2, as Meyer and Mark (1928) had already suggested. The reactions of chitobiose octa-acetate, that is, oxidation with iodine in alkaline conditions followed by hydrolysis led to the conclusion that a 1,4-linkage was likely and X-ray evidence and changes of rotation ($^{20}[\alpha]_D = -14° \rightarrow +56°$) on hydrolysis indicated a β-form, in agreement with the results published by Meyer and Mark (1928) and by Karrer (Karrer and Hofmann, 1929; Karrer and von François, 1929). All the results clearly demonstrated that carbon-1 and carbon-2 of the reducing moiety cannot be involved in the glycosidic linkage. Shortly thereafter, the postulated β-glycosidic linkage was also confirmed by Zechmeister and the same conclusions as Bergmann about the structure and formula of chitin and its similarity to cellulose (Zechmeister et al., 1932; Zechmeister and Tóth, 1933).

4.3 László Zechmeister

László Zechmeister (1890–1972), professor of organic chemistry emeritus, is considered a pioneer of chromatography (Ettre, 1989; Wirth, 2013). He also published a series of works on chitin chemistry (Zechmeister and Tóth, 1931, 1932, 1933, 1934, 1939a,b,c; Tóth and Zechmeister, 1939; Zechmeister et al., 1932, 1939a,b). The isolation of the chitin saccharides only became feasible with the development of chromatographic techniques developed by him.

László Zechmeister was born in 1890 in the city of Györ, a town in the north-west of Hungary. In 1907, he began his studies in chemistry at the *Eidgenössische Technische Hochschule* in Zurich, under the guidance of Richard Willstätter (1872–1942, German chemist, Nobel Prize in chemistry in 1927). Zechmeister received his degree as a technical chemist on 1911 and then worked at the Kaiser Wilhelm Institute for Chemistry in Berlin-Dahlem from 1912 to 1914. During this period, Zechmeister received his doctorate on "*Zur kenntnis der cellulose und des lignins.*" The experiments were conducted first in Zurich (1912) and then in Berlin-Dahlem (1913). During the World War I, László Zechmeister was enlisted for the Hungarian army. He was taken prisoner and sent to a prison in Siberia. During his captivity, he learned the English language himself. In 1919, he was released and first returned to Munich to work with Willstätter and then returned to Hungary. Two years later, Dr. Zechmeister accepted a position as an instructor at the Royal Danish Agriculture and Veterinary Academy in Copenhagen until 1923 under Professor Niels Bjerrum (1879–1953). For his work, Dr. Zechmeister was elected a Foreign Member of the Royal Danish Academy of Science. Then, he became professor of medical chemistry at the University of Pécs, Hungary (he was only 33 years old). In 1935, Professor Zechmeister received the Pasteur Medal of the *Société Française de Biochimie.* (French Society of Biochemistry.) In 1949, he was awarded a Guggenheim fellowship in order to lecture in European universities. Ten years later, Professor Zechmeister became professor emeritus at Caltech (California Institute of Technology). In 1962, he received the Labline Award of the American Chemical Society (of which he was an active member of the society's journal) for his work in chromatography and electrophoresis. László Zechmeister died in Pasadena on February 28, 1972. Wirth (2013) has published a comprehensive biography on the life and scientific career of László Zechmeister.

Like Meyer, Zechmeister agreed with the similarity of chitin and cellulose in many of their properties. He regarded chitin to be a derivative of cellulose, in which the C-2 hydroxyl groups have been replaced by acetamido residues. Like other research groups, Zechmeister was interested in the hydrolysis products of "pure" chitin to demonstrate its structure and configuration. In order to achieve this goal, he was first interested in preparing pure chitin from fungal and animal resources. However, he knew the difficulties of this preparation. Zechmeister and his colleague Géza Tóth then took up the work of Scholl (1908a,b) and improved the purification protocol. They then examined the products of partial acid hydrolysis of their "pure" chitin. In 1931, chitobiose, a disaccharide found by Bergmann (Bergmann et al., 1931a,), was also isolated by Zechmeister and Tóth (1931). Chitobiose, originally isolated from lobster shells, was isolated from beetles, snail radulae, and also fungi (Zechmeister and Tóth, 1934; Tóth and Zechmeister, 1939). This crystalline disaccharide had six oxygen-linked ester groups and two *N*-acetyl groups. Chitobiose had reducing properties toward alkaline cupric and silver reagents. However, the sugar lost these reducing properties when oxidized by sodium hypoiodite to the corresponding chitobionic acid. Treatment of the acid with acetic anhydride led to an unsaturated glycoside in which a double bond was located between the C-2 and C-3 carbons of the lactone ring, suggesting that the formation of the product was due to a β-elimination. The β-linkage in chitobiose was finally proven by X-ray diffraction data during hydrolysis of chitobiose by enzyme preparations from emulsion. Indeed, Zechmeister separated two enzymes from the digestive juice of *Helix pomatia* by chromatography on bauxite. The first, a chitinase, catalyzed the degradation of chitin into chitodextrins, and the second was inactive toward chitin but hydrolyzed chitodextrins. The main product of the reactions was *N*-acetylglucosamine, and the analysis of all the data suggested the β-linkage. However, at that time, the biological role of chitinase in emulsion was controversial. In addition to chitobiose, partial degradation of chitin also yielded a trisaccharide chitotriose from the acetolysis mixture, which had been characterized as its crystalline acetate (chitotriose undeca-acetate) (Zechmeister and Tóth, 1932). A more detailed account about the D-glucosamine (or chitosamine) was then provided. The authors, using X-ray diffraction measurements and rotary change experiments (Zechmeister and Tóth, 1932, 1933), also confirmed the β-linkage, in agreement with studies published by Karrer (Karrer and Hofmann, 1929; Karrer and von François, 1929) and Bergmann et al. (1931a). Zechmeister and Tóth (1934) comprehensively discussed the occurrence of chitin in plants and animals and its characteristics and analysis. A few years later, on the basis of the similarity found between chitin and cellulose in X-ray diffraction studies and on enzymic studies, Zechmeister confirmed the β-linkage (Zechmeister and Tóth, 1939a,b,c; Zechmeister et al., 1939a,b). Both the use of chemical and enzymic hydrolysis can provide an estimation of the chitin content, especially in fungal material. However, enzymic hydrolysis was more interesting because the reaction had the advantage that neither deacetylation nor destruction of the products of hydrolysis occurred. The reaction had also the advantage of confirming the presence of both acetyl groups and glucosamine units in the material, providing unequivocal evidence of the location of acetyl groups. The only problem was the fact that complete enzymic hydrolysis can be performed only under restricted conditions. There was a period of confusion for the nomenclature related to chitobiose. Zechmeister and Tóth first considered chitobiose a disaccharide of *N*-acetylglucosamine (Zechmeister and Tóth, 1931, 1939c; Zechmeister et al., 1932). In another work, the same authors considered it a disaccharide of glucosamine (Zechmeister and Tóth, 1939a). This remained a subject of debate between the different laboratories until the 1960s (Kent and Whitehouse, 1955; Barker et al., 1957, 1958; Berger and Reynolds, 1958; Foster and Webber, 1961).

4.4 Norman Haworth

At the end of the 1920s, the first X-ray diffraction patterns showed considerable similarity to patterns from cellulose and strengthened but did not prove the case in favor of glucosamine, the structural unit of chitin (Herzog, 1924; Gonell, 1926; Meyer and Mark, 1928). Glucosamine was then the object of numerous fundamental studies. Many references to glucosamine can be found at that time on its preparation, characterization, quantitative determination, structure, biochemistry, and chemical modification (Karrer and Smirnoff, 1922; Fränkel and Jellinek, 1927; Bergmann and Zervas, 1931; Elson and Morgan, 1933; Bergmann et al., 1934; Kawabe, 1934; Morgan and Elson, 1934; Cox et al., 1935; Cutler et al., 1937; Cox and Jeffrey, 1939; Cutler and Peat, 1939; Bierry et al., 1939; Freudenberg et al., 1942). However, much controversy and debate especially surrounded the constitution of glucosamine, and in particular the stereochemical position of the amino group (Meyer, 1942). Many reactions indicated that the amino group had the same position as the hydroxyl group in mannose, while other reactions suggested a relationship with glucose.

Unequivocal evidence for the structure and configuration of glucosamine, the amino-hexose from chitin as 2-amino-2-deoxy-D-glucose, was finally provided by the synthesis due to Haworth et al. (1939). Walter Norman Haworth (1883–1950) was a British chemist at the University of Birmingham, known for his work on ascorbic acid and the development of the Haworth projection used in organic chemistry to characterize sugar structures. Haworth studied organic chemistry at the University of Göttingen earning his Ph.D. in Otto Wallach's laboratory. Professor Haworth received the Nobel Prize in chemistry in 1937 (shared with Professor Karrer) for his work on carbohydrates and vitamin C.

In 1939, Haworth, Lake, and Peat proposed a synthetic method which distinguished between the alternative configurations (glucose or mannose) for glucosamine (chitosamine). The fission by alkali such as sodium methoxide of the ethylene oxide ring in anhydro-sugars leads to the formation of two isomeric sugars, consequent upon the independent rupture of the two bonds of the oxide oxygen atom. In each case, ring opening was accompanied by a Walden inversion at the carbon of the epoxide ring at which the attack occurred and to which the amine group, therefore, became attached (Haworth et al., 1939). The method has been used to prepare a dimethyl 2,3-anhydro-methylmannoside which has been shown to give rise, with sodium methoxide or sodium hydroxide, to both a glucose derivative and an altrose derivative. The opening of the anhydro-ring in the dimethyl 2,3-anhydro-methylmannoside was then carried out by the help of ammonia, which gave rise to a derivative of 3-amino-altrose on the one hand and of 2-amino-glucose on the other. The latter was shown to be identical with chitosamine, which might be considered configurationally to be glucosamine and, therefore, related to the parent sugar glucose. A more detailed description can be found in the book by Roberts (1992).

It should be pointed out that two other groups came to similar conclusions (Neuberger and Pitt Rivers, 1939; Cox and Jeffrey, 1939). Indeed, the same year, the carbon-2 configuration was also conclusively established by both Albert Neuberger and Rosalind Pitt Rivers using chemical and physical (optical rotation) methods. On the basis of a comparison of the relative rates of hydrolysis of the methyl α- and β-glycosides of glucosamine with those for methyl α-D-glycosaminide and its β-anomer that the methoxyl group in the methyl α-D-glycosaminide had a *cis* configuration with respect to the carbon-2-amine group, and hence glucosamine must have the same carbon-2 configuration as does D-glucose (Neuberger and Pitt Rivers, 1939). Ernest Gordon Cox and G.A. Jeffrey also determined the crystal structures of α-chitosamine hydrochloride and hydrobromide by X-ray crystallography and showed that the configuration of chitosamine was the same as that of D-glucose, the substituents on carbon-1 and carbon-2 of the α-form being in the *cis* position with respect to each other. Their results confirmed that glucosamine was a glucose and not a mannose derivative (Cox and Jeffrey, 1939).

4.5 Albert Glenn Richards

At the end of the 1940s, the works of Albert Glenn Richards (1909–1993), an American professor of entomology and zoology at the University of Minnesota, on the biology of the arthropod cuticle and the biochemistry of chitin were acknowledged to have made an important contribution (Anderson and Richards Jr., 1942; Richards, 1947a,b, 1949, 1951, 1952, 1958; Richards and Anderson, 1942; Richards and Cutkomp, 1946; Richards and Korda, 1948; Richards and Pipa, 1958).

From 1890 to 1930, the most extensive work on chitin in fungi has been carried out with the genera of *Aspergillus*, *Agaricus*, *Boletus*, and *Phycomyces*. The first and complete list of forms tested in the fungi was only reported in 1951 by Richards in his book on the integument of the arthropods, which really set the benchmark for chitin zoological research (Richards, 1951). This monograph (411 pages and 1800 references) is considered a fundamental guide to the voluminous literature on the subject. Richards also reviewed some histochemical methods for the detection of chitin, although Hackman later reported that histochemical estimations were inaccurate or nonspecific (Hackman, 1984). This monograph was updated 7 years later by Richards in a famous article entitled *The Cuticle of Arthropods* and published in the journal *Ergebnisse der Biologie* (Richards, 1958). Another fundamental book on the biology of the arthropod cuticle was published in 1975 by Professor Anthony C. Neville, University of Bristol, Department of Zoology, Great Britain (Neville, 1975).

4.6 Other studies

Pringsheim and Krüger (1932) were the first to point out the fact that the chemical modification of chitin was difficult because of the general insolubility of the material. According to Pringsheim and Krüger, the isolation of pure chitin from fungi was also "very complicated including multitreatments" by boiling water, alkali, hydrochloric acid and oxalic acid solutions, and by potassium permanganate and alcohol (Pringsheim and Krüger, 1932). By this method, at the mid-1890s, Gilson and Winterstein previously found that chitin constituted 5%–6% of the dry matter, while Pringsheim and Krüger (1932) reported that about 20%–50% of the cell walls of fungi consisted of chitin. The same year, using the diaphänol method, Koch (1932) found 48% chitin in the elytra of *Melolontha* (Odier previously obtained a value of 29%), 64% in the cuticle of *Pieris*, and 38% in the abdominal sclerites of *Periplaneta* (Table 4.1). In a large number of insects, in which the cuticle was treated with 10% potassium hydroxide at 100°C, until the residue was of constant of weight, Lafon (1943a) found an average value of 33% chitin, ranging from 55% in the larva of *Calliphora* to 25% in the hemielytra of *Cercopis*.

Although it has long been known that the chitin of fungi had the same composition as that of arthropods, the final demonstration of this was due to the work by Diehl and van Iterson Jr (1935). These authors demonstrated that the composition of fungal chitin obtained in pure form from the sporangiophores of *Phycomyces blakesleeanus* was $C_8H_{13}O_5N$, which was the

Table 4.1 Examples of amount of chitin in insect cuticle.

Source	**Amount present, in %**	**Reference**
Elytra of *Melolontha*	29	Odier (1823)
Elytra of *Melolontha*	48	Koch (1932)
Abdominal sclerites of *Periplaneta*	35	Campbell (1929)
Abdominal sclerites of *Periplaneta*	38	Koch (1932)
Dorsal abdominal cuticle of *Periplaneta*	37.6	Tauber (1934)
Hind wings of *Periplaneta*	18	Tauber (1934)
Cuticle of *Sarcophaga* larvae: wet weight	20	Dennell (1946)
Cuticle of *Sarcophaga* larvae: dry weight	42	Dennell (1946)
Cuticle of *Sarcophaga* larvae: dry weight	52	Fraenkel and Rudall (1947)
Cuticle of *Pieris*	64	Koch (1932)
Cuticle of *Oryctes*	32.3	Lafon (1943a)
Endocuticle of the cockroach	60	Campbell (1929)
Exocuticle of the cockroach	22	Campbell (1929)
Cuticle of *Tenebrio* larvae	28	Lafon (1943b)
Exuviae of *Tenebrio* larvae	14.6	Lafon (1943b)
Insect cuticle (average)	33	Lafon (1943a,b)

same as animal chitin. In another detailed study, the same group also reported that the physical properties were analogous, and the X-ray diagram was essentially the same as that of crustacean chitin (van Iterson Jr et al., 1936). The authors pointed out the fact that the X-ray method was of great value in comparing chitins from various plant and animal sources. However, considerable cleaning of samples, especially those from fungal sources, is required before satisfactory patterns can be obtained as highlighted by Kreger (1954). Later, Lotmar and Picken (1950) and Hackman (1960) also claimed that the X-ray method must be used to distinguish the different crystalline forms of the polysaccharide. Tracey (1955) has suggested that enzymatic methods can be applied to the qualitative detection of chitin.

In 1936, distinct diagrams were obtained with the cell walls of the sporangiophores of the fungus *Phycomyces* and studied by Anton N.J. Heyn at the University of Utrecht, The Netherlands (Heyn, 1936a,b,c). Heyn's work showed that the fiber diagrams were identical with those obtained from chitin of animal origin (wing of *Periplaneta*), in agreement with the results reported by van Iterson Jr et al. (1936). Heyn made the interesting observation that a higher orientation occurred in the cell wall of the sporangiophores. The plane of the glucose rings was approximately radial. The X-ray diagram of fungal chitin was somewhat poorer in reflections than that of arthropod chitin and, in this respect, might be compared with molluscan chitin. A discussion in the differences in three-dimensional structure between the chitin and cellulose has also been made. Heyn mainly attributed these differences to the presence of the acetamide groups in chitin chains (Heyn, 1936a,b,c). Later, Meyer (1950) reported that a high degree of molecular orientation has also been observed in other fibrous structures, for example, in tendons and in the laminae of beetle elytra. He pointed out the fact that physical methods such as X-ray analysis were very useful in comparing chitins from different sources.

Edward S. Castle, studying the double refraction of chitin at Harvard University, Cambridge, reported that structurally chitin resembled cellulose, the unit of structure being the acetyl-glucosamine residue instead of the glucose residue (Castle, 1936), in agreement with the conclusion by Meyer and Mark (1928). Organized chitin was optically anisotropic and exhibited birefringence, which was ordinarily positive in relation to the principal axis of the structure in which it occurred. Using a series of relatively unreactive liquids and fluid mixtures which permeated the chitin framework, the type of curve relating double refraction, and refractive index of the imbibed fluid was found to depend greatly on the chemical nature of the fluid. Either a positive or a negative residual birefringence may be found, depending on the choice of imbibing liquid (Castle, 1936).

George L. Clark and his pupil Albert F. Smith (University of Illinois) reported that X-ray diffraction analyses of chitin dissolved in solutions of lithium salts showed no evidence of hydrolysis even after several months (Clark and Smith, 1936). The polysaccharide can be recovered from this salt dispersion by dilution by acetone (the speed of precipitation could be controlled by varying the proportion of water). Clark and Smith also published the first X-ray pattern of a well-oriented fiber made from chitosan obtained by chemical treatment of a lobster tendon chitin. The chitosan chains were packed in an orthorhombic unit cell with the dimensions $a=8.9$, $b=10.25$, and $c=17.0$ Å. Sixty years later, Okuyama et al. (1997) reported similar values.

In the mid-1930s, several chitin derivatives have been prepared by Paul Schorigin and co-workers at the *Österreich Institut für Kunstseide*, Moskau, Austria (Schorigin and Hait, 1934, 1935; Schorigin and Makarowa-Semljanskaja, 1935a,b). Schorigin and Hait (1934) obtained a acetylated product with 2.99 acetyl groups per sugar residue by using acetic anhydride, although the authors pointed out that chitin was not readily acetylated. The acetate derivative was soluble in formic acid but insoluble in other organic solvents. At the same period, a similar chitin acetate (2.5 acetyl groups per sugar residue) was obtained by Meyer and Wehrli (1937), but the derivative was strongly degraded in formic acid. In another study, Schorigin and Hait reported that acetylation with the normal reagent cellulose modification was not applicable to chitin because the polymer did not swell in the conventional medium (Schorigin and Hait, 1935). Methylation of chitin was also difficult because of its inability to swell in alkali, and only the monomethyl ether has been prepared (Schorigin and Makarowa-Semljanskaja, 1935b). The product obtained after 45 methylation reactions with dimethyl sulfate-sodium hydroxide still contained appreciably less than 1 methyl group per sugar residue. Later, this reaction was studied in detail by Wolfrom et al. (Wolfrom, 1958; Wolfrom et al., 1958).

von Fürth and Scholl (1907) previously reported that chitin could be nitrated with fuming nitric acid to give a substance partially soluble in acetic acid. Later, Schorigin et al. also reinvestigated this reaction to obtain chitin nitrates (1.5 nitrate groups per sugar residue) using concentrated nitric acid and found substantially the same result (Schorigin and Hait, 1934). However, they pointed out that a considerable portion was soluble in formic acid. Since the nitrogen content of the fractions was the same, the difference in solubility was explained by difference in chain length. This was demonstrated by Clark and Smith (1936) using X-ray diffraction data (in this study, chitin nitrates have been prepared using fuming nitric acid). Meyer and Wehrli (1937), using the Clark and Smith's protocol, also reported that the relatively short-chain products were not homogenous and could be fractioned by the use of formic and acetic acids.

Among possible derivatives of longer-chain lengths, several xanthates have been made by Thor in different forms (filaments, films, etc.), opening new perspectives in various fields (Thor, 1939a,b; Thor and Henderson, 1940). An alkali-chitin was prepared by steeping the polymer in 50% aqueous solution hydroxide for 2 at 25°C. With carbon disulfide, this product gave a chitin xanthate.

At the same period, the observation that the anticoagulant activity of heparin preparations appeared to be related to their ester sulfate content led to the synthesis of a number of sulfated polysaccharides using chlorosulfonic acid in pyridine as suitable sulfating agent (Bergström, 1935, 1936). This was the beginning of numerous studies on the synthesis of amino-polysaccharides containing functional groups used as anticoagulants (Karrer et al., 1943; Astrup et al., 1944; Barsøe and Selsø, 1946; Doczi et al., 1953; Wolfrom et al., 1953; Terayama, 1954). Indeed, chitin derivatives were shown to be synthetic, active anticoagulants. For instance, Karrer et al. (1943) first prepared a phosphoric ester of chitosan and studied its anticoagulant activity. Astrup et al. (1944) observed that chitin disulfates could show up to 20% of the anticoagulant potency of heparin, accompanied by only slight toxicity. Later, Doczi et al. (1953) proposed a highly *N*-sulfated chitosan having approximately half the activity of heparin. Wolfrom et al. (1953) also prepared an *N*-sulfated, *O*-sulfated chitosan of similar activity. Terayama (1954) synthetized an *N*-polymethylated chitosan derivative and reported its antibiotic activity. Sulfated chitins have also been proposed for use as thickeners in pastes, adhesives, and drilling muds (Brimacombe and Webber, 1964).

The bulk of the nonchitinous material in the cuticle was protein and, in the insects, the content varied from 25% to 37% (Lafon, 1943a,b). Trim (1941) previously reported the fundamental presence of protein in the larval cuticle of *Sarcophaga* and in the larva of *Sphinx ligustri*. Stacey (1943, 1946) and Haworth (1946) suggested that the arthropod cuticle might well be regarded as mucoprotein, and the polysaccharide fraction (chitin) might vary in amount in different species or in different regions of the integument. Such a view was consistent with the results reported by Richards (1947a). In 1947, from a study of the X-ray diffraction and swelling properties of the cuticle of the larva of *Sarcophaga*, Fraenkel and Rudall pointed out that within the micellae, chitin was intimately associated with protein (Fraenkel and Rudall, 1947). The chitin/protein ratio in insect cuticle was around 55/45. These authors suggested that the cuticle might consist of alternating monolayers of protein and chitin, the latter served as a fibrous element in biological (composite) materials. It was then accepted that except in some Diatomea, chitin was always associated with proteins (which function as the matrix), minerals (predominantly calcium carbonate in Crustacea), and later with also polyphenols (in insects). One year later, Richards and Korda (1948) proved the presence of chitin-protein complexes using electron microscope studies. Later, Hackman (1955) finally demonstrated the formation of weak bonds between chitin and proteins and the presence of stable complexes (glycoproteins). The association was also stabilized by salt formation between protonated-free amino groups of the biopolymer and carboxylate groups of the proteins (polypeptides). The arthropod cuticle was then considered a laminar structure of these different complexes. Later, Hackmann et al. also showed that in biological materials, chitin fibers had varying orientations, including parallel (rigid materials) and parallel layers arranged in blocks or a helicoidal orientation (Hackman, 1987).

At the end of the 1940s, it was well-accepted that the organic skeletal substance of fungi, arthropods, annelids, and mollusks was chitin, a nitrogen-containing carbohydrate (Dennell, 1947a,b, 1958; Wigglesworth, 1948; Meyer, 1950; Richards, 1951; Whistler and Smart, 1953). Chitin was also found in certain unicellular Protista and some algae. The distribution among living organisms was related to the main function of chitin: it was a supporting material. Cellulose, chitin, and collagen were the three supporting systems in organisms. From numerous X-ray diffraction data, the sharp separation of Phycomycetes into chitin- and cellulose-containing fungi was also confirmed. Like cellulose and starch, chitin was considered a homopolysaccharide, although some authors continued to consider it a mucopolysaccharide (Kulagin, 1939; Meyer, 1938; McLintok, 1945). From a chemical point of view, the equations described in Scheme 4.3 were accepted (Foster, 1949). On hydrolysis with acids, glucosamine and acetic acid were obtained in equivalent quantities and in theoretical yields. Under certain conditions, acetylglucosamine might also be obtained. This product was the structural unit of chitin, just as glucose was the structural unit of cellulose. Under the name chitosan, various more or less deacetylated breakdown products of chitin have been described, which formed crystalline salts with acid (Meyer, 1942).

$$\underset{\text{chitin}}{(C_{32}H_{54}N_4O_{21})x} + 2xH_2O \xrightarrow{\text{KOH}} \underset{\text{chitosan}}{x(C_{28}H_{50}N_4O_{19})} + \underset{\text{acetic acid}}{2xCH_3COOH}$$

$$\underset{\text{chitosan}}{C_{28}H_{50}N_4O_{19}} + 3H_2O \xrightarrow{\text{HCl}} \underset{\text{N-acetylglucosamine}}{2C_6H_{11}O_5(NH\text{-}COCH_3)} + \underset{\text{glucosamine}}{2C6H_{11}O_5(NH_2)}$$

$$\downarrow \text{HCl}$$

$$\underset{\text{glucosamine}}{2C_6H_{11}O_5(NH_2)} + \underset{\text{acetic acid}}{2CH_3COOH}$$

Scheme 4.3 General equations/reactions for hydrolysis of chitin and chitosan.

4.7 Literature review

Relatively few reviews of chitin were published during the period 1930–1950. Among them, the distribution of chitin in a wide range of animals was reviewed in 1932 by Conrad Koch, German biochemist and professor of zoology at the *Universität Rostock* (Koch, 1932); and a year later by Vincent B. Wigglesworth (1899–1994; British entomologist and professor of biology at the University of Cambridge) (Wigglesworth, 1933). Zechmeister and Tóth (1939a) published a review in the journal *Fortschritte der Chemie Organischer Naturstoffe* where they summarized the data on chitin and other polysaccharides containing amino sugars. In 1942, different aspects of the structure and chemistry of chitin were discussed by Meyer (1942). Meyer (1950) also published a chapter entitled "Chitin and other polysaccharides containing amino sugars" in the famous book "*Natural and Synthetic High Polymers*." A comprehensive review of the fine structure of biological systems was published by Laurence E.R. Picken in 1940 (Picken, 1940). In the late 1940s, Wigglesworth updated his overview published in 1933, focusing the subject on insect cuticle (Wigglesworth, 1948).

CHAPTER FIVE

The period of doubt: 1950–1970

Chitin does not occur naturally as a distinct and separate chemical entity in the cuticle but is always found as one component in a chitin-protein complex; then it follows that chitin is no more a natural compound than are the various degradation products that can be prepared from it.

Professor Albert Glenn Richards, 1951,
Professor of Entomology and Zoology, University of Minnesota, United States

During the period of exploration (1930–1950), chitin and chitosan attracted considerable attention with the exploration of natural fibers and the first applications, but lack of adequate manufacturing facilities and mostly cutthroat competition from synthetic polymers restricted their commercial development. From 1950 to 1970, chitin and chitosan entered into a period of doubt, although much progress has also been made on their isolation, production, and fundamentals (Wigglesworth, 1948, 1957; Meyer, 1950). Most of the fundamental information on chitin has been compiled in the 1950s in several comprehensive books and reviews. Table 5.1 reports a selection of fundamental books, monographs, and reviews published during this period. All these remarkable scientific contributions have gathered a significant amount of knowledge and results on chitin and chitosan, which are still valid today.

The first books on chitin were the well-known monographs on the integument of the arthropods by Richards edited in 1951 (Richards, 1951) and on the biochemistry of aminosugars edited in 1955 by Paul Welberry Kent (1923–2017) and Michael Wellesley Whitehouse (Kent and Whitehouse, 1955). A few years later, other comprehensive books were published by Peter Bernfeld on the biogenesis of chitin (Bernfeld, 1963), by Elwyn T. Reese on a discussion of the advances in the enzymic hydrolysis of natural substances (Reese, 1963), and by Charles Jeuniaux (1928–2001) on chitin and its enzymatic breakdown (Jeuniaux, 1963). Jeuniaux reported that the presence or abundance of chitin in animals was depended on the presence or absence of chitin synthase. Numerous studies have focused on chitin synthases that contributed to the virulence of human pathogens. Their inhibition by structural analogs of the chitin biosynthesis substrates proved to be fungistatic and also fungicidal, opening the way to new

https://doi.org/10.1016/B978-0-323-96119-6.00001-3

Table 5.1 A selection of fundamental reviews on chitin published during the period of doubt (1950–1970).

Year	Journal/book	Author(s)	Title	Topics
1950	*Natural and Synthetic High Polymers*	Kurt H. Meyer, University of Geneva	Chitin and other polysaccharides containing amino sugars	Constitution of chitin; crystalline structure; molluscan chitin; fungal chitin; reactions of chitin
1951	*The Integument of Arthropods. The Chemical Components and their Properties, the Anatomy and Development, and the Permeability*	Albert Glenn Richards, University of Minnesota	The integument of arthropods	Glucosamine, *N*-acetylglucosamine, chitobiose and chitin; chitin derivatives and metabolic sources; detection and estimation of chitin; distribution of chitin in the animal and plant kingdoms
1952	*Advances in Carbohydrate Chemistry*	Allan B. Foster and Maurice Stacey, University of Birmingham	The chemistry of the 2-amino sugars	Glucosamine; preparation, chemistry; reactions; chondrosamine; detection and determination
1953	*Polysaccharide Chemistry*	Roy Lester Whistler and Charles Louis Smart, Purdue University, Indiana	Chitin	Occurrence; preparation; composition and structure; properties; derivatives; microbiological decomposition; uses
1955	*Biochemistry of the amino-sugars*	Paul W. Kent and Michael W. Whitehouse, University of Oxford	Biochemistry of the amino-sugars	Amino-sugars in the biological environment; classification and distribution of mucosubstances; enzymic degradation; metabolism of aminosugars; complexes; chemistry of aminosugars and their derivatives

1955	*Modern Methods of Plant Analysis*	M.V. Tracey, Rothamsted Experimental Station, Harpenden, Herts	Chitin	Detection and determination of chitin
1957	*Annual Review of Entomology*	Vincent B. Wigglesworth, Cambridge	The insect cuticle	Composition of the soft cuticle; hardening and darkening of the cuticle; the epicuticle; deposition of the cuticle
1957	*Reviews of Pure and Applied Chemistry*	M.V. Tracey, Rothamsted Experimental Station, Harpenden, Herts	Chitin	Discovery; distribution; preparation; chemical evidence; physical evidence; detection and determination; properties; derivatives and uses;
1958	*Encyclopedia of Plant Physiology*	Allan B. Foster and Maurice Stacey, University of Birmingham	The aminosugars and chitin	Occurrence; biosynthesis; structure and properties; chitinases; role in the cell wall
1951	*Ergebnisse der Biologie*	Albert Glenn Richards, University of Minnesota	The cuticle of arthropods	Chitin chemistry; chemistry of other components; structure of the cuticle; the multiple barriers of the cuticle; ecological aspects of the cuticle
1959	*Advances in Carbohydrate Chemistry*	Allan B. Foster and Derek Horton, University of Birmingham	Aspects of the chemistry of the amino sugars	Chemical syntheses; natural products, chemistry, detection and determination of amino sugars, oligosaccharides
1961	*Advances in Carbohydrate Chemistry*	Allan B. Foster and J.M. Webber, University of Birmingham	Chitin	Occurrence; isolation; properties, derivatives; chitosan; uses

Continued

Table 5.1 A selection of fundamental reviews on chitin published during the period of doubt (1950–1970)—cont'd

Year	Journal/book	Author(s)	Title	Topics
1963	*Advances in Insect Physiology*	K.M. Rudall, University of Leeds	The chitin/protein complexes of insect cuticles	Structure of chitin; chemistry of the proteins associated with chitin; X-ray studies on chitin/protein complexes; optical studies; electron microscopy
1963	*Chitine et Chitinolyse*	Charles Jeuniaux, *Université de Liège*	*Chitine et chitinolyse, un chapitre de la biologie moléculaire*	Distribution; biochemistry; enzymes; determination; reactions
1964	*Comparative Biochemistry: A Comprehensive Treatise*	Paul W. Kent, University of Oxford	Chitin and mucosubtances	Distribution of chitin; molecular structure and function of chitin; biochemistry of chitin; chitin in relation to mucosubstances
1964	*Mucopolysaccharides: Chemical structure, distribution and isolation*	John S. Brimacombe and JM Webber, University of Birmingham	Mucopolysaccharides	Chitin; occurrence; isolation; properties; composition and structure; derivatives; chitosan; biosynthesis
1965	*The Fungi—An Advanced Treatise*	Jerome M. Aronson, University of Detroit	The cell wall	Microscopic structure; physical conformation; chemical composition; functional attributes
1969	*Journal of Polymer Science*	K.M. Rudall, University of Leeds	Chitin and its association with other molecules	Chemistry; X-ray diffraction; association of chitin with protein; electron microscopy; analogies with lysozyme; chitin in fungi

applications. Different aspects of the structure, properties and function of chitin have been reviewed in 1953 by Roy Lester Whistler and Charles Louis Smart in their famous book "*Polysaccharide Chemistry*" (Whistler and Smart, 1953). In 1955, M.V. Tracey published a review on the detection and determination of chitin in plant materials and the quantitative analytical methods (Tracey, 1955). Two years later, Tracey also published a detailed review entitled "*Chitin*" (Tracey, 1957). In 1958, Allan B. Foster and Maurice Stacey (1907–1994, a student of Professor Haworth) published a detailed chapter on the aminosugars and chitin in *Encyclopedia of Plant Physiology* (Foster and Stacey, 1958). Biological aspects of chitin were summarized in general reviews by Richards (1958), Picken (1960), Jeuniaux (1963), and K.M. Rudall (Rudall, 1963); chemical aspects by Foster and Webber (1961); and its production by James N. BeMiller (BeMiller, 1965). Later, K.M. Rudall also addressed the concept of the chitin-protein. He detailed the chitin-protein complexes and their conformation in two comprehensive reviews (Rudall, 1967, 1969). From crystallographic and electron microscopic studies, Rudall pointed out a close association of chitin with protein, suggesting a quantitative relationship between the components, although this did not necessarily imply covalent bonding. In animals, the principal skeletal structure was the collagen system, but chitin was a prominent supporting material in invertebrates. The role of collagen was as an internal connective tissue supporting organs, and chitin was organized as a cuticle at one surface of an epithelium. Their structures were also different. Chitin-protein systems were formed by a fibrous framework of polysaccharide (chitin) reinforced by a protein matrix. Collagenous connective tissue presented a fibrous framework of protein reinforced by a polysaccharide matrix (Rudall, 1967, 1969; Attwood and Zola, 1967). In 1964, Paul W. Kent (1923–2017) published a comprehensive review on chitin and mucosubstances containing 223 references (Kent, 1964). The same year, John S. Brimacombe and J.M. Webber discussed the chemical structure, distribution, and isolation of mucopolysaccharides (Brimacombe and Webber, 1964). The progress on chitin orientation in the cuticle and its control were summarized by Neville (1967). Three chapters by Friedman (1970), Honke and Scheer (1970), and Jeuniaux (1971) were devoted to the zoological importance of chitin and its role in biochemical evolution. The first interdisciplinary book on chitin was only published in 1973 (Muzzarelli, 1973).

During the period of doubt between 1950 and 1970, numerous works were published on all scientific aspects of chitin and chitosan, including their production, chemistry, biology, and potential applications. From this period, I choose to highlight the following studies.

Chitin was present in green algae but not in bacteria and *Actinomycetes*. There was still controversy about its presence in yeasts. Using micro-chemical methods of detection, Roelofsen and Hoette (1951) found "pure" chitin in a variety of filamentous and nonfilamentous yeasts, thus confirming earlier observations published by Schmidt (1936). Roelofsen and Hoette (1951) modified van Wisselingh's protocol by first detecting chitosan by acid extraction followed by the sulfate test. However, a year earlier, Frey (1950) had failed to detect chitin in some *Saccharomycetes* and imperfect yeasts using X-ray data. These results called into question the micro-chemical methods of detection and suggested that X-ray diffraction was more appropriate.

Until the beginning of the 1950s, the methods for isolating pure chitin by decalcification with acids and deproteinization with alkalis proposed in the 19th century by Odier, Children, Gilson and Winterstein, and later by other colleagues such as Scholl have continued to be virtually unchanged (Kent, 1964). Between 1950 and 1960, "new" methods for the isolation of chitin were proposed by Blumberg (Blumberg et al., 1951), Hackman (Hackman, 1953a,b, 1954, 1959, 1960, 1962; Foster and Hackman, 1957), Roseman (Horowitz et al., 1957; Blumenthal and Roseman, 1957), BeMiller (1965), and Broussignac (1968). Methods for the isolation of a series of oligosaccharides (Barker et al., 1957, 1958) and for the production of chitosan (Wolfrom, 1958; Wolfrom et al., 1958; Wolfrom and Han, 1959; Horowitz et al., 1957; Broussignac, 1968) from chitin have also been proposed. All these methods were detailed in the book by Muzzarelli (1973).

The most common source of chitin for laboratory purposes was crustacean shells, and a typical isolation procedure (considered a "standard" method) was that of Hackman. This author found a 6.6% conversion to 2-amino-2-deoxy-D-glucose on treatment of chitin with 2 N hydrochloric acid at 100°C for 7 h and removal of only 1.3% of the acetyl groups by the action of N aqueous alkali at 100°C for 48 h (Hackman, 1954). Hackman pointed out the fact that drastic methods used for chitin isolation can cause degradation. Drastic methods were necessary because chitin is closely associated with other materials in nature (Hackman, 1954; Muzzarelli, 1973). Three years later, Horowitz et al. (1957) proposed a "modified method" for chitin isolation. For most purposes, using the standard (Hackman, 1954) or modified (Horowitz et al., 1957) method, chitin is obtained as an amorphous solid, insoluble in water, dilute acids, dilute and concentrated alkalis, and organic solvents, in accordance with all the results published in the last century. Chitin can be dissolved in concentrated hydrochloric acid or in concentrated sulfuric acid, but dissolution is accompanied by degradation. Hackman (1962) reported a rapid decrease in the molecular weight of chitin following its dissolution in concentrated mineral acids at 20°C. In hydrochloric acid, chitin initially had $[\alpha]_D^{20} = -14°$ which slowly changed to +56° because of hydrolysis, in agreement with the previous observations by Irvine (1909) and later by Clark and Smith (1936). These authors indicated that for a disaccharide comprising two D-sugar residues, the rotary (optical) change was indicative of β-D-linkages. Subsequently, this approach of determining the nature of the two glycosidic linkages by optical rotation was also used for trisaccharides (Neuberger and Pitt Rivers, 1939). Hackman's group then used this relatively simple and rapid method to characterize the prepared saccharides and oligosaccharides, in conjunction with other methods (X-ray, infrared spectroscopy).

In 1946, Purchase and Braun pointed out the fact that chitin disaccharide contained two amino sugar residues joined by a (1 → 4)-glycosidic linkage by examination of its octaacetate (Purchase and Braun, 1946). These authors also showed that 2-amino-2-deoxy-D-glucose can be isolated in yields of 60%–70% of crab chitin with concentrated hydrochloric acid (yields were lower from mold chitin). Later, for complete hydrolysis of chitin, heating for several hours at 100°C with 5.7 N hydrochloric acid has been recommended by Hackman (1962). In 1950, Jeanloz, studying periodate oxidation of chitin and chitosan, demonstrated that only (1 → 4)-linkages were present in these polysaccharides (Jeanloz, 1950; Jeanloz and Forchielli, 1950, 1951). Later, this was confirmed by Barker et al. using infrared studies (Barker et al., 1957, 1958). The chitin saccharides, and chitin itself, all showed absorption at 884–890 cm^{-1}, the region indicative of the β-linkage. The same year, Darmon and Rudall (1950), studying the infrared spectra of chitin, were able to identify the principal bands associated with the amido and hydroxyl groups vibrations and observed their dichroism. The authors correlated the infrared absorption spectrum with the X-ray data in order to give a detailed structure for chitin. The IR spectrum of α-chitin also showed two absorption bands at 1655 and 1625 cm^{-1}, characteristic of hydrogen-bonded amide groups. Later, Pearson et al. (1960) also reported these two characteristic bands attributed to the C=O stretching region of the amide moiety and especially interesting for distinguishing α-chitin and β-chitin. In the latter case, the amide I band is unique, whereas for α-chitin, the band is split. In contrast, the amide II band is unique in both chitin allomorphs. Darmon and Rudall (1950) also studied the infrared spectra of deacetylated chitin (they observed the disappearance of these bands during deacetylation) and chitin nitrate. It is interesting to point out that the use of these bands for the determination of the degree of *N*-acetylation was only reported in the mid-1970s (Muzzarelli, 1977). During the preparation of chitosan with aqueous alkali, half the acetyl groups were readily removed to give a mono-acetyl-di-glucose-amine structure, confirming the suggestion of Löwy (1909), whereas, on prolonged treatment, the remaining acetyl groups were slowly removed. In this latter case, the structure approached that of a poly(2-amino-2-deoxy-D-glucose), as suggested by Meyer and Wehrli (1937). Later, Pearson et al. (1960) commented that "this behavior may be attributable to the acetamido groups having different degrees of accessibility as a result of rotational isomerism."

In 1954, Hackman, using an enzyme preparation from the intestine of the snail *Helix aspersa*, obtained a 44% yield of crystalline 2-acetamido-2-deoxy-D-glucose and only 0.5% (colorimetric determination) of 2-amino-2-deoxy-D-glucose from chitin (Hackman, 1954). The results were in agreement with those obtained from lobster chitin (Karrer and Hofmann, 1929) and mushroom chitin (Karrer and von François, 1929) treated by *Helix pomatia*. In 1958, Giles et al. reported that chitin was a polymer of the amino sugar acetylglucosamine (Giles et al., 1958). However, in a small percentage of residues, the acetyl group may be missing, leaving glucosamine, so that the chain as a whole may be positively charged. This could be important in chitin-protein cross-linking (Hackman, 1960) and as one of the factors involved in orientation of the chitin chains (Attwood and Zola, 1967; Neville, 1967). At the beginning of the 1960s, polysaccharides were regarded as condensation polymers of monosaccharides resulting from the formation of glycosidic linkages. It was evident that chitin, considered exclusively a homopolysaccharide, was not normally found, and in situ it was associated with other substances, notably proteins, by hydrogen bonds and covalent linkages (Whistler and Smart, 1953; Hackman, 1954; Rudall, 1963, 1965; Kent, 1964; Brimacombe and Webber, 1964). A convenient method of distinction proposed by Hackman (1960) reserved the name chitin for the chemically purified material and native chitin for the complex in which it is involved in tissues. The chemical structure of chitin was agreed to be that of a long unbranched polysaccharide in which *N*-acetyl-D-glucosamine (2-acetamido-2-deoxy-D-glucopyranose) residues were linked in the β-(1 → 4) positions. At the end of the 1960s, it was accepted that three different molecular systems occurred as the chief skeletal support of living organisms, namely, the cellulose systems present in plants, collagenous systems present in animals, and chitinous system (Rudall, 1965, 1967, 1969; Rudall and Kenchington, 1973). Chitin occurred as an alternative to cellulose in plants and as an alternative to collagen in animals. Significant progress was made in the chemistry and the production of chitin (Brimacombe and Webber, 1964; Conrad, 1966). However, many questions continued to be raised in studies using X-ray diffraction and infrared spectroscopy such as the association between chitin and other molecules (Rudall, 1969).

Although initiated in 1920s, an important contribution was made on the structure of "purified" chitin and its polymorphism during the period 1950–1970. Indeed, this was the subject of hot debate between the different laboratories (Kreger, 1954; Rudall, 1955). The first important milestone was the fact that X-ray fiber diagrams of oriented chitin samples finally showed an obvious similarity to those of cellulose. The next was its polymorphism. Indeed, comparison of the X-ray data for chitin from different sources had also revealed the existence in nature of more than one polymorphic form, namely, α-chitin (Meyer and Mark, 1928), β-chitin (Lotmar and Picken, 1950), and γ-chitin (Rudall, 1963). The structure of α-chitin has been investigated more extensively than that of either the β- or γ-form because it was the more common polymorphic form. The three forms have been found in different parts of the same organisms, suggesting that these forms were relevant to the different functions and not to animal grouping. However, it was difficult to assign a physiological role to these crystalline forms.

As already mentioned, the earliest X-ray investigation was that of Gonell (1926) whose results formed the basis for the discussion of Meyer and Mark (1928) on α-chitin. In this article, a structure analogous to that of cellulose was first proposed. They established the unit cell as orthorhombic, and the proposed unit cell contained eight 2-acetamido-2-deoxy-D-glucose residues. The crystal structure of α-chitin was confirmed by Meyer and Pankow (1935), by Clark and Smith (1936), and by Lotmar and Picken (1950). Its structure was then modified by Darmon and Rudall (1950). Later, the first detailed structure analysis was that of Carlström (1957) who concurred with an orthorhombic unit cell but obtained different dimensions (Table 5.2). Carlström also rejected a straight arrangement of the residues in the disaccharide units, as originally proposed by Meyer for chitin and also cellulose, in favor of a bent conformation arising by rotation around the glycosidic linkages. Dweltz (1960), using the optical transform method, also arrived at a new structure.

Evidence for the existence of a second crystalline form of chitin was first obtained by Lotmar and Picken (1950). These authors observed a new X-ray pattern for deproteinized pens from the squid *Loligo*. This type form which apparently had a unit cell of dimensions $a=9.32$ Å, $b=10.17$ Å and $c=22.15$ Å, was named β-chitin to distinguish it from the much more common α-chitin. Density measurements indicated that the unit cell dimensions

Table 5.2 Unit cell dimensions (in Å) for α-chitin, β-chitin, and γ-chitin.

Chitin	*a*	*b*	*c*	Reference
α	11.58	10.4	19.42	Gonell (1926)
α	9.40	10.46	19.25	Meyer and Pankow (1935)
α	9.25	10.46	19.25	Clark and Smith (1936)
α	9.40	10.26	19.25	Lotmar and Picken (1950)
α	4.76	10.28	18.85	Carlström (1957)
α	4.69	10.43	19.13	Dweltz (1960)
α	4.74±0.01	10.32±0.02	18.86±0.01	Minke and Blackwell (1978)
α	4.71	10.30	18.78	Sikorski et al. (2009)
β	4.7	10.3	10.5	Dweltz (1961)
β	4.85	10.38	9.26	Blackwell et al. (1967)
β (anhydrous)	4.85	10.38	9.26	Blackwell (1969)
β				(monohydrate)
4.8	10.4	10.5	Blackwell (1969)	
β (dehydrate)	4.8	10.4	11.1	Blackwell (1969)
γ	4.7	10.3	28.4	Walton and Blackwell (1973)

could not result from the normal cell by swelling along the c axis. Later, Dweltz (1961) also reported the presence of a very different crystallographic type of the polysaccharide chitin (also from the pen of *Loligo*), but he has suggested a smaller unit cell (Table 5.2). A similar chitin was found in annelid chaetae (*Aphrodite aculeate*), in the brachiopod *Lingula*, and in the skeletal pen of squids (*Loligo*). In arthropods and fungi, only α-chitin appeared to occur. The squid *Loligo* was of particular interest since α-chitin was present in the beak, radula, and linings of the stomach and esophagus of this animal, while β-chitin was restricted to the pen. Lotmar and Picken (1950) also noted the extra X-ray reflections in the pattern of intact *Aphrodite chaetae* and other tanned material and suggested that they may be due to ordered protein sequence. In a previous work, Fraenkel and Rudall (1940), studying the X-ray characterization of chitin-protein complexes, pointed out the differences between the fiber diagrams of intact and deproteinized insect cuticles. They suggested that the differences were due to modification of the chitin structure due to the presence of complexing protein. Later, this was confirmed by Rudall (1955) who, studying the distribution of α- and β-chitins among the various types of lower animals, concluded that the presence of β-chitin was associated with collagen-type cuticles. A full discussion on this subject was made by Richards 3 years later in a review entitled The *Cuticle of Arthropods* (Richards, 1958).

Rudall showed that the X-ray patterns of the intact complexes gave many more layer lines than the purified chitin component (Rudall, 1955, 1963, 1967). Rudall first remarked that β-chitin was found associated with collagen, whereas α-chitin occurred alone or in association with a non-collagenous protein such as arthropodin (Rudall, 1955). He advanced reasons for supposing that "the production of α-chitin and collagen may be mutually exclusive." Chitin from parts of Loligo (beaks, radulae) other than the pen gave the α-chitin pattern. After recovery from formic acid solution, the pen chitin also gave the α-pattern (Rudall, 1955). The pattern of the beak chitin was unaffected by this procedure. Treatment with fuming nitric acid or cold hydrochloric acid also converted β-chitin to α-chitin. Differences between the two forms of chitin were evidently slight. The infrared spectra were also essentially similar, though there were differences indicating less tight bonding β-chitin (Rudall, 1955). There was also evidence from the X-ray diffraction pattern that a third modification of chitin might occur in coelenterates. It was probable that the α- and β-forms did not differ in any essential chemical manner since both were readily hydrolyzed by chitinases from a number of sources. All these observations were also reported by Hackman (1960) and by Dweltz (1961) using both X-ray diffraction and hydrolysis studies.

Detailed crystallographic investigations have also been reported for the α- and β-forms of chitin by Dweltz (Dweltz, 1960, 1961; Dweltz and Anand, 1961). Dweltz proposed new crystal structures and comprehensively discussed the spatial configuration of the polymer chain. In both cases, the structures were based upon backbones consisting of straight polysaccharide chains. Basic to the proposed structure for the two systems was the presence of sheets of parallel chains linked by $C{=}O\cdots H{-}N$ hydrogen bonds through the amide groups. The forms differed in the sense of the chains in successive sheets. In β-chitin, the sheets were all arranged in a parallel manner, whereas in the α-form, successive sheets were antiparallel. For α-chitin ($[C_8H_{13}O_5N]_n$), the unit cell contained two polysaccharide chains running in opposite directions and four asymmetric *N*-acetyl-glucosamine units. This structure was in agreement with infrared absorption data (Dweltz, 1960, 1961). In 1960, Dweltz obtained a unit cell having dimensions ($a = 4.69$ Å, $b = 19.13$ Å and $c = 10.43$ Å) similar to those of Carlström ($a = 4.76$ Å, $b = 18.85$ Å and $c = 10.28$ Å) (Carlström, 1957), but postulated a straight arrangement of the sugar residues (Dweltz, 1960). The main feature was intermolecular hydrogen bonding between the hydroxymethyl groups of adjacent chains. However, the same year, Pearson et al. (1960), using infrared studies, were unable to confirm this. For β-chitin, obtained from the conversion of α-chitin by treatment with formic acid or fuming nitric acid using the experimental protocol published by Lotmar and Picken (1950) and Rudall (1955), Dweltz reported the unit cell to be approximately half that of α-chitin (Table 5.2). β-Chitin must be considered to be a monohydrate in the dry state with the chemical formula $[C_8H_{13}O_5N{\cdot}H_2O]_n$, and the sugar (glucosamine) in β-chitin was the same as that in α-chitin (Dweltz, 1961). As with the α-chitin structure, C=O ··· H—N bonding was permitted along a axis. It was further postulated that introduction of more water molecules could lead to separation of the chains and the variable b repeat distance observed with changes in humidity.

In 1962, Diego Carlström has criticized the structures of α- and β-chitin proposed by Dweltz. He has emphasized that stereochemical requirements have been ignored (Carlström, 1962). Indeed, the main criticism was that the polysaccharide chain configuration was stereochemically unsatisfactory on the basis of unfavorable H···H contacts at the carbon atoms C-1 and C-4 adjacent to the glycosidic oxygen. A straight-chain configuration consisting of glucose units linked together by 1,4-β-glucosidic bonds was sterically impossible because of the bulky acetamido side-chains (Carlström, 1962). First, Carlström, using the optical transform method, arrived at a new structure for α-chitin from apodemes of the lobster *Homarus americanus* (Carlström, 1957). An orthorhombic unit cell with $a = 4.76$ Å, $b = 10.28$ Å (fiber axis), and $c = 18.85$ Å was proposed. There were two chitobiose units (di-*N*-acetylchitobiose residues) per unit cell in two polymer chains running in opposite direction. The repeating period along the fiber axis was the same as that of native cellulose. The chain was in the bent form, also similar to that proposed for cellulose and for cellobiose (Carlström, 1962). This was a major difference in the structure compared with model originally proposed by Meyer (Meyer and Mark, 1928; Meyer and Pankow, 1935). The —$NHCOCH_3$ groups were also assumed to be planar and to be predominantly perpendicular to the fiber axis. This was previously suggested by Darmon and Rudall (1950) using infrared spectroscopic results. An intramolecular hydrogen bond was formed between the carbon-3 hydroxyl group and the ring oxygen of the next acetylglucosamine residue (Carlström, 1957). Carlström (1962) proposed a scheme of full intermolecular hydrogen bonding of the C=O ··· H—N groups along the direction of the a axis. He noted that the hydroxyl group attached to carbon-6 was found to have some rotational freedom. Each glucose residue had a distance between the connecting oxygens of about 5.45 Å, and this was supported by the excellent crystal structure determination of cellobiose. Carlström finally reported that optically derived Fourier transforms based on his proposed structure had intensity distributions similar to the observed X-ray intensities. The same year, Carlström's structure was partially confirmed by Ramachandran and Ramakrishnan (1962). These authors suggested that the data obtained by Dweltz were re-estimated due to differences in the observed intensities. Re-measurement of the repeat spacing along the fiber axis of the specimen used by Dweltz provided a value of 10.3 Å and not 10.43 Å as reported by Dweltz (Carlström's value was 10.28 ± 0.03 Å). Ramachandran and Ramakrishnan concluded that "it was not possible to say that one structure was superior to the other." The complete intermolecular C=O···H—N hydrogen bonding scheme, originally proposed by Carlström from X-ray data, was also in agreement with the polarized infrared measurements on various chitin preparations published by Pearson et al. (1960). These authors also reported that there were no free OH and NH groups nor any C=O···H—O bonds in the chitin crystal. With the aid of a scale model of the chitin unit cell, a number of hydrogen-bonding schemes involving the primary hydroxyl groups were proposed. The C-6 hydroxyl groups were intramolecularly hydrogen bonded both to the nitrogen of the adjacent residue and to the bridge oxygen. Several attempts were also made over the following 20 years to improve on Carlström's model (Bouligand, 1965; Blackwell et al., 1965, 1967, 1980; Blackwell, 1969; Neville and Luke, 1969a,b; Neville, 1970; Ramakrishnan and Prasad, 1972; Gardner and Blackwell, 1975; Minke and Blackwell, 1978).

Another milestone in the discovery of chitin's structure and arrangement was made by Bouligand in 1965 through extensive ultra-structure analyses of crustacean cuticles (Bouligand, 1965; Berezina, 2016). Bouligand discovered that chitin adopted a stereotypic arrangement (helicoid structure) in arthropods. Thus, three types were also found: α-, β-, and γ-chitin. In 1969, Neville and Luke also found that chitin in insect cuticles adopted the Bouligand arrangement as well (Neville and Luke, 1969a,b; Neville, 1970). They suggested that cuticle was arranged in non-lamellate and lamellate systems. The original discovery of cuticle deposition by Bouligand was extensively reviewed and discussed by Neville in 1975 (Neville, 1975; Neville et al., 1976).

The research group of John Blackwell reinvestigated the structure of chitin (Blackwell, 1969; Blackwell et al., 1965, 1967, 1980; Gardner and Blackwell, 1975; Minke and Blackwell, 1978). Blackwell and co-workers, studying the crystalline forms of the chitin, claimed that there were different regular crystalline modifications of polysaccharides based on chain direction with differences in the packing and polarity of the adjacent chain. α-Chitin was the most abundant form, and it was found in certain fungi and in arthropod cuticles. α-Chitin was the tightly compacted, most crystalline polymorphic form where the chains were arranged in anti-parallel orientation. Sheets of chains were arranged in stacks along the a axis, the sheets being linked by C=O···H—N hydrogen bonds approximately parallel to the a axis. The strong inter- and intramolecular bonding leads to the formation of long microfibrils (Blackwell, 1969; Blackwell et al., 1965, 1967). Later, Blackwell and et al. also pointed out a number of deficiencies in the model proposed by Carlström (Minke and Blackwell, 1978; Blackwell et al., 1980). For α-chitin, their two main criticisms were the following: (1) the CH_2OH side chains were not hydrogen-bonded, although infrared spectroscopic studies showed that all the hydroxyl groups formed donor hydrogen bonds and (2) the presence of two amide I peaks suggested that the arrangement of the amide groups cannot be correct. Marchessault and Sarko (1967) previously reported similar conclusions. The structure of β-chitin from pogonophore tubes and from the spines of marine diatoms (*Thalassiosira fluviatilis*, *Cyclotella cryptica*) has been refined by rigid-body least-squares methods by Blackwell and et al. The β-chitin was the form where the chains were parallel. The unit cell was monoclinic, in agreement with Dweltz' model, but different cell dimensions were found (Table 5.2). The increase in the values along the a axis, compared with that for α-chitin, indicated a greater spacing between the chains in this direction. The structure consisted of an array of poly-*N*-acetyl-D-glucosamine chains all having the same environment, which were also linked together in sheets by C=O···H—N hydrogen bonding of the amide groups. The structure proposed was consistent with the swelling properties of β-chitin and can be seen to be analogous to that of native cellulose (Gardner and Blackwell, 1975). Both structures contained extended parallel chains and can be visualized as an array of hydrogen-bonded sheets. β-Chitin swelled extensively in water and has been shown to form a series of crystalline hydrates. β-Chitin from Polychaetae, when precipitated from acids, also assumed the α-form. In 1988, Blackwell suggested that γ-chitin may be "a distorted version of either α- or β-chitin, rather than a true third polymorphic form" (Blackwell, 1988).

CHAPTER SIX

The period of application: From 1970 until now

> *Aujourd'hui, une nouvelle place s'est développée pour la chitine dans l'intérêt de l'humanité : la chitine est entrée dans l'ère industrielle. Certaines technologies modernes peuvent désormais s'appuyer sur l'utilisation de la chitine et de certains de ses dérivés, comme le montre un grand nombre de brevets, dans le traitement des déchets, la transformation des aliments, l'agriculture et les produits pharmaceutiques.*
>
> *Today, a new place has developed for chitin in the interest of mankind: chitin has entered the industrial age. Some modern technologies may now rely on the use of chitin and of some of its derivatives, as shown by a large number of patents, in waste treatment, food processing, agriculture and pharmaceuticals.*
>
> ***Professor Charles Jeuniaux, 1976, Professor of Zoology and Entomology, Université de Liège, Belgium***

Despite the significant progress that was made between 1930 and 1970, few researchers at that time believed in the potential that chitin and chitosan had. "Re-discovery" and revived interest in the 1970s encouraged the need to better utilize biowastes from marine crustaceans due to the introduction of regulation on the dumping of untreated shellfish wastes into the oceans (Muzzarelli, 1977; Hirano, 1989; Roberts, 1992; Goosen, 1997; Kurita, 1998; Ravi Kumar, 2000; Khor, 2001; Crini et al., 2009). From 1970 to 1980, chitin and chitosan entered the period when they reached maturity, with their industrial production and the marketing of products containing chitosan. Indeed, this biopolymer was very rapidly considered a "new" biopolymer with strong development potential. It was recognized that its main characteristic was its cationic nature (unique in nature), allowing it a particular polyectrolyte behavior in solution. Most of the industrial applications of chitosan were related to these properties (Muzzarelli and Muzzarelli, 2005; Rinaudo, 2006). Chitosan was produced industrially for the first time in Japan in 1971 (Hirano, 1989). This industrialization of the production has also contributed enormously to its development. Another reason could be mentioned. The majority of commercial polymers and products were derived from petroleum-based raw materials using processing chemistry that

Chitin and Chitosan
https://doi.org/10.1016/B978-0-323-96119-6.00006-2

was not always safe or environmental-friendly. In this period, there was growing interest in developing natural low-cost and renouvelable alternatives to synthetic polymers. Therefore, chitosan was an ideal candidate to prepare innovative materials. The other reasons for this growing interest were its intrinsic, unique, and multiple physicochemical, biological, and technological properties.

6.1 The first applications of chitin and chitosan

The expansion of the two polysaccharides was made possible by the first concrete applications in cosmetology, pharmacy, personal care uses, food uses, agriculture, biotechnology, clarification and waste management (sludge dewatering), dentistry, and medicine (Feofilova, 1984; Zikakis, 1984; Hirano, 1989; Tsugita, 1990; Krajewska, 1991; Roberts, 1992; Goosen, 1997; Dodane and Vilivalam, 1998; Kurita, 1998; Ravi Kumar, 2000; Khor, 2001). Several patents also claimed the use of chitosan and its derivatives in papermaking (Slagel and Sinkovitz, 1973a,b; Plisko et al., 1974). Research on chitin synthesis, dormant for many years, has also been revived at the end of the 1970s due to an unexpected discovery relative to the insecticidal properties of certain benzoylphenyl ureas (Verloop and Ferrell, 1977; Hajjar and Casida, 1979; Zoebelein et al., 1980). This insecticidal action has generated great interest in insect chitin biosynthesis, in particular in order to contribute to a better understanding of the functional organization of the chitin synthase within the integument and its intricate regulation and to develop environmentally acceptable pesticides (Hackman, 1984, 1987; Cohen, 1987a). Two steps in arthropod chitin synthesis have been identified, one sensitive to benzoylphenyl ureas and another to tunicamycin. This clearly discriminated chitin formation in animals from that in fungi and yeasts and demonstrated that different targets for interference besides chitin synthase itself can be used successfully (Roberts, 1992; Goosen, 1997). At the same time, the potent ability of chitin to accelerate wound healing is discovered (Prudden et al., 1970). Subsequently, many works attempted to implement this discovery in many fields (Dumitriu, 2005; Kim, 2011, 2014). Another interesting example concerns the production of chitin fiber and chitosan. At the end of 1930s, Kunike was the first to try to produce fibers from chitin (Kunike, 1926a,b,c), although the major difficulty was to find a suitable solvent. Later, Rigby (1936a,b) also patented work on making films and fibers from chitosan. However, although industrial applications have been proposed, they have not been developed because of the difficult operating conditions. The discovery of new solvents in the 1970s stimulated new interest in this topic.

6.2 Books on chitin and chitosan

In 1973, Riccardo A.A. Muzzarelli edited the first interdisciplinary reference book on polysaccharides entitled "*Natural chelating polymers*" including three chapters on chitin, chitosan, and their analytical applications (Muzzarelli, 1973). This book also discusses the different sources of chitin, their properties, and structures, including a section on polymorphism and the three forms of chitin, α-, β-, and γ-chitin. The same year, Walton and Blackwell (1973) and Brimacombe (1973) comprehensively discussed the structural aspects of chitin and its chemistry. In 1977, Muzzarelli edited his famous chitin sourcebook (Muzzarelli, 1977). In the foreword, Jeuniaux wrote: "the publication of the book marks a turning point in the history of chitin and in the place occupied by this biopolymer in the chemical world of mankind." Muzzarelli's book is a remarkable compilation of data on the presence of chitin in many aquatic and terrestrial organisms and certain microorganisms: crustaceans (shrimps and crabs) and shellfish, mollusks (gastropods, cephalopods, lamellibranchs, and polyplacophora), nematodes, insects (ants), ringed worms, fungi (Ascomycetes, Basidiomycetes, and Phycomycetes), some yeasts, bacteria, and some algae (chlorophyceous algae and marine diatoms). Muzzarelli also pointed that the main sources for the isolation of commercial chitin are shellfish. More than 45 years later, this waste from the fishing industry is still the main source of chitin and chitosan on an industrial scale. It is important to note that in addition to the publication of Muzzarelli's book in 1977, this year was also a milestone with the organization of the first international conference on chitin and chitosan. A few years later, in 1992, a second reference book entitled "*Chitin chemistry*" was published by George A.F. Roberts. In the foreword, Marguerite Rinaudo wrote: "the book is a very useful source of knowledge for everyone involved with chitin and its developments." All aspects of chitin and chitosan are covered in this book, from extraction of the biopolymers to their characterization, structure, properties, and applications. Roberts' book is always considered a fundamental reference (Roberts, 1992). In 2009, Grégorio Crini, Pierre-Marie Badot, and Eric Guibal published the first book in French (Crini et al., 2009), considered by Roberts to be "a very useful addition to the literature on chitin and chitosan." Table 6.1 shows a selection of books on chitin and chitosan published in the last two decades.

Table 6.1 Selected books on chitin and chitosan published during the last two decades.

Book title	Topics covered	Reference
Applications of chitin and chitosan	Properties, applications, structure and activity, solution properties, reactivity characteristics, modifications, graft copolymers, food additives, plant protection, medicine, wastewater treatment	Goosen (1997)
Chitin: Fulfilling a biomaterial promise	Production, purity and quality, cost, medicine and biomedicine, nanotechnology	Khor (2001)
Chitine et chitosane—Du biopolymère à l'application	Preparation, properties, complexes, hypocholesterolaemic agent, food applications, biofilms, packaging materials, aerogels, biomedical applications, oenology, wastewater treatment	Crini et al. (2009)
Chitin, chitosan, oligosaccharides and their derivatives: biological activities and applications	Sources and production, derivatives, structural modifications, biological activities, biomedical applications, biotechnology, plant protection, beverage industry, food industry, wastewater treatment	Kim (2011)
Chitosan: Manufacture, properties, and usage	Preparation, properties, uses, complexes, chitosan from fungi, biomedical applications, drug delivery, nanoparticles, nanocarriers	Davis (2011)
Chitosan for biomaterials	Nanoparticles, drug delivery, medicine and biomedicine	Jayakumar et al. (2011)
Focus on chitosan research	Chemical and biological properties, biological activity, complexes, food industry, cosmetology, medicine, pharmaceutical applications	Ferguson and O'Neill (2011)

Table 6.1 Selected books on chitin and chitosan published during the last two decades—cont'd

Book title	Topics covered	Reference
Chitosan-based hydrogels: Functions and applications	Derivatives, chemical properties, biological activities, hydrogels, drug delivery, tissue engineering, enzyme immobilization	Yao et al. (2012)
Chitosan-based systems for biopharmaceuticals	Physical, chemical, and technological properties, modification, nanoparticles, medical and pharmaceutical applications, quality control, regulatory status	Sarmento and das Neves (2012)
Chitin and chitosan derivatives. Advances in drug discovery and developments	Preparation, characterization, modification, biological activities, drug delivery, gene delivery, biomedicine, nanomedicine	Kim (2014)
Advances in marine chitin and chitosan	Sources, production, characterization, modification, biomedical applications	Sashiwa and Harding (2015)
Chitin and chitosan for regenerative medicine	Hydrogels, composites, tissue engineering, cartilage regeneration, cancer chemotherapy, wound healing	Dutta (2016)
Chitosan based biomaterials. Fundamentals	Chemical and physical modifications, hydrogels, composites, sponges, nanoparticles, antibacterial properties, biomedicine	Amber Jennings and Bumgardner (2017)
Chitosan—Derivatives, composites and applications	Production, derivatives, textile industry, food applications, medicine and biomedicine, agrochemistry, water treatment	Ahmed and Ikram (2017)
Chitosan based materials and its applications	Sources, production, characterization, modification, nanofibers, nanocapsules, environmental applications, drug delivery, packaging, membranes	Dotto et al. (2017)

Continued

Table 6.1 Selected books on chitin and chitosan published during the last two decades—cont'd

Book title	Topics covered	Reference
Chitin: Properties, applications and research	Properties, modifications, electrospinning, nanofibers, biomedical applications	Phillips (2017)
Chitin-chitosan. Myriad functionalities in science and technology	Modification, food industry, environmental applications, wound dressings, biomedicine, agriculture	Dongre (2018)
Chitosan-based adsorbents for wastewater treatment	Structure, properties, nanocomposites, wastewater treatment, dye removal	Nasar (2018)
Oligosaccharides of chitin and chitosan: bio-manufacture and applications	Production, modification, characterization, biomaterials, medical and pharmaceutical applications, beverage industry	Liming (2019)
Targeting chitin-containing organisms	Structure, chemistry, biology, biodegradation, enzymes, plant protection	Qing and Tamo (2019)
Functional chitosan. Drug delivery and biomedical applications	Oral drug delivery, transdermal delivery, gene delivery, tissue engineering, tissue regeneration, antibacterial activity	Jana and Jana (2019)
Advances in chitin/chitosan characterization and applications	Characterization, applications, agriculture, drug delivery, medicine, papermaking industry, aerogels	Rinaudo and Goycoolea (2019)
Oligosaccharides of chitin and chitosan. Bio-manufacture and applications	History, preparation and characterization of chitin oligosaccharides, applications in food industry, biomaterials	Zhao (2019)
Chitin and chitosan: History, fundamentals and innovations	Discovery, sources, production, characterization, modification, enzyme immobilization, 3D printing and bioprinting, medical and biomedical applications, textile industry	Crini and Lichtfouse (2019a)

Table 6.1 Selected books on chitin and chitosan published during the last two decades—cont'd

Book title	Topics covered	Reference
Chitin and chitosan: Applications in food, agriculture, pharmacy, medicine and wastewater treatment	Food industry, seafood processing and preservation, packaging materials, agriculture, tissue engineering, drug delivery, bioflocculation, wastewater treatment	Crini and Lichtfouse (2019b)
Chitin- and chitosan-based biocomposites for food packaging applications	Nanocomposites, characterization, packaging industry, food applications, biomaterials, composites, biodegradation, cytotoxicity studies	Jacob et al. (2020)
Handbook of chitin and chitosan. Preparation and properties	Sources, isolation, properties, solubility, derivatives, nanomaterials, electrospinning, aerogels, applications	Thomas et al. (2020)
Chitin and chitosan: properties and applications	Sources, isolation, physicochemical properties, modifications, hydrogels, antimicrobial properties, drug delivery, biomedicine, food applications, packaging	van den Broek and Boeriu (2020)
Chitosan in biomedical applications	Chitosan chemistry, properties, chemical modifications, characterization, materials, applications, biomedical and healthcare fields, tissue engineering, mechanism of biological interaction, regulatory aspects	Nayak et al. (2021)
Chitin and Chitosan. Physicochemical properties and industrial applications	Extraction, physicochemical properties, modification, characterization, applications, biomedical sciences, drug delivery systems, superabsorbents, nanotechnology, textiles, biotechnology	Berrada (2021)

Continued

Table 6.1 Selected books on chitin and chitosan published during the last two decades—cont'd

Book title	Topics covered	Reference
Chitin and chitosans in the bioeconomy	Occurrence of chitin, biosynthesis, production, structure, properties, biodegradation, materials, valorization, bioeconomy, market, applications	Wertz and Perez (2021)
Handbook of chitin and chitosan. Preparation and properties	Sources, isolation, extraction, properties, applications, food industry, edible films, agriculture, beverage industry, biomedical applications	Cassie (2022)
Chitooligosaccharides. Prevention and control of diseases	Bioproduction of chitooligosaccharides, enzymatic production, isolation, insect enzymes, deacetylation, biomedical applications, drug delivery systems, antiinflammatory activity, immunomodulatory effects, anticancer effects, wound healing agent, neuroprotective properties, nutraceuticals	Kim (2022)
Chitosan applications in food systems	Properties, food industry, food preservation, nutritional aspects, packaging, edible films, nanoparticles	Savvaidis (2022)
Role of chitosan and chitosan-based nanomaterials in plant sciences	Nanoparticles, morphology, properties, nanotechnology, plant sciences, toxicity and human exposure, tolerance mechanism, biochemical pathways	Kumar and Madihally (2022)

6.3 Comprehensive reviews and book chapters on chitin and chitosan

Abundant scientific literature has been built up since the 1970s. Between 1970 and 2020, 31,873 and 98,295 publications, including articles and generalist reviews, were published on chitin and chitosan, respectively

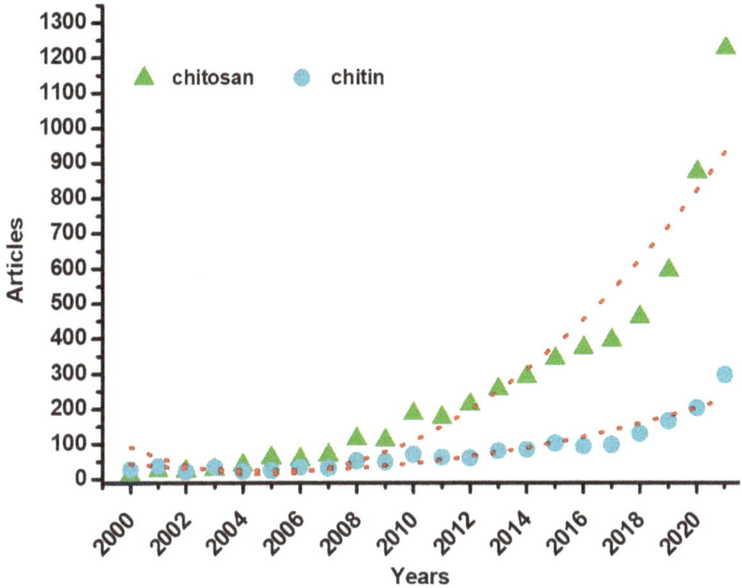

Fig. 6.1 Number of chitin- and chitosan-related reviews published from 1970 to 2021. *(Source: Web of Science Core Collection; the analysis was performed using the terms "chitin" AND "review" and "chitosan" AND "review" in the topic category, on December 31, 2021.)*

(source: Web of Science Core Collection). As an example, Fig. 6.1 shows how the number of reviews on chitin and chitosan increased between 2000 and 2021. During this period, a total of 1809 and 5968 reviews were published for chitin and chitosan, respectively. The vast majority of these publications are focused on applications. Table 6.2 compares the number of reviews published for several highly studied polysaccharides in the literature since the early 2000s. As indicated, after cellulose and starch, chitosan is the biopolymer of most interest to the scientific community. A large number of book chapters have been published, and numerous patents have also been filed on practically all the aspects of chitin and chitosan, so many that it would not be possible to list them all. Table 6.3 presents a series of selected

Table 6.2 Comparison of the number of reviews published between 2000 and 2021 for several polysaccharides (updated on December 31, 2021).

Polysaccharide	Number of reviews
Cellulose	8513
Starch	6213
Chitosan	5968
Cyclodextrin	2955
Alginate	2431
Chitin	1809

Table 6.3 Reviews on different topics related to chitin and chitosan and their derivatives published in the last 50 years (selected papers).

Topics	Keywords	References
Agriculture	Abiotic stress, agrochemistry, antimicrobial agent, antioxidant, bacteriostatic agent, biochemicals, biochemical pathways, biocides, bioeconomy, biopesticides, biostimulation, biotic stress, chitooligosaccharides, coating effects, edible coatings, edible films, elicitor, enzymes, fertilizers, films, fruit coating, horticulture, human exposure, hydrogels, ingredient to improve microbial communities, low molecular weight chitosans, moisturizing agent, morphology, nanochitosan, nutrient, oligochitosan, pathogens, postharvest, plant defense systems, plant protection, pest control, pesticide formulations, preharvest treatment, respiration rate, seed pretreatment, seed soaking agent, soil conditioning agent, soil enrichment, storage ability, structure-activity relationship, and stress resistance	Goosen (1997), Rabea et al. (2003), Hayes et al. (2008b), El Hadrami et al. (2010), Yin and Du (2011), Sharp (2013), Muñoz-Bonilla et al. (2014), Katiyar et al. (2014), Xing et al. (2015), Badawy and Rabea (2016, 2017), Bautista-Baños et al. (2016), Hadwiger (2017), Ippólito et al. (2017), Orzali et al. (2017), Divya and Jisha (2018), Grande-Tovar et al. (2018), Sharif et al. (2018), Betchem et al. (2019), Crini and Lichtfouse (2019a), Morin-Crini et al. (2019), Yuan et al. (2019), Bandara et al. (2020), Maluin and Hussein (2020), Malerba and Cerana (2019, 2020), Mujtaba et al. (2020), Qu and Luo (2020), Shamshina et al. (2020), Adiletta et al. (2021), Dave et al. (2021), Shahrajabian et al. (2021), Singh et al. (2021a), Yu et al. (2021), Mahari et al. (2022), and Prathibhani et al. (2022)
Aquaculture	Antimicrobial, antioxidant, biochemical, bioeconomy, controlled release of compounds, drug delivery, functional foods, immune-stimulant, nutrition, probiotics, quality water, supplements, treatment of seafood effluents, and vaccines	Tsugita (1990), Chung et al. (2005), Chung (2006), Borgogna et al. (2011), Cerezuela et al. (2011), Alishahi and Aïder (2012), Harikrishnan et al. (2012), Niu et al. (2013), Zaki et al. (2015), Lian et al. (2016), Bernardi et al. (2018), Vinay et al. (2018), Crini and Lichtfouse (2019a), Morin-Crini et al. (2019), Yuan et al. (2019), Yin et al. (2021), El-Naggar et al. (2022), and Mahari et al. (2022)

Beverage industry	Acidity-adjusting agent, alcoholic fermentation, antimicrobial agent, antioxidant agent, beer industry, beverages, chelation, clarification, clarifying agent, enology, extending the shelf-life, enzyme immobilization, filtration, fining agent, flocculation, functional beverage, haze prevention, immobilization, inhibition of unwanted microbial growth, metal reduction, packaging, protein haze formation, protein removal, tartrate precipitation, removal of ochratoxin A, removal of volatile compounds, sulfite-free wines, and wines	Kim (2011), Gassara et al. (2015), Rocha et al. (2017), Crini and Lichtfouse (2019a), Morin-Crini et al. (2019), Castro Marín et al. (2020), and Cosme and Vilela (2021)
Biochemistry	β-*N*-Acetylglucosaminidase, biosynthesis, chitin synthase, chitin synthesis, chitinases, chitosomes, hydrolysis, inhibition, molecular mechanism, structure, and synthesis	Jeuniaux (1971, 1982), Brimacombe (1973), Rudall and Kenchington (1973), Neville (1975), Hackman (1984, 1987), Cohen (1987a,b, 2010), Lezica and Quesada-Allué (1990), Horst et al. (1993), and Horst et al. (1993)
Biological properties	Bioactivities properties (antiacid, antibacterial, anticoagulant, antifungal, antimicrobial, antioxidant, antitumor), bioadhesivity, biocompatibility, biodegradability, Gram-negative bacteria, Gram-positive bacteria, hydrating agent, microencapsulation, nontoxicity, regulatory status, structure and activity	Brimacombe (1973), Muzzarelli (1973, 1977), Neville (1975), Cohen (1987a,b), Kurita (1998, 2006), de Alvarenga (2011), Croisier and Jérôme (2013), Younes and Rinaudo (2015), Sahariah and Másson (2017), Crini and Lichtfouse (2019b), Jangid et al. (2019), Kucharska et al. (2019), Rinaudo and Pérez (2019), El-Hack et al. (2020), Santos et al. (2020), Amirthalingam et al. (2021), Ardean et al. (2021), Ashok et al. (2021), Dave et al. (2021), Ke et al. (2021), Joseph et al. (2021), Muthu et al. (2021), Kou et al. (2022), Muñoz-Núñez et al. (2022), and Yu et al. (2022)

Continued

Table 6.3 Reviews on different topics related to chitin and chitosan and their derivatives published in the last 50 years (selected papers)—cont'd

Topics	Keywords	References
Biotechnology	Bioactivity, biocatalysis, biomembranes, biosensors, bioseparation, biosynthesis, biowastes, cell immobilization, cell recovery, chinolytic organisms, gel permeation chromatography, hydrolysate, electronic devices, enzyme immobilization, enzyme technology, enzymology, metabolic analysis	Krajewska (1991, 2005), Hirano (1996), Kim et al. (1999), Hayes et al. (2008b), Dash et al. (2011), Wang (2012), Suginta et al. (2013), Philibert et al. (2017), Grifoll-Romero et al. (2018), Morin-Crini et al. (2019), Verma et al. (2020), Stephen et al. (2021), and Yang et al. (2021)
Characterization	Crystallinity, crystallography, degree of deacetylation, distribution of *N*-acetyl groups, hydrolysis, linear amino-polysaccharides, molecular weight, NMR, purity, rheological behavior, spectroscopy, titration, water retention, X-ray diffraction	Blackwell et al. (1980), Blackwell (1988), No and Meyers (1995), Kurita (2001, 2006), Vårum and Smidsrød (2004a), Rinaudo (2006), Hayes et al. (2008a), de Alvarenga (2011), Sahoo and Nayak (2011), Teng (2012), Nwe et al. (2014), Younes and Rinaudo (2015), Berezina (2016), Dima et al. (2017), Crini and Lichtfouse (2019b), Rinaudo and Pérez (2019), Feng et al. (2020), Kumari and Kishor (2020), Wu et al. (2020), Hossin et al. (2021), Joseph et al. (2021), Achinivu et al. (2022), and Ma et al. (2022)
Chemical applications	Adhesives, analytical chemistry, biocatalysis, biosensors, catalysis, chiral separations, chromatography, click chemistry, CO_2 removal, dialysis, deep eutectic solvents, electrochemistry, electrochemical sensing, gas permeation, green chemistry, green solvents, ionic liquids, membranes, pervaporation, polymer science, reverse osmosis, solvent separation, thin layer chromatography, ultrafiltration	Sharon (1980), Tsugita (1990), Kurita (1998), Kim et al. (1999), Ravi Kumar (2000), Ravi Kumar et al. (2004), Phan et al. (2012), Suginta et al. (2013), Mati-Baouche et al. (2014), Carneiro et al. (2015), Salehi et al. (2016), Shen et al. (2016), Thakur and Voicu (2016), Osman and Arof (2017), Silva et al. (2017), Argüelles-Monal et al. (2018), Galiano et al. (2018), Hong et al. (2018), Marpu and Benton

		(2018), Zdanowicz et al. (2018), Crini and Lichtfouse (2019a), Kadokawa (2019), Morin-Crini et al. (2019), Shamshina (2019), Shamshina and Berton (2020), Shamshina et al. (2019), Xie and Yuan (2019), Jaworska et al. (2020), Boominathan and Sivaramakrishna (2021), Peter et al. (2021), Özel and Elibol (2021), Roy et al. (2021), Vinodh et al. (2021), Yin et al. (2021), Khajavian et al. (2022), Prabhu et al. (2022), Muñoz-Núñez et al. (2022), Wang et al. (2022), and Zhang et al. (2022)
Chemistry (properties, behavior, modification)	Acylation, aggregation, carbohydrate-branched chitosans, carboxyalkylation, carboxymethylation, chain conformation, chemical modification, chitosan sulfates, complexation with polyanions, complexing properties, composites, conductive polymers, cross-linking, depolymerization, derivatization, dissolution, electrospinning, electrostatic properties, gelation, gels, gel state, grafting, hydrogels, ionic conductivity, layer-by-layer, lyophilization, microgels, nanotechnology, pH-sensitive gels, phosphorylation, physical modification, polyelectrolytes, precipitation, quaternized derivatives, reactivity, salts, self-assembly, solubility, structure and activity	Brimacombe (1973), Muzzarelli (1973, 1977), Walton and Blackwell (1973), Whistler (1973), Sharon (1980), Peter (1995), Hudson and Smith (1998), Kurita (1998, 2001, 2006), Vårum and Smidsrød (2004a), Muzzarelli and Muzzarelli (2005), Kurita (2006), Rinaudo (2006), Peniche et al. (2008), de Alvarenga (2011), Teng (2012), Nwe et al. (2014), Younes and Rinaudo (2015), Bonecco et al. (2017), Nezakati et al. (2018), Crini and Lichtfouse (2019b), Morin-Crini et al. (2019), Rinaudo and Pérez (2019), El Knidri et al. (2020), Feng et al. (2020), Kumari and Kishor (2020), Hossin et al. (2021), Joseph et al. (2021), and Torkaman et al. (2021)

Continued

Table 6.3 Reviews on different topics related to chitin and chitosan and their derivatives published in the last 50 years (selected papers)—cont'd

Topics	Keywords	References
Cosmetics	Additives, antiaging, antiperspirants, bacteriostatic, coloring products, cosmeceuticals, creams, deodorizant products, gums, encapsulating agents, essential oils, enzymes, fragrances, hair care, hydrogels, hygiene, moisturizers, oral care, personal care, multifunctional ingredient, shampoos, skin care, thickening agent, toiletry, anti-UV rays	Muñoz et al. (2012), Lima et al. (2012), Senevirathne et al. (2012), Chalongsuk and Sribundit (2013), Jimtaisong and Saewan (2014), Ahmed and Ikram (2017), Bonecco et al. (2017), Costa and Santos (2017), Nechita (2017), Philibert et al. (2017), Serrano-Castañeda et al. (2018), Casadidio et al. (2019), Crini and Lichtfouse (2019a,b), Morin-Crini et al. (2019), El-Hack et al. (2020), Triunfo et al. (2021), and van den Broek and Boeriu (2020)
Cosmetotextiles	Advanced properties, antimicrobial textiles, antiodor textiles, deodorizing textiles, encapsulation of fragrances, intelligent textiles, microencapsulation, moisturizing agent, perfuming textiles, UV-protective textiles	Morin-Crini et al. (2019), Crini and Lichtfouse (2019a), Han et al. (2020), and Raza et al. (2020)
Dermatology	Acne, bacteriostatic, coated textiles, cosmeceuticals, gels, essential oils, moisturizers, nanofibers, nanoparticles, nanostructures, skin care, water-resistant adhesives; self-healing; wound-healing	Bernkop-Schnürch and Dünnhaupt (2012), Ghannoum et al. (2015), Lopes et al. (2015), Elieh-Ali-Komi and Hamblin (2016), Badawy and Rabea (2017), Krishnaswami et al. (2018), Zarrintaj and Saeb (2018), and Tangkijngamvong et al. (2020)
Dentistry	Adhesives, antibacterial effect, biocomposites, carboxymethyl chitin, caries, chewing gums, chitooligosaccharides, composites, conservative dentistry, controlled delivery of fluoride, dental alignments, dental composites, dental surgery, dental	Keegan et al. (2012), Hayashi et al. (2013), Queiroz et al. (2015), Kmiec et al. (2017), Aranaz et al. (2018), Navarro-Suarez et al. (2018), Morin-Crini et al. (2019), Chen et al. (2021), and Sharifianjazi et al. (2022)

	therapy, drug delivery, endodontics, erosive lesions, gels, implants, ingredients, nanochitosans, oral healthcare, oral hygiene, orthodontics, periodontology, plaque inhibitor, preventive dentistry, prosthodontics, salivary secretion, surgery, toothpastes, wound healing	
Derivatives	Aerogels, biocomposites, biofilms, carboxymethylchitin, chitosan derivatives, chitooligomers, chitooligosaccharides, click chemistry, composites, controlled functionalization, cross-linked gels, dendrimers, enzymatic modifications, fibers, films, foams, functional aerogels, gels, gelation, graft copolymers, glucosamines, glycol chitosan, hierarchical structures, hybrid materials, hydrogels, innovative molecular architectures, membranes, microgels, monofilament fibers, nanobiocomposites, nanocomposites, nanocrystals, nanofibers, nanofibrils, nanoparticles, nanoscaffolds, nanowhishkers, oligosaccharides, optical tweezers, pH-sensitive gels, sponges	Kurita (2001, 2006), Nilsen-Nygaard et al. (2015), Duarte et al. (2018), Liaqat and Eltem (2018), Yu et al. (2018), Brasselet et al. (2019), Kaczmarek et al. (2019), Yang et al. (2019), Yuan et al. (2019), Zhang et al. (2019a), El Knidri et al. (2020), Feng et al. (2020), Lang et al. (2020), Peers et al. (2020), Boominathan and Sivaramakrishna (2021), Boroumand et al. (2021), Lima et al. (2021), Singh et al. (2021b), Tabassum et al. (2021), and Muñoz-Núñez et al. (2022)
Environmental chemistry	Adsorption, antifouling agent, biological denitrification, biocoagulation, biofiltration, bioflocculation, biosorbents, biosorption, chelating resins, composite membranes, cross-linked gels, cyanobacterial blooms, dehydration agent, drilling	Kaseamchochoung et al. (2006), Crini and Badot (2008), Renault et al. (2009), Liu and Bai (2014), Vakili et al. (2014), Zemmouri et al. (2014), Boamah et al. (2015), Yong et al. (2015), Azarova et al. (2016), Barbusinski et al. (2016), Kos (2016), Salehi

Continued

Table 6.3 Reviews on different topics related to chitin and chitosan and their derivatives published in the last 50 years (selected papers)—cont'd

Topics	Keywords	References
	muds, dye removal, emerging pollutants, eutrophication, films, filtration, fluoride removal, hydrogels, imprinted membranes, membranes, metal removal, nanochitosan, nanotechnology, pesticides, odors, polymer-assisted ultrafiltration, recovery of precious metals, remediation, sand filtration, sludge treatment, solid-phase extraction, structure-activity relationship, suspended solid removal, turbidity reduction, ultrafiltration, uranium, wastewater treatment, water clarification, water purification	et al. (2016), Zhang et al. (2016), Ahmed and Ikram (2017), Anastopoulos et al. (2017), Crini et al. (2017, 2019), Kyzas et al. (2017), Nechita (2017), Sudha et al. (2017), Alaba et al. (2018), de Andrade et al. (2018), Desbrières and Guibal (2018), El Halah et al. (2018), Kasiri (2018), Pakdel and Peighambardoust (2018), Van Tran et al. (2018), Wei et al. (2018), Boulaiche et al. (2019), Lichtfouse et al. (2019), Lin et al. (2019), Morin-Crini et al. (2019), Samoila et al. (2019), Sarode et al. (2019), Shi et al. (2019), Zhang et al. (2019b), Abhinaya et al. (2021), Begum et al. (2021), da Silva Alves et al. (2021), Dave et al. (2021), Karimi-Maleh et al. (2021), Pal et al. (2021), Sadiq et al. (2021), Saheed et al. (2021), Sheth et al. (2021), Sirajudheen et al. (2021), Spoială et al. (2021), Kong et al. (2022), and Musarurwa and Tavengwa (2022)
Food Industry	Additives, alicament, antimicrobial activity, antioxidant, biocomposites, biodegradable films, bio-inks, biotechnology, dietary fiber, edible film, encapsulating agent, encapsulation, essential oils, feed additives, flavors, food preservation, food protection, food technology, functional foods, high-value products, hydrogels, packaging, microbial films, nanofilms, nutraceutical ingredients, nutraceuticals, nutrition, regulations, removal of target substances, superfoods	Winterowd and Sandford (1995), Shahidi et al. (1999), Hayes et al. (2008b), Vårum and Smidsrød (2004b), Gallo et al. (2016), Gutiérrez (2017), Han et al. (2018), Manigandan et al. (2018), Cazon and Vasquez (2019), Crini and Lichtfouse (2019a), Morin-Crini et al. (2019), Priyadarshi and Rhim (2020), Qu and Luo (2020), Adiletta et al. (2021), Amirthalingam et al. (2021), Bose et al. (2021), Muthu et al. (2021), Singh et al. (2021b), Wang et al. (2021), Yang et al. (2021), Maleki et al. (2022), Savvaidis (2022), and Yang et al. (2022)

Medicine	Bandages, biomedical adhesives, biomedical engineering, bio-imaging, bio-printing, bio-sensing, cancer nanotechnology, cancer therapy, cell biology, chemotherapy, drug delivery, gene therapy, hydrogels, immunology, implants, low molecular weight chitosans, medical devices, nanofibers, nanoparticles, nanotechnology, oncotherapy, orthopedics, radiation therapy, regenerative medicine, restoration of cutaneous tissue, scaffolds, skin-regeneration agent, structure-activity relationship, therapeutic domains, tissue engineering, vaginal infections, wound dressing	Burkatovskaya et al. (2006), Rauh and Dornish (2006), Minagawa et al. (2007), Hayes et al. (2008b), Muzzarelli and Muzzarelli (2009), Lahiji et al. (2000), Ribeiro et al. (2009), Dash et al. (2011), Jayakumar et al. (2011), Kean and Thanou (2011), Wang et al. (2012), Croisier and Jérôme (2013), Anitha et al. (2014), Elieh-Ali-Komi and Hamblin (2016), Guarino et al. (2016), LogithKumar et al. (2016), Ahmed and Ikram (2017), Amber Jennings and Bumgardner (2017), Sahariah and Másson (2017), Singh et al. (2017), Al Aboud (2018), Akbar and Shakeel (2018), Ahmed et al. (2018), Aljohani et al. (2018), Dimassi et al. (2018), Liaqat and Eltem (2018), Nezakati et al. (2018), Pellá et al. (2018), Zhao et al. (2018), Crini and Lichtfouse (2019a), Jana and Jana (2019), Jangid et al. (2019), Massella et al. (2019), Morin-Crini et al. (2019), Zhang et al. (2019a), Chivere et al. (2020), Lang et al. (2020), Satitsri and Muanprasat (2020), Zamboulis et al. (2020), Amirthalingam et al. (2021), Ardean et al. (2021), Araujo et al. (2021), Boroumand et al. (2021), Dave et al. (2021), Ke et al. (2021), Kumara et al. (2021), Lima et al. (2021), Mohammadi et al. (2021), Nayak et al. (2021), Rezaei et al. (2021), Torkaman et al. (2021), Valachová and Šoltés (2021), Gao and Wu (2022), Sánchez-Machado et al. (2022), and Yu et al. (2022)

Continued

Table 6.3 Reviews on different topics related to chitin and chitosan and their derivatives published in the last 50 years (selected papers)—cont'd

Topics	Keywords	References
Nutraceuticals	Additives, alicaments, allergy risks, anticholesterol, chitooligosaccharides, diabetes, dietary fiber, essential oils, fats, flavors, food technology, functional foods, glucosamine, high-value products, ingredients, lipids, nutrition, obesity, regulations, sugars, super foods, supplements, vitamins	Kumar and Tharanathan (2004), Hayes et al. (2008b), Kim and Karadeniz (2013), Karadeniz and Kim (2014), Je and Kim (2012a,b), Nasri et al. (2014), Patti et al. (2015), Nasri et al. (2014), Morin-Crini et al. (2019), and Akbari-Alavijeh et al. (2020)
Ophthalmology	Antimicrobial activity, carriers, colloidal carriers, coatings, contact lenses, eyes, gels, liposomes, micelles, mucoadhesion, nanocarriers, nanoparticles, ocular bandages, ocular drug delivery	Felt et al. (1999), Alonso and Sánchez (2003), Bernkop-Schnürch and Dünnhaupt (2012) Elieh-Ali-Komi and Hamblin (2016), Badawy and Rabea (2017), Krishnaswami et al. (2018), Paliwal et al. (2019), and Zamboulis et al. (2020)
Packaging	3D printing, active films, additives, biocomposites, biodegradable films, biodegradable nanomaterials, biotechnology, coatings, composites, edible coatings, edible films, essential oils, environmental impact, films, food preservation, food protection, hydrogels, metal oxides, microbial films, nanocomposites, nanofibers, nanocrystals, nanoemulsions, nanofillers, nanoscale reinforcements, nanotechnology, plasma, plasticized chitosan films, preservatives, pickering stabilizers, reduction of pesticide use, safe and ecofriendly materials, thermoplastic films	Nakano et al. (2007), Muñoz-Bonilla et al. (2014), van den Broek et al. (2015), Han et al. (2018), Wang et al. (2018), Cazon and Vazquez (2019), Jacob et al. (2020), Priyadarshi and Rhim (2020), Souza et al. (2020), Adiletta et al. (2021), Amirthalingam et al. (2021), Bose et al. (2021), Dutta et al. (2021), Wang et al. (2021), Yang et al. (2021), Muñoz-Núñez et al. (2022), Oladzadabbasabadi et al. (2022), and Savvaidis (2022)

Pharmacy	Biopharmacy, cosmeceuticals, dermatology, drug delivery, drug release, excipients, ocular delivery formulations, nanotechnology, nutraceuticals, ophthalmology, oral delivery carriers, oral delivery platforms, polyelectrolyte complexes, vaccines	Felt et al. (1998), Illum (1998), Dodane and Vilivalam (1998), Kim et al. (1999), Ravi Kumar (2000), Singla and Chawla (2001), Kato et al. (2003), Illum and Davis (2004), Ravi Kumar et al. (2004), Vårum and Smidsrød (2004a,b), Hayes et al. (2008b), Kean and Thanou (2010), Dash et al. (2011), Bernkop-Schnürch and Dünnhaupt (2012), Croisier and Jérôme (2013), Mati-Baouche et al. (2014), Badwan et al. (2015), Younes and Rinaudo (2015), Berezina (2016), Amber Jennings and Bumgardner (2017), Sahariah and Másson (2017), Akbar and Shakeel (2018), Crini and Lichtfouse (2019a), Jana and Jana (2019), Jangid et al. (2019), Morin-Crini et al. (2019), Shariatinia (2019), Lang et al. (2020), Peers et al. (2020), van den Broek and Boeriu (2020), Wu et al. (2020), Zamboulis et al. (2020), Ardean et al. (2021), and Singh et al. (2021a)
Pulp and paper industry	Biodegradable packaging, chromatography paper, decontamination of wastewaters, paperboard, papermaking industry, packaging, paper sizing, photographic paper, pulp and paper, retention and drainage agents, surface coating application, toilet paper, water treatment, wood wrapping	Ravi Kumar (2000), Crini et al. (2009), Phillips (2017), Samyn et al. (2018), Song et al. (2018), and Morin-Crini et al. (2019)

Continued

Table 6.3 Reviews on different topics related to chitin and chitosan and their derivatives published in the last 50 years (selected papers)—cont'd

Topics	Keywords	References
Production	Aquatic crustaceans, bleaching, bioactive molecules, bioeconomy, biomass, biosynthesis, biowastes, chitin deacetylases, chitinolytic enzymes, chitooligosaccharides, conversion, cost, cost-purity relationship, deacetylation, decoloration, demineralization, deproteinization, extraction (chemical, biological, enzymatic), fish waste, fungi, high molecular weight, high-value products, industrial production, isolation, ionic liquids, insects, lipids, low molecular weight, microorganisms, microwave-assisted methods, molecular weight, proteins, properties, purification, purity, quality control, regulatory status, solubility, sources, structure, valorization	Hirano (1989), No and Meyers (1995), Winterowd and Sandford (1995), Hudson and Smith (1998), Kurita (2001, 2006), Rinaudo (2006), Peniche et al. (2008), Kean and Thanou (2011), Nwe et al. (2011a,b, 2013, 2014), Sahoo and Nayak (2011), Teng (2012), Nwe et al. (2014), Younes and Rinaudo (2015), Berezina (2016), Hamed et al. (2016), Bonecco et al. (2017), Dima et al. (2017), Philibert et al. (2017), Hong et al. (2018), Sieber et al. (2018), Bastiaens et al. (2019), Kadokawa (2019), Hahn et al. (2020), Santos et al. (2020), Sebastian et al. (2020), van den Broek and Boeriu (2020), Abhinaya et al. (2021), Kou et al. (2021), Hossin et al. (2021), Pakizeh et al. (2021), Stephen et al. (2021), Achinivu et al. (2022), Huq et al. (2022), Iber et al. (2022), Khajavian et al. (2022), Ma et al. (2022), Oladzadabbasabadi et al. (2022), and Wang et al. (2022)
Other applications	Adhesives, antifouling paints, biodiesel production, biolubricants, biomimetic photonic materials, bioprinting, cement industry, cigarette industry, CO_2 removal, composites, deep eutectic solvents, detergents, leather, liquid crystals, lithium batteries,	Dutta et al. (2004), Heuser et al. (2009), Pelletier et al. (2009), Banerjee et al. (2011), Cheba (2011), Dash et al. (2011), Ifuku and Saimoto (2012), Heuser and Cárdenas (2014), Aljohani et al. (2018), Galiano et al. (2018), Marpu and Benton (2018), Morin-Crini

	optical materials, paints, petroleum industry, photography, photonics, photonic nanohybrids, plastics, printing, quantum dots, reinforcement agent, semiconductors, sensors, surfactants, wood industry, wood protection	et al. (2019), Lizundia et al. (2021), Negi et al. (2021), Özel and Elibol (2021), Peter et al. (2021), Roy et al. (2021), Zhang et al. (2021), Vinodh et al. (2021), and Willson et al. (2021)
Structure	Crystallinity, morphology, block-type copolymer, random-type copolymer, low-molecular weight macromolecules, structure-property relationship	Muzzarelli (1973, 1977), Vårum and Smidsrød (2004a), Rinaudo (2006), Peniche et al. (2008), Dash et al. (2011), Teng (2012), Younes and Rinaudo (2015), Hossin et al. (2021), Joseph et al. (2021), and Achinivu et al. (2022)
Textiles	Active compounds, deodorizing property, cosmetotextiles, functional textiles, medical textiles, microcapsules, nonallergenic fibers, sanitary fibrous products, surgical threads, regulations	Ravi Kumar (2000), Giri Dev et al. (2005), Enescu (2008), Ummu Habeeba et al. (2007), Crini et al. (2009), Francesko et al. (2010), Islam et al. (2013), Şahan and Demir (2014), Hamed et al. (2016), Voncina et al. (2016), Gutiérrez (2017), Roy et al. (2017b), Kucharska et al. (2019), Massella et al. (2019), and Morin-Crini et al. (2019)
Toxicology	Biodegradation, cytotoxicity, regulations, toxicity	Kurita (1998), Peniche et al. (2008), Nwe et al. (2014), Berezina (2016), Bonecco et al. (2017), Dima et al. (2017), and Matica et al. (2017)
Technological properties	Adhesivity, adsorption properties, coagulating agent, film-forming ability, flocculation properties, polyelectrolytes, synthesis-structure relationship, versatility	Nwe et al. (2014), Philibert et al. (2017), Verma et al. (2020), Ashok et al. (2021), and Joseph et al. (2021)

Continued

Table 6.3 Reviews on different topics related to chitin and chitosan and their derivatives published in the last 50 years (selected papers)—cont'd

Topics	Keywords	References
Veterinary Medicine	Adjuvant, antibiotics, antiparasitics, body-care products, bone regeneration, drug delivery, hemostatic products, hemostatic and antibacterial barrier, mucosal immunization, nutrition, nutritional supplement, probiotics, sprays, vaccines, regeneration medicine, shampoos, supplements, tissue engineering, veterinary dermatology, wound healing	Şenel and McClure (2004) Şenel (2011), Underwood and van Eps (2012) Drewnowska et al. (2013), Gerdts et al. (2013), Tonda-Turo et al. (2016), Morin-Crini et al. (2019), and Maldonado-Cabrera et al. (2021)

reviews on different topics related to chitin and chitosan and their derivatives published in the last 50 years, especially reviews and book chapters on chitosan applications used to argue the discussion in the next chapter.

6.4 The first international chitin and chitosan conference

In 1977, the first International Conference on Chitin and Chitosan (ICCC) was held in Boston in April (11th–13th) organized by Vincent LoCicero of the Massachusetts Science and Technology Foundation and hosted jointly by the MIT Sea Grant Program and the Massachusetts Science and Technology Foundation. This symposium was a great success, with participants coming from all over the world, for example, Arthur Glenn Richards, Minesota; Charles Jeuniaux, Belgium, the Chairman Riccardo A.A. Muzzarelli, Italy, José Ruiz-Herrera, and México. The proceedings of this conference were published by Muzzarelli and Ernst R. Pariser, Massachusetts Institute of Technology, in 1978 (Muzzarelli and Pariser, 1978). Since the 2nd ICCC at Sapporo, Japan, in 1982, the conference has been held every 3 years. Japan hosted the conference a second time in 2000 in Yamaguchi. The other countries that hosted the conference are Italy in 1985, Norway in 1988, United States in 1991, Poland in 1994, France in 1997 and 2006, Canada in 2003, Taiwan in 2009, Brazil in 2012, and Germany in 2015. The 14th and most recent ICCC was organized by the Japanese Society for Chitin and Chitosan in Osaka, Japan, in August 27–30, 2018.

There are also several national groups that are very active such as in Japan, which enables an excellent quality of communication and exchange on current research into chitin/chitosan in each country. The first Japanese Chitin & Chitosan Symposium was held in Osaka in 1981 (Hirano, 1989). Following the 2nd and 3rd symposia in 1985 (Tottori) and 1988 (Tokyo), the Japanese Society for Chitin and Chitosan was founded in 1989 with the objective of contributing to the development of chitin and chitosan research and their applications. Since 1990, this scientific society has been holding an annual meeting each summer in Japan. The society also publishes the journal Chitin and Chitosan Research (two issues/year). Due to the 2019 coronavirus outbreak, many national and international events have been postponed.

6.5 The European Chitin Society

The European Chitin Society (EUCHIS) as a nonprofit organization was founded on March 27, 1992, in Villeurbanne, France. The first President was Alain Domard (1993–1997) and, in 1995, the first General Assembly was held at the 1st EUCHIS conference in Brest, France. Domard was followed by Riccardo Muzzarelli (Italy), Martin Peter (Germany), Kjell Vårum (Norway), Sevda Şenel (Turkey), Angeles Heras (Spain), and Bruno M. Moerschbacher (Germany, actual President). The main objective of EUCHIS is "to encourage basic and applied scientific studies of all aspects of chitin and chitosan, including their derivatives, and related enzymes." EUCHIS conferences are organized usually on a 2-year schedule (Table 6.4).

Table 6.4 History of the International Conference on the European Chitin Society.

	City/country	Date
1st	Brest/France	September 11–13, 1995
2nd	Lyon/France	September 3–5, 1997
3rd	Potsdam/Germany	August 31–September 3, 1999
4th	Senigallia/Italy	May 6–10, 2001
5th	Trondheim/Norway	June 26–28, 2002
6th	Poznan/Poland	August 31–September 3, 2004
7th	Montpellier/France	September 6–9, 2006
8th	Antalya/France	September 8–11, 2007
9th	San Servolo—Venice/Italy	May 23–26, 2009
10th	St Petersburg/Russia	May 20–24, 2011
11th	Porto/Portugal	May 5–8, 2013
12th	Münster/Germany	August 30–September 2, 2015
13th	Seville/Spain	June 1–4, 2017
14th	Kazan/Russia	Postponed

CHAPTER SEVEN

Chitin and chitosan: Production, properties, and applications

Over the last 30 years, the scientific research on chitin and chitosan has been increasing exponentially, many potential end-uses have been identified, and several of these are now commercially important.

Professor George A.F. Roberts, 2009, Professor in Biopolymers, Emeritus Professor of Textile Science, University of Nottingham, United Kingdom

This last chapter summarizes the production, fundamental properties, current applications (illustrated by examples of existing products on the market), recent advances, and challenges of chitin and chitosan. The main objective of this bibliographic synthesis is to give a general overview of these two biopolymers, supported by both historical and more recent bibliographic references. Any reader wishing to obtain additional information on the applications (customized products with advantages and disadvantages, costs, trends, etc.) and on the scientific issues (e.g., the role of the intrinsic parameters of chitin and chitosan on the performance of the products, the scientific problems specific to each research theme, the state of progress of knowledge, future research and challenges, etc.) can refer to the numerous recent publications cited in Table 6.3 (Chapter 6) The following recent books can also be consulted: Dongre (2018), Nasar (2018), Liming (2019), Qing and Tamo (2019), Jana and Jana (2019), Rinaudo and Goycoolea (2019), Zhao (2019), Crini and Lichtfouse (2019a,b), Jacob et al. (2020), Thomas et al. (2020), van den Broek and Boeriu (2020), Berrada (2021), Wertz and Perez (2021), and Nayak et al. (2021). However, this list of the selected work is not exhaustive as the bibliography is vast.

7.1 Production

Currently, marine crustaceans (especially shrimps, crabs, and lobsters) are the main source of commercial chitin extraction because a significant of wastes is available as low-cost and renewable by-products of the fishing industry in many places around the world. Other sources of chitin are other

Chitin and Chitosan
https://doi.org/10.1016/B978-0-323-96119-6.00004-9

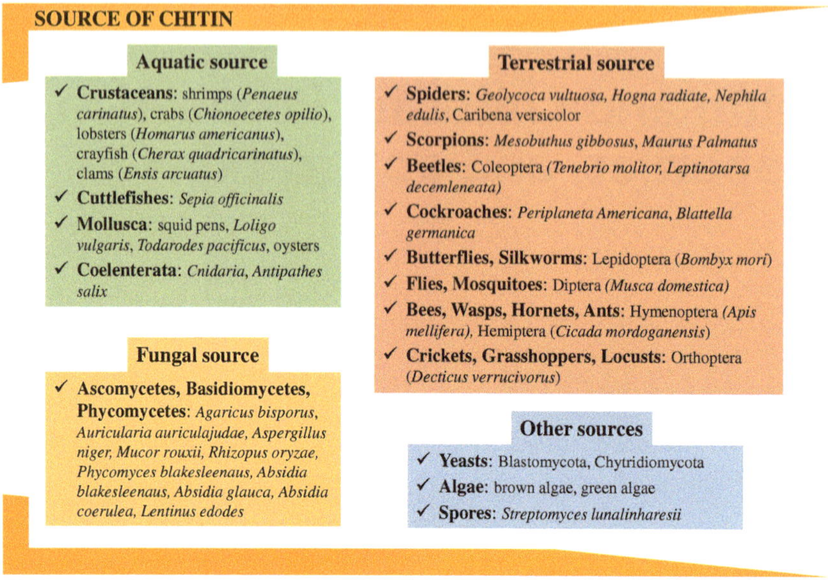

Fig. 7.1 Simplified classification of the main sources of chitin.

sea animals (squids, cuttlefishes, and mollusca), fungi and mushrooms, insects (scorpions, spiders, beetles, mosquitoes), other microorganisms (algae, yeast, spores), and other terrestrial crustaceans (nematodes). A simplified classification of the main sources of chitin is represented in Fig. 7.1. Some researchers separate the fungal sources of chitin into fungi and mushrooms (Table 7.1).

The main sites where chitin is present are as follows:

- the exoskeleton, cuticle, and the membrane between the segments for arthropods (crustaceans, insects, arachnids, and myriapods); Fig. 7.2 shows the hierarchical structure of chitin in the exoskeleton of a shrimp;
- the tubes for pogonophores;
- the shells, teeth, stomach plates, and feathers for mollusks (gastropods; cephalopods);

Table 7.1 Occurrence of chitin and chitosan in fungi and mushrooms.

Organisms	Species
Fungi	*Aspergillus niger*, *Absidia caerulea*, *Absidia glauca*, *Cunninghamella blakesleeana*, *Mucor rouxii*, *Mortierella isabelina*, *Rhizopus oryzae*, *Rhizopus delemar*, *Candida albicans*, and *Candida auris*
Mushrooms	*Agaricus bisporus*, *Auricularia auricula-judae*, *Trametes versicolor*, *Armillaria mellea*, *Lentinula edodes*, *Pleurotus ostreatus*, *Pleurotus sajo-caju*, and *Pleurotus eryngii*

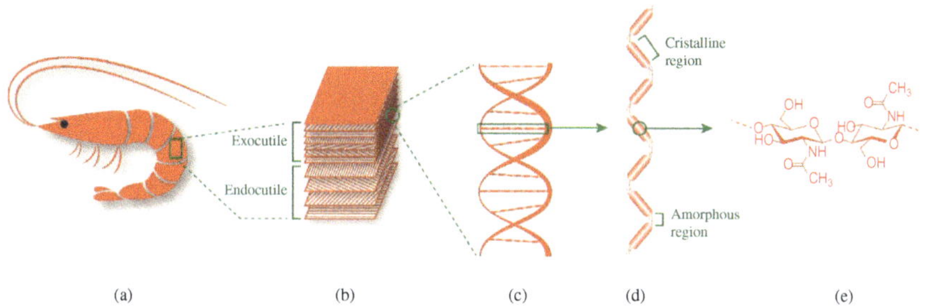

Fig. 7.2 Hierarchical structure of chitin in (A) exoskeleton of a shrimp; (B) macrostructure of the exoskeleton comprising layered chains, protein and calcium carbonate; (C) helicoidal arrangements of chains; (D) chitin nanofibrillar bundles; and (E) chemical composition of chitin.

- the eggshells and internal and median membranes for corals (Cnidaria; Coelenterata);
- the setae for annelids;
- the cell wall, mycelium, stem, and spores for fungi (Ascomycetes, Basidiomycetes, and Phycomycetes);
- the cuticles and shells for brachiopods;
- the eggshell and pharynx for nematodes;
- the cell wall for algae (Chlorophyceae; marine diatoms).

The content of chitin varies according to its source, animal or fungal, the species, and the sites of presence. For example, the chitin present in the exoskeleton of crustaceans, the cuticle of insects, and the cell walls of fungi varies; for example, its content is 30%–40% in shrimps, 15%–30% in crabs, 20%–30% in krills, 20%–40% in squids, 3%–6% in clams, 3%–6% in oysters, 5%–25% in insects, and 10%–25% in fungi (Rudall and Kenchington, 1973; Muzzarelli, 1977; Muzzarelli et al., 1986; Skjåk-Braek et al., 1989; Roberts, 1992; Horst et al., 1993; Felt et al., 1998; Kurita, 2006; Rinaudo, 2006; Nwe et al., 2011a,b, 2013, 2014). The shells of the marine crustaceans (shrimp, crab, lobster, krill, and squid wastes) contain not only chitin (15%–40%) but also two other major components, namely, proteins (20%–40%) and calcium carbonate (20%–50%). Other minor components are pigments, lipids, and other minerals including metals and salts. These percentages vary considerably with the species used (quality and freshness of shell), the season, and the geography (Nwe et al., 2013, 2014; Berezina, 2016; Hamed et al., 2016; Bonecco et al., 2017; Dima et al., 2017; Philibert et al., 2017; Santos et al., 2020; Kou et al., 2021; Sieber et al., 2018; Bastiaens et al., 2019; Hahn et al., 2020; van den Broek and Boeriu, 2020; Hossin et al., 2021; Pakizeh et al., 2021; Ma et al., 2022).

The other important point is the structural organization of chitin depending on the species. For example, in an exoskeleton or a carapace, chitin forms crystalline microfibrils together with proteins, which are organized in planes into well-defined hierarchical levels (Fig. 7.2). As in most fibrillar materials, the internal structure of chitin is, indeed, designed in a hierarchical arrangement formed by the self-assembly of its fundamental building blocks into higher-order fibrous structures, which are stabilized by noncovalent interactions. This hierarchical organization has always interested the scientific community because it is useful for understanding chitin biosynthesis, which is a highly complex and sequential process, with variations between species (it is now known that the cellular machinery for chitin synthesis is consistent in fungi and insects). Based on the understanding of these molecular architectures, three different forms of chitin were identified due to the arrangement of the macromolecular chains.

Chitins found in the nature differ in their more or less crystalline state and the origin of these polymorphs from the biopolymer source itself. In the solid state, the macromolecular chains are in the form of a helix and each repeating unit consists of two residues. This is the reason why there are various arrangements of the chains in their structure. Three distinct forms are recognized. The first is α-chitin which is the most abundant and stable; it exists in crustaceans, for example, in the exoskeletons of shrimps and crabs, coelenterates, annelids, fungi, and algae. The second is β-chitin, which is less abundant than α-chitin and is found in the feathers of squid, cuttlefish, annelids, algae, and in the cocoons of the figwort weevils *Cionus* and *Cleopus*. β-Chitin is a softer material with higher affinity for solvents and higher reactivity. The last one is γ-chitin which has been identified in the cuticle of inarticulate brachiopods, in cocoon *Ptinus* fibers in beetles (the Australian spider beetle), and in diatoms (Bacillariophyceae). Its abundance is medium to low, depending on the source of chitin. γ-Chitin is considered to be a mixture or an intermediate form of α-chitin and β-chitin. A general rule follows that α-chitin is found in hard structures, while β- and γ-chitin are found in soft structures.

All structures involve hydrogen bonding arrangements of the macromolecular chains through the amide group, the essential difference being the way in which these chains stack with respect to each other, as shown in Fig. 7.3. This figure schematically describes the differences between the three polymorphs of chitin. It shows how the macromolecular chains are stacked by hydrogen bonds and how the direction of the chains changes as they are stacked. The structure of the three polymorphs can be simply described as follows: the macromolecular chains of α-chitin are arranged

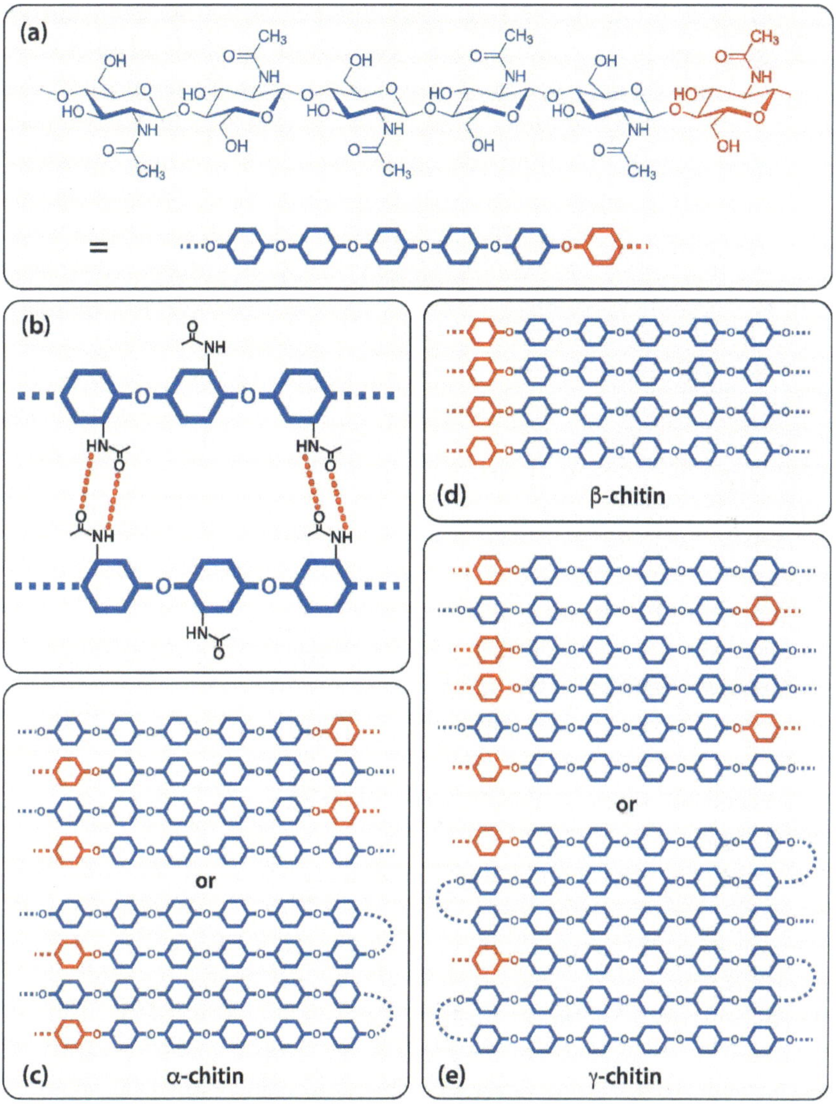

Fig. 7.3 Schematic representation describing the differences between the three polymorphs of chitin. (A) Structure of a macromolecular unit of chitin; (B) stacking of chains via hydrogen bonds; (C–E) change of direction of the chains during stacking according to each of the three forms of chitin.

antiparallel and crystallize in an orthorhombic system containing two molecules; β-chitin has chains stacked parallel to each other in a monoclinic mesh containing only one molecule, while γ-chitin has a structure consisting of two parallel chains to one antiparallel chain (Fig. 7.3).

To isolate chitin, for example, from the shells of shrimps and crabs and bone plates of squids, three main steps using chemical treatments are required, namely, deproteinization to dissolve proteins, demineralization to dissolve calcium carbonate, and decolorization to remove pigments and obtain a colorless chitin. The decolorized final product corresponds to the commercial chitin. To produce chitosan, a fourth deacetylation step is necessary. The final characteristics of commercial chitin and chitosan, such as quality (purity, protein, and mineral content, etc.), degree of acetylation (DA) or degree of deacetylation (DD), molecular weight (MW), and polydispersity vary depending on the source of the raw material used for chitin extraction. All these characteristics are also highly dependent on the conditions used in the three chitin extraction steps and the deacetylation step (Acosta et al., 1993; No and Meyers, 1995; Kurita, 1998; Percot et al., 2003; Hayes et al., 2008a; Peniche et al., 2008; Nwe et al., 2011a,b, 2013, 2014; Younes and Rinaudo, 2015; Dima et al., 2017; Philibert et al., 2017; Santos et al., 2020; Kou et al., 2021; Sieber et al., 2018; Bastiaens et al., 2019; Hahn et al., 2020; van den Broek and Boeriu, 2020; Hossin et al., 2021; Kou et al., 2021; Pakizeh et al., 2021).

For chitin extraction, the wastes are first crushed to the appropriate size and washed thoroughly with water to remove any organic material adhering to their surface (Nwe et al., 2014). The chemical protocol then consists of using alkalis (e.g., NaOH, Na_2CO_3, or KOH) to extract proteins (deproteinization step) and acids (HCl, HNO_3) to remove inorganic compounds such as calcium carbonate and calcium phosphate (demineralization step). Small amounts of pigments and lipids are also removed during these two steps. The deproteinization step is difficult due to the strength of the chemical bonds between chitin and proteins. Nevertheless, the complete removal of proteins is particularly important for biomedical applications to avoid any risk of allergy to shellfish. This step invariably leads to depolymerization of the biopolymer by hydrolysis, which lowers its molecular weight, anomerization, and partial deacetylation of chitin. The chemical demineralization to remove minerals is easier because of the easy decomposition of calcium carbonate into water-soluble calcium salts with the release of CO_2, the salts being filtered out. Other minerals are also removed by salt formation. However, it is difficult to remove all minerals, so the reaction conditions must be modified (larger volume and more concentrated acid, temperature, and processing time), depending on the desired quality. These changes in reaction conditions can also affect the quality of the chitin. For enzymatic treatments, the order of the two steps (deproteinization and demineralization) is reversed, whereas for microbial fermentation, these two steps are processed simultaneously. An additional step of decolorization using oxidants (bleaching agents) is applied to remove residual pigments. Inorganic salts, proteins, carotenoids, and pigments can be recovered as high-value by-products, lowering the manufacturing costs of chitin production. For example, proteins and carotenoids can be used as high-quality feed additives for livestock and fish. Pigments may also be recovered as high-value side products (Pakizeh et al., 2021).

Chitin is then converted to chitosan by chemical hydrolysis of acetamide groups of chitin using NaOH solutions through homogenous or heterogenous processes. This is called chemical deacetylation or alkali deacetylation. In the heterogenous chemical method, chitin is treated with a hot concentrated NaOH solution for a few hours, whereas in the homogenous method, alkaline chitin is prepared after dispersion of chitin in concentrated NaOH at room temperature for several hours, followed by dissolution in crushed ice. In general, the latter method results in chitosan with a lower average degree of deacetylation, but the acetyl groups are evenly distributed along the biopolymer chains. The chemical protocols for chitosan production are well-known, and again, the DA and MW characteristics of chitosan depend on the chitin source and on the deacetylation conditions used. These parameters have a significant effect on the physical, chemical, biological, and technological properties of chitosan (No and Meyers, 1995; Kurita, 1998; Jo et al., 2011; Nwe et al., 2014; Younes and Rinaudo, 2015; Hamed et al., 2016; Bonecco et al., 2017; Dima et al., 2017; Philibert et al., 2017; Santos et al., 2020; Kou et al., 2021; Sieber et al., 2018; Hahn et al., 2020; Sebastian et al., 2020; van den Broek and Boeriu, 2020; Pakizeh et al., 2021). Another important feature is the distribution of acetyl groups along the macromolecular chitosan chains as this feature, together with MW and polydispersity, can influence chitosan solubility, viscosity, and gelling capacity (Younes and Rinaudo, 2015). Thus, each commercial batch of chitosan and also chitin is characterized by their DA/DD, MW, purity, and other characteristics such as crystallinity, viscosity, ash content, and metal content. Chitosan purity is also an important characteristic, and three grades are proposed: a technical grade ($70\% < DD < 85\%$) for applications in wastewater treatment, for example, a pure grade ($85\% < DD < 95\%$) for food industry and an ultrapure grade ($90\% < DD < 95\%$) for medical applications (Figs. 7.4 and 7.5). The ultrapure grades are obviously the most difficult to obtain, the rarest but the most valuable. Few production plants are able to provide these types of purity grades. Of course, the prices ranging

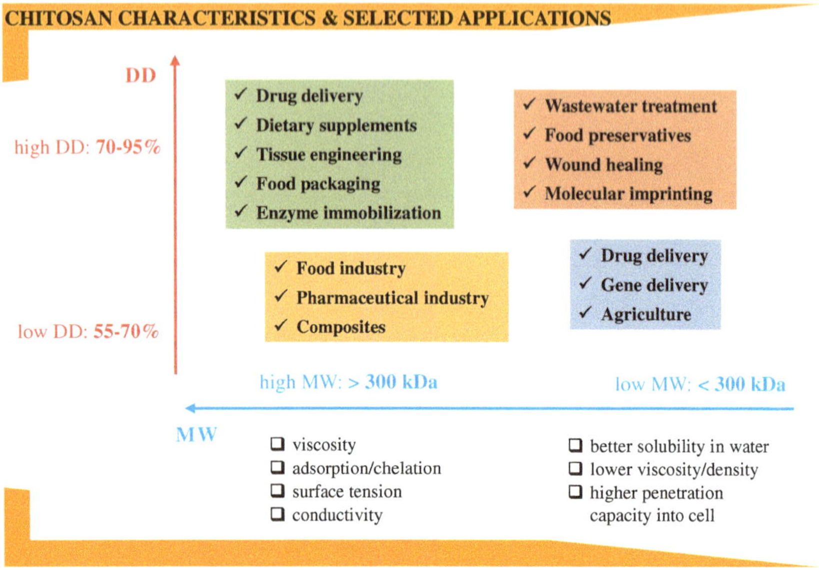

Fig. 7.4 Relationship between chitosan characteristics' (*MW*, molecular weight and *DD*, degree of deacetylation) and its applications.

Fig. 7.5 Relationship between chitosan quality and volumes and the application market.

from 15 to US$ 20,000/kg depend on purity and other characteristics (DD, MW). Low, medium, and high molecular weight products are prevalent on the market. Products with desired characteristics can be acquired or prepared in the laboratory. Nevertheless, the high price of the highly purified chitosans hinders their development in other industrial sectors than the medical one (Sebastian et al., 2020).

From an industrial production point of view, two other commercial sources of chitin and chitosan and their derivatives can be used. Indeed, fungi are the second main source of chitin and are an alternative, vegan source of both biopolymers. Zygomycetous fungi (e.g., *Mucor rouxii*, *Rhizopus oryzae*) have been studied extensively. Other organisms such as *Aspergillus niger* and *Agaricus bisporus* and wine yeasts have also been proposed. In general, the extraction is carried out under milder alkaline and acidic treatment conditions than those used in the chemical method used for the extraction of crustacean chitin (Ghormade et al., 2017; Philibert et al., 2017; de Lima Batista et al., 2018; Abo Elsoud and El Kady, 2019; Sebastian et al., 2020; Iber et al., 2022). In addition, the demineralization step is not necessary because the calcium level is lower than that of marine waste. The process is also more ecological. However, commercial applicability of fungal chitosan has not yet started, although locally in some countries (China, India, Canada, Norway, and France), there are initiatives of micro-enterprises (Morin-Crini et al., 2019).

Insects represent a third promising and sustainable source of chitin and chitosan that is both ecological and economic (Abidin et al., 2020; Hahn et al., 2020; Mohan et al., 2020; Kou et al., 2021; Ma et al., 2022). Insect farming, which is widespread in Asia, is developing considerably in Europe for the sustainable production of animal feed or for human food applications. The insect production industry is in full development, and the waste from this industry represents a highly valuable biomass. The availability of large quantities of insect biomass and chitin-rich by-products, such as exoskeletons, is increasing and is due to the fact that insects can be easily reared due to their high fertility and reproductive rate. Unlike crustaceans, they are not subject to seasonality (Ma et al., 2022). In addition, other benefits are cited: A comparison of the DD and MW parameters of chitin and chitosan from insects and crustaceans, their stability, crystallinity, and surface structure showed great similarity in the products obtained (Philibert et al., 2017; Hahn et al., 2020). Although it is more difficult to purify insect chitin, it seems that it has a greater degree of purity and higher macromolecular chain lengths, which may be advantages for certain areas of application (biomedical, cosmetics). However, the chitin content of whole insects is generally lower than that of crustaceans and depends on the species, type of feed, and of life cycle stage. Although local initiatives exist (in France, a micro-company offers chitin and chitosan from the larvae of the fly *Hermetia illucens*), the production of chitin and chitosan from insects has not yet been extended to an industrial scale (Abidin et al., 2020; Ma et al., 2022).

For several years, there are environmental and societal questions and scientific challenges (especially on the purity and polydispersity of chitin) on the production of products from waste from the fishing industry. Indeed, the availability of the exoskeleton of crustaceans is limited by geography and is very seasonal as commercial shellfish fisheries begin in the spring, after the spawning period. The sustainability of shellfish farming is currently under debate. The chemical methods for chitin extraction or for chitosan production by deacetylation are often criticized because they have several disadvantages (Crini et al., 2009): a significant consumption of water and energy; the use of hazardous and environmentally unfriendly chemicals (hydrochloric acid, sodium hydroxide, and bleaching agents); the generation of large volumes of concentrated alkaline and acidic solutions and wastes, which must be treated; the difficulty to purify chitin (especially to remove proteins and fats, which can pose problems for applications in the biomedical field); and especially the difficulty to control the characteristics of commercial products (Philibert et al., 2017; Santos et al., 2020; Kou et al., 2021; Sieber et al., 2018; Hahn et al., 2020; Mutreja et al., 2020; Oyatogun et al., 2020; van den Broek and Boeriu, 2020; Iber et al., 2022). Indeed, chemical extraction processes may denature the structure of biopolymers. So, the concentration of the reagents, reaction time, and temperature must be carefully controlled to minimize depolymerization and deacetylation of chitin and prevent thermal degradation.

Similarly, the DD and MW of chitosan also depend on the conditions used in the deacetylation step. This is the reason why it is difficult to obtain similar commercial batches with the same characteristics. To overcome these drawbacks, studies suggest the use of more environmentally friendly chitin extraction or enzymatic deacetylation processes called bioextraction processes. Indeed, since the 2000s, in addition to the production of chitin from microorganisms, biotechnological approaches such as fermentation and enzymatic pathways have been proposed as alternatives to chemical extraction. The process of biotechnological fermentation of chitin from shellfish waste is feasible (with little or no risk to handlers), involves less energy, water and chemical consumption, and is environmentally friendly. However, proteins and minerals are not completely removed by biological processing, and commercial chitin also has a lower quality and higher cost. Reaction times for production by the microbial biotechnology route are also much longer than with the chemical route, and the yield of purified chitin and chitosan is also lower than that of shellfish sources. In addition, bioextraction processes are not yet widespread, although local initiatives exist (Nwe et al., 2013, 2014; Younes and Rinaudo, 2015; Berezina, 2016; Hahn et al., 2020; Mutreja et al., 2020; Oyatogun et al., 2020; Kou et al., 2021; Huq et al., 2022).

There are many companies around the world that offer chitin and chitosan and their derivatives (low and high molecular weight chitosans, chitooligosaccharides), mainly of animal origin (shellfish waste). It is difficult to give an exact number, but it is estimated that there are between 50 and 70 industrial sites in the world producing chitin and chitosan, mainly in Asia and North America. The main producing, exporting, and consuming countries are Japan, China, India, Canada, and the United States, far ahead of the European market which is mainly a consumer (Mutreja et al., 2020; Oyatogun et al., 2020). It should be noted, however, that there is no standardized supply chain, each company using its own production protocol. For chitin extraction and chitosan production, there is no standardized protocol because the reaction conditions (concentrations, temperature, and treatment time) vary considerably from one protocol to another. In addition, the order of the steps can differ, and the decoloration step is not systematic. The characteristics such as DD, MW, purity, and polydispersity of these biopolymers can, therefore, vary from one company to another and even for different batches from the same company. This is the most important problem of the chitin industry. The price of biopolymers is strongly dependent on these characteristics. There are also companies that specialize in the production of plant-based chitin (for example obtained from mushroom and *Aspergillus niger*), which is referred to as plant chitin/chitosan or fungal chitin/chitosan obtained using biological technologies. These products have a much higher purity than animal-based biopolymers, but a higher price. The other advantages of vegetable/fungal chitosan are as follows: a guaranteed 100% natural nonanimal production and a constant traceable quality; a product suitable for vegans, vegetarians, and people allergic to shellfish; a gluten-free and nonallergenic product (e.g., free of tropomyosin, the allergenic crustacean protein); GMO-free and also from renewable and inexhaustible resources (Pakizeh et al., 2021; Huq et al., 2022; Ma et al., 2022).

Customization of DD and MW parameters of biopolymers is more easily controlled, and their quality is more consistent from batch to batch. Fungal chitosan with a DD of over 98% is suitable for medical applications due to a lower dispersity of the macromolecular chains, which improves both performance and reproducibility of the results for a given application. There is a growing market, especially in the food and cosmetic sectors, due to the high demand from people for products that do not contain animal ingredients. As for the biopolymers of animal origin, particular attention must be paid to their intrinsic characteristics (Berezina, 2016; Hamed et al., 2016; Bonecco et al., 2017; Dima et al., 2017; Philibert et al., 2017; Santos et al., 2020; Kou et al., 2021; Sieber et al., 2018; Hahn et al., 2020; Mutreja et al., 2020; Oyatogun et al., 2020; van den Broek and Boeriu, 2020; Huq et al., 2022; Iber et al., 2022).

7.2 Basic and intrinsic properties

Chitin (CAS Number 1398-61-4) is a natural polymer of monosaccharide comprising β-(1 → 4)-2-acetamido-2-deoxy-β-D-glucopyranose units. It is generally represented as a linear long-chain homo-polymer comprising *N*-acetyl glucosamine (GlcNAc) units, [poly(*N*-acetyl-β-D-glucosamine)]. Chitosan (CAS Number 9012-76-4), obtained by partial deacetylation of chitin, is an aminopolysaccharide comprising macromolecules of D-glucosamine (GlcN) and GlcNAc. It is a copolymer of β-(1 → 4)-linked GlcNAc units and 2-amino-2-deoxy-β-D-glucose and refers to products having different proportions of GlcN and GlcNAc units and of varying chain lengths. Indeed, the fully deacetylated product is rarely obtained due to the risks of side reactions and chain depolymerization. Thus, each commercial chitosan is characterized by its average percentage of DA (see Fig. 7.6A) or by its DD. DA is defined as the fraction (F_A) of *N*-acetylated glycosidic units in chitosan or chitin biopolymer (with $DD = 1 - F_A$ or = 100-DA percent).

Table 7.2 describes the differences between chitin and chitosan, the main one being the much greater presence of amine functions in the macromolecular chains of chitosan. In general, the chitosan "label" corresponds to commercial products with less than 25%–40% *N*-acetyl groups (Fig. 7.6A). It is also recognized that when the content of these groups is higher than 50%, chitin is considered, while for lower valuer, chitosan is considered. In fact, in the latter case, another parameter is used to distinguish them, that of solubility and their differences in physicochemical and

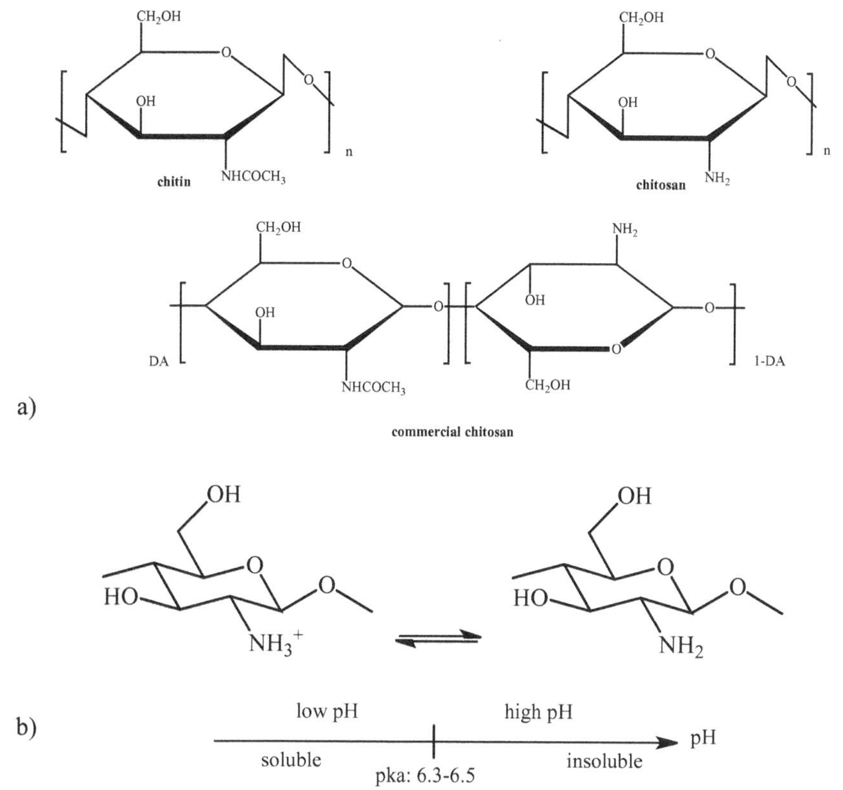

Fig. 7.6 (A) Schematic representation of completely acetylated chitin, completely deacetylated chitosan, and commercial chitosan, a copolymer characterized by its average degree of acetylation (DA) percent (the chitosan "label" generally corresponds to commercial products with less than 25%–40% *N*-acetyl groups; it is also recognized that when the content of these groups is higher than 50%, chitin is considered, while for lower valuer, chitosan is considered); (B) Chitosan's versatility in aqueous solution (at low pH, less than pKa, chitosan's amines are protonated, conferring polycationic behavior to macromolecules; at higher pH, above pKa, chitosan's amines are deprotonated).

biological behavior. Both biopolymers have common chemical and biological properties such as non-toxicity, biocompatibility, biodegradability, vulnerability to enzymatic hydrolysis by lysosymes or adsorption properties, and other different properties. For example, chitosan is soluble in acidic aqueous media, is an antibacterial, and highly reactive and hydrolyzable by lipases, unlike chitin. Their technological properties such as versatility (e.g., implementation of materials) are also different (Brimacombe, 1973; Muzzarelli, 1973, 1977; Neville, 1975; Blackwell et al., 1980; Cohen, 1987a,b; Hirano, 1989, 1996; Kurita, 1998).

Table 7.2 Main common properties and differences between chitin and chitosan (*DD*, degree of deacetylation; *GlcNAc*, *N*-acetyl glucosamine units; *GlcN*, D-glucosamine).

Chitin	Chitosan
Very common in the animal and plant kingdoms	Very rare in nature
Natural substance (biopolymer) often associated with organic substances (proteins) and impregnated with minerals (calcium salts)	Biopolymer obtained from chitin by chemical reaction
Three different forms: chitin-α, chitin-β, and chitin-γ	Coarse ground flakes and powder
Powder form or in granules (or sheets)	White to pale yellow products
White to off-white products	Linear polysaccharide comprising two subunits: GLcN and GlcNAc
Linear polysaccharide comprising two subunits: GlcNAc and GLcN	High deacetylation degree: DD < 50%
High acetylation degree: DD > 50%	Number of GlcN units higher than 50%
Number of GlcNAc units higher than 50%	$(C_6H_{13}NO_5)_n$ with a molecular weight of 161 g/mol for the repetitive unit
$(C_8H_{13}NO_5)_n$ with a molecular weight of 221 g/mol for the repetitive unit	Molecular mass: 200,000 g/mol
Molecular mass: from 80,000 to 10^6 g/mol	Melting point: 88–102°C
Melting point: >300°C	Reactivity of amino and hydroxyl groups
Practically inert substance	A unique natural polycationic polysaccharide with high charge density
Very stable chemical properties	Adhesion to negatively charged surfaces
Reactivity of hydroxyl groups	Properties strongly depend on the pH of the solution/medium
Biodegradable and nontoxic	pKa: 6.3–6.5
Hydrophobic (depending on the presence of blocks of *N*-acetylated units)	Stable
Insoluble in water, alkaline solutions, and most organic solvents	Hydrophilic
Soluble in concentrated formic acid and methane sulfonic acid	Biodegradable, nontoxic, and biocompatible
Soluble in dimethylformamide-lithium chloride and dimethylacetamide-lithium chloride mixtures	Weak base, powerful nucleophile
Easily hydrolyzed by lysozyme and by chitinases	A very active chemistry allowing the synthesis of numerous derivatives
Weak antibacterial properties	Unlike chitin and cellulose, the presence of amine groups allows to perform chemical reactions specific to this function
Film-forming properties	Insoluble in water and alkali
Capable of triggering defense mechanisms in plants against infections and parasitic aggressions	Soluble in dilute aqueous acid (pH < 6.5) and in organic acids: formic acid, acetic acid, benzoic acid, naphthenic acid
	Used both in the solid state and in solution
	Film-forming properties
	Recognized as a food additive (Japan, Korea)

Table 7.2 Main common properties and differences between chitin and chitosan (*DD*, degree of deacetylation; *GlcNAc*, *N*-acetyl glucosamine units; *GlcN*, D-glucosamine)—cont'd

Chitin	Chitosan
Specific applications: cosmetics, cosmeceuticals, treatment of chronic wounds, ulcers and bleeding, beverage industry, wastewater treatment	Recognized as safe (US Food and Drug Administration) Antibacterial Dietary fiber Hydrolyzed by lipases (human fluids) Antioxidant Antitumor activities Capable of triggering defense mechanisms in plants against infections and parasitic aggressions A versatile material/biomaterial A wide range of different applications: food and nutrition, nutraceuticals, agriculture, pharmacy, medicine and biomedicine, cosmetics, cosmeceuticals, biotechnology, membranes, pulp and paper, textile industry, wastewater treatment, etc.

Another definition to distinguish chitin from chitosan is based on their difference of solubility in water. Chitin is, indeed, an inert material, insoluble in water (because its hydrophobic, stable macromolecule chains have a low content of amino groups) and in the vast majority of common solvents, while chitosan as hydrophilic biopolymer is soluble in aqueous acetic acid. Chitin is also insoluble in diluted acids, diluted and concentrated bases, and organic solvents. It is only soluble in high concentrations of hydrochloric, sulfuric, and phosphoric acid solutions. Other examples of solvents for chitin are dimethylformamide-lithium chloride and dimethylacetamide-lithium chloride mixtures, hexafluoroisoacetone sesquihydrate, hexafluoroisopropanol, $2H_2O$-saturated methanol, and other organic salts (LiSCN, alkali metal/urea system).

Although its structure is rigid and crystalline through intra and intermolecular hydrogen bonding, the presence of free amino groups on the C-2 position of the GlcN unit along the macromolecule chains allows chitosan to dissolve in dilute aqueous acidic solvents through the protonation of these groups and the formation of the corresponding polymer salt. The biopolymer chitosan is then converted to a polyelectrolyte in acidic media, which is unique among natural polymers. So, chitosan is considered a high charge density polyamine whose properties are highly dependent on the pH of the solution/medium. Fig. 7.6B, indeed, illustrates the versatility of chitosan in aqueous solution. In general, the intrinsic pKa of commercial chitosans is close to 6.3. At low pH, less than pKa, chitosan's amines are protonated conferring polycationic behavior to macromolecules. Chitosan is then soluble in water. This property is interesting, for example, in water treatment to interact with anionic species such as dyes (flocculation processes), fatty acids and proteins (cosmetology), or in the preparation of membranes by forming polycation/polyanion complexes. At a higher pH, higher than the pKa, chitosan precipitates because its amine groups are deprotonated and its structure then presents chelating properties (non-ionized amine function but carrying a free electron doublet) and very effective adsorbing properties to complex metals, for example. At higher pH, chitosan's amines are also reactive, and the presence of amine functions along the macromolecular chain allows, unlike chitin and cellulose, to perform chemical reactions (useful for derivatization, substitution, grafting, cross-linking) specific to this function. In the latter case, chitosan can undergo associations that can lead to networks such as gels and films to fibers.

This behavior and its chemical reactivity present a great commercial interest in various fields of application of chitosan in a wide range of forms, soluble (solutions and suspensions) or insoluble (flakes, beads, sponges, foams, films, membranes, or fibers) (see the following seminal papers: Muzzarelli, 1973, 1977; Cohen, 1987a,b; Hirano, 1989, 1996; Peter, 1995; Hudson and Smith, 1998; Kurita, 1998, 2006; Kim et al., 1999; Vårum and Smidsrød, 2004a; Muzzarelli and Muzzarelli, 2005; Rinaudo, 2006). This potential for use under different forms is particularly appreciated in the pharmaceutical, medical, and cosmetic fields. For example, solution forms are used to prepare bacteriostatic, homeostatic and cosmetic formulations, powders for surgical glove and enzyme immobilization, gels for delivery vehicle, membranes in dialysis applications, contact lenses and wound dressing, and sponges in water treatment.

Chitosan also has many other intrinsic characteristics and properties, as described in Fig. 7.7 (Muzzarelli and Muzzarelli, 2005; Rinaudo, 2006). Besides its cationic character, another important characteristic is its antibacterial activity, which is why chitosan is used as a bactericide or bacteriostatic agent acting on various common bacteria. It is recognized that the

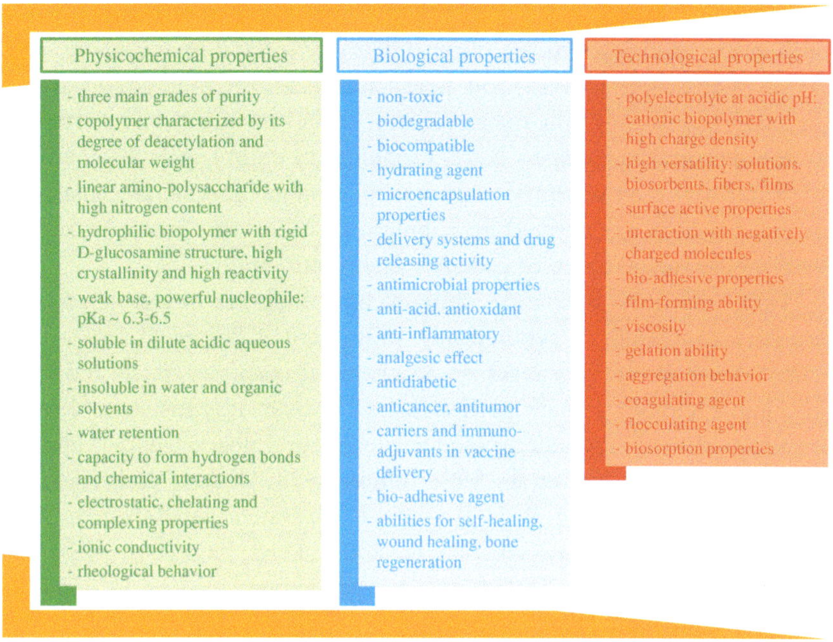

Fig. 7.7 Detailed physicochemical, biological and technological properties of chitosan.

presence of the amino groups on macromolecular chains also explains most of other biological properties such as mucoadhesivity, hemostatic activity, and analgesic effects. Chitosan is also of great interest for other reasons: it is nontoxic, biocompatible, and biodegradable, for example, used as excipient (to date, chitosan and chitosan hydrochloride are only accepted as excipients by regulatory agencies and not as drugs for the treatment of disease) and as biomaterial (it is present on the GRAS—generally recognized as safe—list) for use in wound dressings, bandages, and hemostatic agents (Amirthalingam et al., 2021). Chitosan is considered a food additive, for example, by the Korean Food and Drug Administration, as a dietary fiber and approved also as a biomaterial (but not for pharmaceutical use) for food and dietary applications, for example, in weight loss products. Since 2009, the European Parliament regulation N° 1107/2009 has approved chitosan hydrochloride as a dietary supplement additive (last directive: 2021/1446, September 03, 2021). In Europe, for organic farming, chitosan has been authorized since 2008 as an elicitor of plant defense mechanisms under specific conditions (CE 889/2008), and 3 years later as a clarifying agent in oenology (CE 53/2011). Since 2018, chitosan extracted from *Aspergillus niger* is authorized as a wine stabilizer (CE 2018/1584). However, the applications of chitin products in food, pharmaceuticals, and therapeutics are still limited by regulations. The main regulatory issues are structural heterogeneity of the products, batch-to-batch variability in biopolymer characteristics and thus properties, potential impurities in samples, and the fact that commercial products are derived from potentially allergenic materials such as shellfish, the main source of production of chitin and chitosan.

As an aminopolysaccharide, chitosan has a high percentage of nitrogen (6%–7%) compared to synthetically substituted cellulose (1.25%), these amine groups being very reactive. Its chemical reactivity, facile modification, and technological properties make chitosan an ideal candidate for the preparation of biomaterials used in medical and pharmaceutical applications (Peter, 1995; Felt et al., 1998; Illum, 1998; Dodane and Vilivalam, 1998; Kim et al., 1999; Ravi Kumar, 2000; Singla and Chawla, 2001; Kato et al., 2003; Illum and Davis, 2004; Ravi Kumar et al., 2004; Vårum and Smidsrød, 2004a,b; Kean and Thanou, 2010; Dash et al., 2011; Croisier and Jérôme, 2013; Badwan et al., 2015; Younes and Rinaudo, 2015; Berezina, 2016; Sahariah and Másson, 2017; Akbar and Shakeel, 2018; Crini, 2019; Crini and Lichtfouse, 2019a; Jana and Jana, 2019; Jangid et al., 2019; Morin-Crini et al., 2019; Lang et al., 2020; Peers et al., 2020; Mutreja et al., 2020; Oyatogun et al., 2020; van den Broek and Boeriu, 2020; Wu et al., 2020; Zamboulis et al., 2020; Ardean et al., 2021; Singh et al., 2021a). It is essential to emphasize that depending on the intended use, specific polymer forms (purity) with targeted properties (MW and DD) are required as shown in Figs. 7.4 and 7.5. It is also important to note that depending on the application, only a part of the properties of each biopolymer is put forward to explain their efficiency or their advantage. For example, for pharmaceutical and therapeutic applications, the main recognized properties/activities of chitosan are as follows: antimicrobial (the most important and most often cited activity in the literature), antifungal, mucoadhesive, antioxidant, antiinflammatory, hemostatic, analgesic, antihyperglycemic, antitumor, and wound healing. But usually, only two or three of these properties are pointed out for the intended application. For agriculture, it is often the antifungal property that is mainly pointed out to explain the performance of chitosan in the fight against pathogens, while for environmental applications (treatment of polluted water), it is the complexing and chelating properties of the amine groups of the macromolecular chains of chitosan.

Among the main commercial derivatives of chitosan are oligosaccharides (chitooligosaccharides, glucosamines) and low molecular weight chitosans, obtained from high molecular weight chitosans. These derivatives are particularly important for the agricultural and medical fields and are produced by physical, mechanical, chemical, and enzymatic processes (Nwe et al., 2011a,b, 2013, 2014; Brasselet et al., 2019; Kim, 2022). The main methods for their production are as follows:

- physical technique: gamma-ray or microwave irradiation;
- mechanical technique: sonication;
- chemical treatment: HCl, $NaNO_2$, H_2O_2, HNO_2, and H_3PO_4;
- enzymatic method: cellulase, lysozyme, chitinase, lipase, papain, pectinase, and protease.

Finally, the presence of the amine function (Fig. 7.6A) along the macromolecular chain of chitosan allows, contrary to chitin or cellulose, to carry out specific chemical reactions to this function. Indeed, the reactive groups in the chitosan backbone macromolecule are the primary amino group at the C-2 position and the primary and secondary hydroxyl groups at the C-6 and C-3 positions, respectively. The other functional groups are the glycosidic bonds and the acetamido group. These functional groups allow a large number of possible modifications, resulting in new polymers with targeted properties and behaviors. However, it is recognized that the reactive amino (mostly) and hydroxyl groups of the macromolecular chain are the main functional groups that facilitate the chemical modification of chitosan. All physical (e.g., polymer blending), chemical, and biological processes do not alter the basic structure of the chitosan macromolecule, but its characteristics (MD, DD, solubility, zeta potential, behavior, and rheology). In addition, any modification affects the chemical, biological, and technological properties of chitosan and chitin.

The main chemical modifications are hydroxyalkylation producing *N*-hydroxyalkyl or *O*-hydroxyalkyl chitosans or a combination of both, carboxyalkylation (e.g., carboxylmethyl chitosan), phosphorylation, thiolation, sulfation (*N*- and *O*-sulfated derivatives), quaternization, graft copolymerization, and hydrolysis (chemical or enzymatic). For example, the sulfation of chitosan results in a biopolymer (called heparinized chitosan) mimicking heparins, opening up new challenges for medical products (as anticoagulants, for tissue engineering and cancer therapy) and devices (films). Another type of modification is the reaction of chitosan with enzymes. Thus, the derivatives of chitosan are much more important than those of chitin, which is one of the reasons why the applications of chitosan are more important and varied than those of chitin. Many articles, reviews, and books on the functionalization of chitosan can be consulted on being published (Table 6.3; see for example: Kim, 2014; Amber Jennings and Bumgardner, 2017; Ahmed and Ikram, 2017; Dimassi et al., 2018; Brasselet et al., 2019; Liming, 2019; Jacob et al., 2020; Amirthalingam et al., 2021).

7.3 Current industrial applications of chitin and chitosan: An overview

There are more than 2000 applications of chitin and chitosan with practical applications in all industrial fields (Fig. 7.8), and it is a fast growing world market. It is estimated that the demand for chitin is twice as great as its industrial production, mainly concentrated in Asia (Kim, 2022), but this production is, indeed, limited, while the world demand is increasingly strong. Chitin and chitosan as biopolymers are products of the future in an existing (since the 1970s) and future market with multiple promising and constantly growing applications.

Japan as a marine nation is the country of chitin and chitosan, whose production, marketing, and use began in the 1970s (Hirano, 1989, 1996). Japanese annual production of chitin is estimated at 3000 tons/year, the main use being the production of chitosan. Currently, in Japan, as elsewhere in Asia, all industrial sectors use these biopolymers, the fishery industry being an important industrial sector. The main markets for chitosan are food, cosmetics, and water treatment (Badawy and Rabea, 2017; Bonecco et al., 2017; Philibert et al., 2017; Crini, 2019; Morin-Crini et al., 2019; Mutreja et al., 2020; Oyatogun et al., 2020; Kim, 2022). By 2030, China will be the leading consumer country for chitosan, with the food industry and nutraceutical market developing rapidly (Manigandan et al., 2018; Morin-Crini et al., 2019; Mutreja et al., 2020). In 2019, China imported 285,900 tons of shrimp, which was 186% more than the 100,000 tons imported in 2018, which represents an important source of waste potentially recoverable to produce chitin (source: The Food and Agriculture Organization of the United Nations). In the United States, marketing began in the early 1990s with the approval of the sale of chitin-based products by the Food Drug Administration. The current North American market for chitin and its derivatives also concerns the medical, agriculture, and wellness sector, including dietary supplements and anticholesterolemic drugs (Morin-Crini et al., 2019; van den Broek and Boeriu, 2020; Triunfo et al., 2021). The

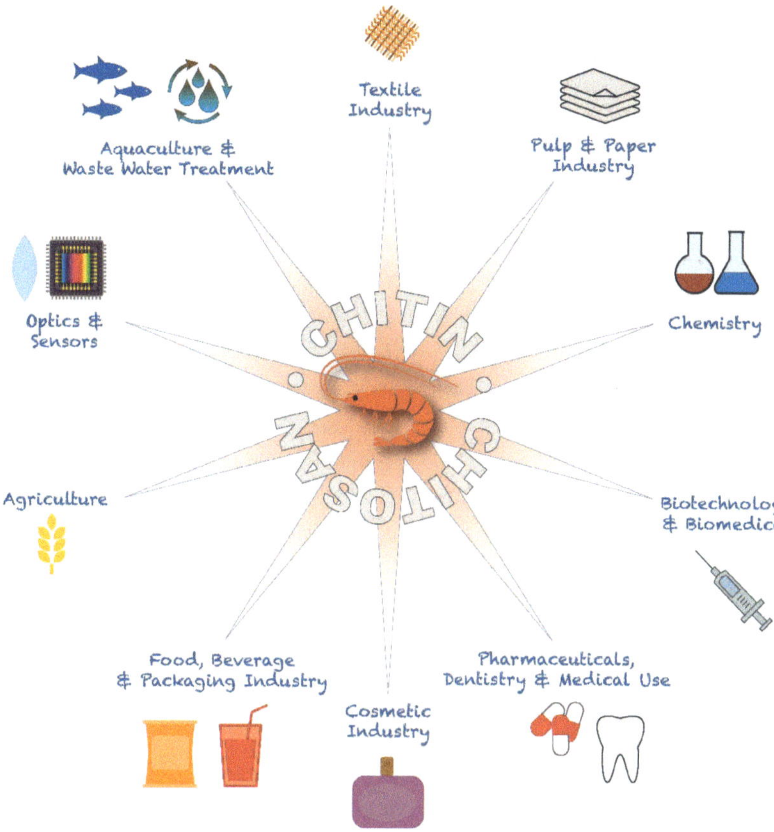

Fig. 7.8 The main applications and markets of chitin and chitosan.

European market has started to develop since the 2000s. High-quality natural products are becoming increasingly popular in the cosmetics and wellness sector (Ahmed and Ikram, 2017; Bonecco et al., 2017; Nechita, 2017; Philibert et al., 2017; Serrano-Castañeda et al., 2018; Crini and Lichtfouse, 2019a,b; Morin-Crini et al., 2019; Mutreja et al., 2020; Oyatogun et al., 2020; van den Broek and Boeriu, 2020; Triunfo et al., 2021; Kim, 2022).

It is important to note that the applications of chitin, chitosan, and their numerous derivatives, including carboxymethylated (carboxymethylchitin, carboxymethylchitosan) and phosphorylated products, other high-value products (glucosamine) and oligosaccharides (chitooligosaccharides) concern the same industrial sectors, mainly those of food (including beverage),

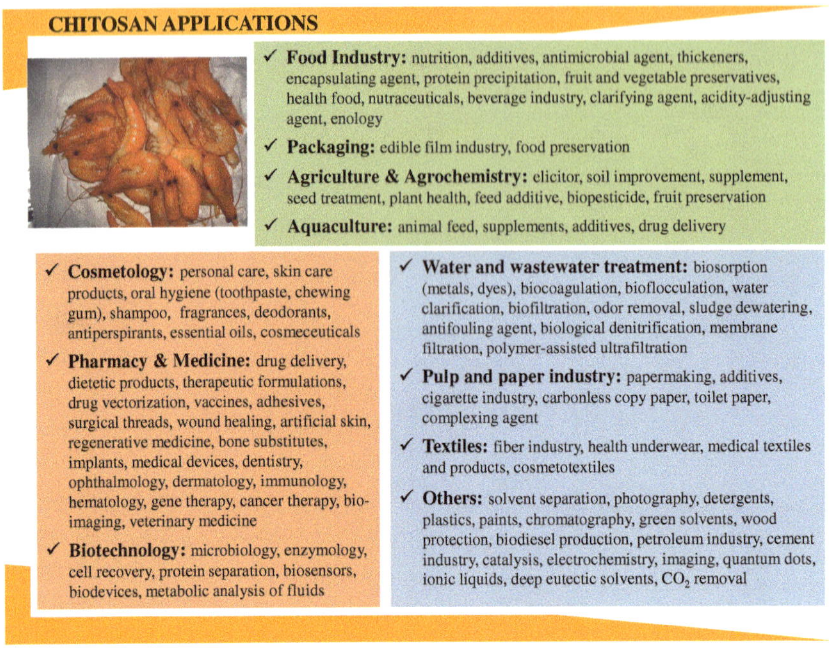

Fig. 7.9 The main applications of chitosan.

cosmetics, medical, and environment (water and wastewater treatment). Nevertheless, the applications of chitosan are much more numerous and varied (Fig. 7.9), which can be explained by its greater aqueous solubility and versatility, and also by the important number of its chemical (polycation) and biological (antibacterial activity) properties, compared to chitin (Fig. 7.10) (Philibert et al., 2017; Phillips, 2017; Santos et al., 2020; Kou et al., 2021; Sieber et al., 2018; Hahn et al., 2020; Joseph et al., 2021). Each application of chitosan is intimately related to one or more of its intrinsic properties (Table 7.3). There is a relationship between its structure, properties, and behavior for given pharmaceutical and therapeutic applications (Dash et al., 2011; Croisier and Jérôme, 2013; Younes and Rinaudo, 2015; Sahariah and Másson, 2017; Ardean et al., 2021; Ashok et al., 2021).

7.4 Food and beverage industries

Chitosan from shellfish waste is recognized as a safe food additive, functional ingredient, and dietary fiber (hypocholesterolemic effect) for

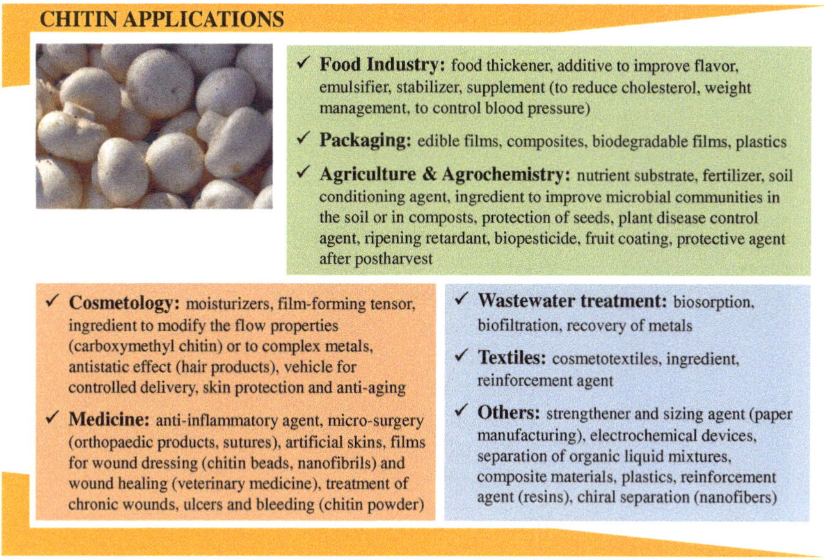

Fig. 7.10 The main applications of chitin.

Table 7.3 Relationship between the application of chitosan in a given industrial field and its intrinsic properties.

Application	Examples of propertie(s)
Food and Bevery Industry	Functional ingredient
	Antioxidant activity
	Antimicrobial activity
	Antifungal
	Encapsulation
	Color stabilizer
	Flocculation and filtration properties
Agriculture	Polycation
	Interaction with negatively charged species
	Plant growth regulator
	Elicitors of plant defense
	Regulation of resistance
	Seed conservation
	Stimulation of chitinase-producing soil bacteria
	Antifungal agent
	Antinematodals

Continued

Table 7.3 Relationship between the application of chitosan in a given industrial field and its intrinsic properties—cont'd

Application	Examples of propertie(s)
Cosmetology	Polycation
	Gel formation
	Film formation
	Interfacial properties
	Antibacterial
	Compatibility with other ingredients
	Polyelectrolyte complexation
Pharmacy	Polyelectrolyte
	Gel formation
	Drug delivery
	Mucoadhesivity
	Analgesic effects
Medicine	Polycation
	Gel formation
	Self-assembly
	Tissue adhesion
	Antibacterial
	Anticoagulant activity
	Hemostatic activity
	Biological activity on cells
	Lipid complexation
Biotechnology	Chemical functionality
	Encapsulation
	Plant culture medium supplement
Paper industry	Complex formation with polysaccharides and polyanions
	Wet strength agent
	Adsorption and flocculation
Textiles	Polycation
	Antibacterial properties
	Film formation
Wastewater treatment	Polycation
	Pollutant complexation
	Surface activity
	Flocculation
	Resistance in extreme medium conditions
	Biodegradability
Chemistry	Polycation
	Gel formation
	Interfacial properties
	Surface activity
	Self-assembly
	Film formation
	Emulsification properties
	Encapsulating agent
	Stabilizing agent

the consumer and it is approved in Japan, Korea, United States, and Europe (Gutiérrez, 2017; Mutreja et al., 2020; Oyatogun et al., 2020). Fungal chitosan extracted from *Aspergillus niger* is also approved as a food additive by the European Food Safety Authority (Philibert et al., 2017), as a clarifying and coagulating agent for fruit juices and nectars (Codex Alimentarius, 2013), and as a treatment to prevent protein haze in grape must and wine by the International Organization of Vine and Wine (Oenological Codex, 2009). According to the Japan Food Chemical Research Foundation and the Korea Food & Drug Administration, derivatives such as chitooligosaccharides and glucosamine are also considered safe for use in dietary supplements (Philibert et al., 2017; Brasselet et al., 2019; Cao et al., 2022; Kim, 2022).

7.4.1 Food applications

The food sector (including nutrition and beverage industry) is the main user of chitosan. Philibert et al. (2017) reported that, in the United States, 2288 metric tons of chitosan were used in 2018 in food and beverages. The main products obtained by enzymatic degradation of interest for the food industry are dimers and oligomers with a depolymerization degree of 2–3 using chitosanase as enzyme, dimers, trimers, tetramers, pentamers, and hexamers using hemicellulose, glucosamine, and *N*-acetylglucosamine oligomers with a depolymerization degree of 2–6 using pepsine or papain, and other oligomers with high depolymerization degree using lipase.

In food industry, chitosan and its derivatives (chitooligosaccharides) are used:

- as additives: texture controlling agent, emulsifying agent, gelling agent, flavor extender, color stabilizer, thickener, and stabilizer for sauces (soybean sauce with fermented chitosan), shelf-life extension;
- as nutritional ingredients to improve nutritional quality: Japanese noodles with chitosan;
- as antioxidant and antimicrobial agents: food protection; for example, Chitoseen™ F is used as an antimicrobial agent in food industry and to control moisture transfer between food and environment;
- as dietary fibers: hypocholesterolemic activity, reduction of fats/lipids, reduction of blood sugar; decrease of blood pressure;
- as prebiotic ingredients (LongLife® Chitosan, Solaray®, Lipidol®): stimulation of the growth and activity of beneficial enteric bacteria and their antimicrobial and antitumor characteristics;

- as encapsulating and bioactive agents: nutraceuticals, to improve calcium absorption (Chitosan PLUS Complex, TIENS Chitosan);
- as preservative agents: preservation of fruits and vegetables from microbial deterioration (coatings, e.g., strawberries coated with chitosan-based formulation); biodegradable and microbial films; food wrapping;
- for the removal of substances: suspended solids, clarification, and dyes.

Many other examples can be found in the following references: Winterowd and Sandford (1995), Shahidi et al. (1999), Vårum and Smidsrød (2004b), Gallo et al. (2016), Gutiérrez (2017); Han et al. (2018), Kasiri (2018), Manigandan et al. (2018), Cazon and Vasquez (2019), Crini and Lichtfouse (2019a), Morin-Crini et al., 2019, Samoila et al. (2019), Priyadarshi and Rhim (2020), Qu and Luo (2020); Cosme and Vilela (2021), Dutta et al. (2021), Muthu et al. (2021), Shahrajabian et al. (2021), Singh et al. (2021b), Wang et al. (2021), Yang et al. (2021), and Maleki et al. (2022). Chitosan products also offer benefits as ingredients of animal feeds. They have beneficial nutritional properties, and they can control the release of feed additives in animals. For example, chitosan is a food supplement for dogs and cats (for example, Epakitin™ formulation limits the absorption of phosphorus and contributes to healthy kidney function by binding to it), an alternative to antibiotic feed additives in poultry and pig production, and its presence in ruminant diets modifies ruminal fermentation, thereby reducing gas emissions (Jiménez-Ocampo et al., 2019). There are also dog food products enriched with chitosan and highly digestible proteins (Veterinary HPM®), particularly suitable for dogs of breeds prone to digestive sensitivity.

7.4.2 Packaging

Chitin and chitosan play an important role in the food industry not only as functional additives but also as edible films or coatings instead of plastics due to their many biological properties. Thus, they are proposed as raw functional materials to form biodegradable films, for example, to preserve food from microbial deterioration. Indeed, chitosan is also an alternative as a coating material for reducing the application of pesticides in food protection (Muñoz-Bonilla et al., 2014; van den Broek et al., 2015; Han et al., 2018; Wang et al., 2018; Adiletta et al., 2021; Bose et al., 2021; Dutta et al., 2021; Oladzadabbasabadi et al., 2022). The interests often cited are as follows: raw materials from waste or biomass, suitability for food contact (for example as packaging for the protection of fruit, fresh vegetables, or meat), antimicrobial effectiveness, efficient barrier properties to grease and oxygen, very varied potential applications, not only in food packaging but also for special papers, medical products, sanitary papers, and filters (Han et al., 2018; Cazon and Vazquez, 2019; Jacob et al., 2020; Priyadarshi and Rhim, 2020; Souza et al., 2020; Wang et al., 2021; Yang et al., 2021). From the point of view of their mechanical characteristics, chitin films cannot undergo elongation, while chitosan-based films can be stretched only twice. In the case of chitosan, these properties can be improved by chemical modification of the biopolymer or by formulation adjustments (Nakano et al., 2007). However, due to their poor mechanical properties, lower performance, and high cost compared to traditional packaging, the applications at the industrial scale are yet rare. The laboratory results are convincing, but the application has not yet reached the stage of industrialization (the price of chitosan is still high for the packaging sector). Nevertheless, it is an important area of current research (Adiletta et al., 2021; Bose et al., 2021; Oladzadabbasabadi et al., 2022).

7.4.3 Beverage industry

The interest of using chitinous compounds including chitin-glucan in oenology and beverage industry has been previously reviewed by Kim (2011). However, the only chitosan authorized in wine industry is fungal chitosan extracted from *Aspergillus niger* (Oenological Codex, 2009, OIV-OENO 368-2009). Its applications include clarification of beverages, stabilization of white wines, prevention of oxidation of the wine, and the removal of undesired substances, for example, the complexation of ochratoxin A (a carcinogen nephrotoxin produced by several molds), metals (iron, lead, and cadmium) and excess of tannin. Chitosan is also used to extend the quality and shelf-life of beverages. It is used as a bioflocculant for the clarification of fruit juices (grapefruit juice and nectars) and other beverages (beer, coffee, tea, and herbal teas), as a flavor enhancer, as an acidity adjusting agent, and as a preservative (due to its antimicrobial and antioxidant activities). It permits also to stabilize color, prevent oxidation, and mask of undesirable taste. As a fining agent, fungal chitosan not only facilitates the settling and clarification of wines but also prevents protein haze. For example, chitosan is able to coagulate the anionic compounds (proteins, pectin) present in tea infusion and clarify infusions without effect on the composition of the phenolic compounds. All the studies highlight the important role of amine functions of chitosan and its cationic character in acidic media, useful for example in the clarification process. Chitin and chitosan are also used as biomaterial for the immobilization of enzymes and microorganisms (Cosme and Vilela, 2021). Fungal chitosan's antimicrobial properties make it a potential substitute for sulfites to control yeast (Brettanomyces, often abridged Brett). Indeed, this type of chitosan, authorized by the European Union in 2011 and included in the specifications for organic winemaking since 2018, is mainly used as an antiseptic and as a substitute for sulfites (Petrova et al., 2016). Chitosan can also be used in active packaging which contains encapsulated bioactive compounds and in water purification (Kim, 2011; Gassara et al., 2015; Rocha et al., 2017; Crini and Lichtfouse, 2019a; Morin-Crini et al., 2019; Castro Marín et al., 2020; Cosme and Vilela, 2021). The challenges of the beverage industry are to prevent the alteration of products, to maintain their color, to increase their shelf-life, and to avoid aromatic defects by limiting oxidation (case of wines for example). This is a promising market for chitin, chitosan, and their derivatives, although at the moment, few of these products are allowed for this application.

7.4.4 Nutraceuticals

Today, consumers demand safe foods with new functionalities (this is the area of nutraceuticals or "borderline products") and longer shelf life and environmentally friendly packaging. This is forcing the industry to innovate and develop new-food processing and packaging strategies. Chitosan fits perfectly into these challenges. The term nutraceuticals, derived from the terms nutrition (the food component) and pharmaceutics (the drug), refers to food products that have additional health benefits beyond the basic nutritional value of the food. Another term used is "alicament," which comes from the contraction of the words food and medicine and refers to foods that have a beneficial effect on health, with effects superior to those of other foods. The terms "health food" or "superfood," which are also foods that combine the concept of food and medicine, are also used. Food supplements, functional foods, food additives, dietary supplements, and herbal products are among these products. These products are widely consumed but are not regulated (Nasri et al., 2014). Indeed, the concept of "borderline products," which includes not only neutraceuticals but also cosmetoceuticals and cosmetotextiles, has no real status. Neutraceuticals are certainly at the forefront of food innovation, but regulations must evolve in order to market these products not only on fashionable benefits but also on real scientific data.

China is expected to be the world's largest consumer of nutraceuticals by 2030 (Morin-Crini et al., 2019; Mutreja et al., 2020). Food supplements based on chitosan, chitooligosaccharides, and glucosamine to absorb fats, cholesterol, and sugars are on the market, for example, Cellulox®, LipoBIND®, ALTISA® Chitosvelt, ChitoClear®, FORTEX, OligOcaps, MicroChitosan™, NaturTierra, etc. All these products are marketed as fat reducers, cholesterol-lowering agents, and as antioxidant agents. They claim to provide important support in maintaining a healthy skin balance owing to the many biological activities of chitosan (Hayes et al., 2008b; Je and Kim, 2012a,b; Nasri et al., 2014; Patti et al., 2015; Morin-Crini et al., 2019; Akbari-Alavijeh et al., 2020; Mutreja et al., 2020). For example, Cellulox®, containing chitosan, vitamin C, and tartaric acid is indicated to contrast the imperfections of cellulite, contribute to the normal formation of collagen for the function of blood vessels, and regulate the functionality of female microcirculation. The claimed properties are based on the antioxidant power of the neutraceuticals and their capacity to make people lose weight. Chitosan acts as an intestinal film and plays a role of a natural fiber that binds dietary fats (this affinity becomes more consistent in the presence of vitamin C and tartaric acid). It would inhibit the absorption of lipids and sugars (Ylitalo et al., 2002), or would directly activate the destocking of lipids (Helgason et al., 2009). However, the mechanisms are not yet fully elucidated. LipoBIND® reduces caloric intake during a diet or a heavy meal. Here again, it is the role of the dietary fiber of chitosan that is highlighted. The presence of chitosan in formulations allows for the progressive delivery of bioactive ingredients, while also improving their bioavailability and stability. ALTISA® Chitosvelt helps maintain normal cholesterol levels. The oligomers of chitosan also seem to prevent the risks of diabetes (Kumar and Tharanathan, 2004; Kim and Karadeniz, 2013; Karadeniz and Kim, 2014). ChitoClear® also has potential in weight management and obesity treatment. Although numerous studies have been published, prospective clinical trials are needed to confirm such benefits (Patti et al., 2015). Indeed, it is important to note that some authors had previously claimed that chitin-based supplements were safe and provided many health benefits (Muzzarelli, 2010), while others had reported allergy risks due to the presence of substances such as metabisulfites, biogenic amines, and protein residues in purified biopolymers (Reese et al., 2007; Kim, 2012). Chitin also triggered immune responses that could lead to allergic reactions. One study also related the role of chitosan against inflammation related to allergic reactions and asthma (Brinchmann et al., 2011). Liao et al. (2007) showed that the consumption of chitosan limited the absorption of calcium in contrast to chitooligosaccharides. There is also a debate surrounding the effectiveness of chitosan at blocking fat absorption. Studies suggest using chitin and chitosan of fungal origin to avoid allergy problems (Reese et al., 2007; Kim, 2012; Hahn et al., 2020; Mohan et al., 2020; Kou et al., 2021; Ma et al., 2022). In any case, further scientific studies are needed to validate the beneficial effects of neutraceuticals and their potential dangers.

7.5 Pharmaceutical and therapeutic applications

In the pharmaceutical and therapeutic applications, chitosan and its derivatives are used in the form of solutions, suspensions, gels/hydrogels, tablets, capsules, microspheres, (nano)particles, sponges, foams, membranes, films, (nano)fibers, microneedles, microscopic threads, coated liposomes, adhesives, (nano)composites, and 2D- and 3D-scaffolds, which allows uses in pharmacy (drug delivery), dermatology, ophthalmology, medicine (vaccines), biomedicine and nanomedicine, dentistry, and veterinary medicine. Consequently, chitosan and its derivatives may be used in oral, ocular, nasal, vaginal, buccal, parenteral, intravesical, and transdermal administration and as bandages and implants (see numerous references cited in Table 6.3).

7.5.1 Pharmacy

The examples of applications of chitosan in pharmacy are numerous and varied:

- as an excipient with active properties: safe excipients; tablets are the likely most favorable dosage form since they provide an accurate dosage, are easy to manufacture and handle, and are favored by patients;
- as a carrier for the controlled delivery of sensitive drugs via encapsulation (to increase and modulate drug release rate): for example, antiinflammatory (diclofenac, aspirin), anticancer (5-fluorouracil, tamoxifen, doxorubicin), bronchodilator (salbutamol), anesthetic (lidocaine), hormones (testosterone, insulin), proteins and peptides (bocine serum albumin), serine protease (trypsin), and alkaloids (caffeine);
- for gene delivery: nucleic acid;
- as a raw ingredient to prepare vaccines, bandages and biological adhesives, coated textiles, in oral hygiene products, and in the cosmeceutical and nutraceutical (body weight management) formulations;
- for wound healing (self-healing);
- as a bacteriostatic, hemostatic, or anticoagulant agent.

Other properties of chitosan that are of interest to the pharmaceutical field are mucoadhesive (useful for buccal delivery), gelling, transfection, and permeation properties (Illum and Davis, 2004; Bernkop-Schnürch and Dünnhaupt, 2012; Amber Jennings and Bumgardner, 2017). Many other examples can be found in the references given in Table 6.3.

7.5.2 Dermatology and ophthalmology

In dermatology, chitosan is used as a moisturizing and bacteriostatic agent and in wound healing (self-healing). It is also used to prepare water-resistant adhesives (Hamedi et al., 2022). Incorporated in dermatological formulations or in textiles, it is used to treat acne and psoriasis (cyclosporine A) (Bernkop-Schnürch and Dünnhaupt, 2012; Ghannoum et al., 2015; Lopes et al., 2015; Elieh-Ali-Komi and Hamblin, 2016; Badawy and Rabea, 2017; Krishnaswami et al., 2018; Zarrintaj and Saeb, 2018; Tangkijngamvong et al., 2020). Lopes et al. (2015) showed that textiles coated with chitosan can fight atopic eczema by modulating the staphylococcal profile of the skin. Ghannoum et al. (2015) showed that hydroxypropyl chitosan as a topical medication can be used as a nail polish to guard against dermatophyte infections. Tangkijngamvong et al. (2020) indicated that due to their antifungal effect, chitosan particles showed an interesting anti-sebum effect, useful for promising applications to combat seborrheic dermatitis, a common and chronic dermatosis (infection of the skin and mucous membranes).

In ophthalmology, chitosans with a homogenous and controlled composition and structure such as Viscosan™ are particularly in demand. Chitosan is then used in ocular drug delivery systems due to its nontoxicty and permeation-enhancing properties (Felt et al., 1999; Alonso and Sánchez, 2003; Bernkop-Schnürch and Dünnhaupt, 2012; Nilsen-Nygaard et al., 2015; Elieh-Ali-Komi and Hamblin, 2016; Badawy and Rabea, 2017; Krishnaswami et al., 2018; Paliwal et al., 2019; Zamboulis et al., 2020). However, its main use is in the production of contact lenses because this biopolymer had important characteristics required for making an ideal contact lens such as optical clarity, mechanical stability, optical correction, gas permeability, wettability, and immunological compatibility. Contact lenses are also high value-added products for the industry despite the fact that the price of chitosan is high because of its high purity. In ophthalmic surgery, contact lenses without optical properties are used for the healing of damaged corneas. The antimicrobial activity, film-forming ability, and wound healing (self-healing) properties also make chitosan suitable for development of ocular bandage-lenses for traumatic injuries (Elieh-Ali-Komi and Hamblin, 2016; Badawy and Rabea, 2017).

7.5.3 Medicine, biomedicine, and nanomedicine

The main biomedical applications of chitosan are in wound healing, combining two important properties of chitosan, its biocompatibility and antimicrobial behavior. The main fields concerned are regenerative medicine (including cellular biology, cell therapy, dermatology, ophthalmology, dentistry, and veterinary medicine), tissue engineering (biomaterials, bone substitutes, regeneration of ligaments, cartilage, neurons, and skin), and 3D printing (medical devices, implants). Many products have been on the market since the early 1990s, mainly in North America and Asia, and more recently in Europe. Other medical, biomedical, and nanomedical applications are particularly numerous and vast: therapeutic formulations, drug delivery and drug vectorization (antibiotics, antiinflammatory, antihypertensive, and anticancer drugs, peptides, proteins, vaccines, growth factor), hematology (bood interactions), microbiology (antimicrobial, antiinfectious), gene therapy, immunology, cancer therapy, bio-imaging, and magnetic resonance imaging. However, most of these applications have not yet reached the market. Like for pharmaceutical applications, chitosan can be easily processed into different forms (solutions, gels, sponges, nanoparticles, films, and fibers) for target medical applications (Croisier and Jérôme, 2013; Elieh-Ali-Komi and Hamblin, 2016; Dutta, 2016; Ahmed and Ikram, 2017; Amber Jennings and Bumgardner, 2017; Liaqat and Eltem, 2018; Chivere et al., 2020; Satitsri and Muanprasat, 2020; Zamboulis et al., 2020; Amirthalingam et al., 2021; Ardean et al., 2021; Araujo et al., 2021; Boroumand et al., 2021; Dave et al., 2021; Khan and Alamry, 2021; Ke et al., 2021; Kumara et al., 2021; Lima et al., 2021; Mohammadi et al., 2021; Rezaei et al., 2021; Torkaman et al., 2021; Valachová and Šoltés, 2021; Gao and Wu, 2022; Kim, 2022; Yu et al., 2022).

Due to its healing activity, chitosan is used to make dressings and bandages and products are on the market (e.g., HemCon® Bandage, ChitoGauze® PRO, Chitopack C®, ChitoFlex® PRO, ChitoSam™, Syvek-Patch®, Tegasorb®, Chitoderm® Plus, Axiostat®, Celox™) (Burkatovskaya et al., 2006; Minagawa et al., 2007; Lahihi et al., 2000; Ribeiro et al., 2009; Singh et al., 2017). Studies show that the healing performance is improved if the degree of deacetylation of chitosan increases.

HemCon® Bandage is a sterile, foldable chitosan acetate dressing used as a hemostatic control dressing to stop severe external bleeding after trauma or penetrating injury. The hemostatic dressing gives the patient/injured the time needed to reach care or to gain the critical time needed for clotting. It also provides an antibacterial barrier against a wide range of microorganisms. This dressing has been used on the battlefield by the US Army since 2002 (Singh et al., 2017). Tegasorb®, a chitosan formulation in gel or hydrocolloid form, is offered for the healing of large internal wounds. Chitoderm® Plus is a chitosan-based hydrogel dressing that promotes healing in a moist environment, while reducing the risk of infection through antibacterial action. Other biomedical products on the market include, for example, Reaxon® (a chitosan-based nerve conduit to protect and promote regrowth of severed nerves) and ChitoSeat™ (a family of chitosan-based hemostatic sealants suitable for hard and soft tissue surgical bleeding). A comprehensive review on the use of many chitinous derivatives in bone tissue engineering can be found in the references Anitha et al. (2014), LogithKumar et al. (2016), and Ahmed et al. (2018). Badawy and Rabea (2017) and Khan and Alamry (2021) discussed the applications in protein, peptide, and gene delivery and gene therapy. The beneficial effects of chitosan in other applications (e.g., anti-Alzheimer's disease, anti-atherosclerosis, kidney injury alleviation, alleviation of obesity, anticancer activity) have been reported by Satitsri and Muanprasat (2020).

Chitin is also used as biomaterial for sutures because of its nontoxicity, biocompatibility, and low immunogenicity. It accelerates wound healing in spray, gel, and gauze, and it can be used to control drug release. Beschitin W®, an artificial skin made of chitin (chitin-coated gauze), is recommended for the rapid healing of burns, skin abrasions, postoperative wounds, and ulcers. Chitopack S® is also a nonwoven chitin dressing designed to repair activity on burned skin. As with chitosan, it is the healing properties of the biopolymer due to its biostimulant activity that are used in this application. This artificial skin also has analgesic properties. Published data show that the healing of burns and the reconstitution of the epidermis is much faster and less painful than with the products normally used. Moreover, this skin does not present any risk of rejection. However, compared to the medical applications of chitosan, those of chitin are more restricted because its insolubility is a major problem in the development of biomaterials and processes using chitin. Nevertheless, it is an important field of academic and industrial research (Kou et al., 2021; Sieber et al., 2018; Hahn et al., 2020; van den Broek and Boeriu, 2020; Sánchez-Machado et al., 2022).

7.5.4 Dentistry

In dentistry, chitosan is proposed in preventive dentistry (oral healthcare, controlled delivery of fluoride), conservative dentistry, endodontics, surgery, periodontology, prosthodontics, and orthodontics (Keegan et al., 2012; Hayashi et al., 2013; Queiroz et al., 2015; Kmiec et al., 2017; Aranaz et al., 2018; Navarro-Suarez et al., 2018; Morin-Crini et al., 2019; Chen et al., 2021; Sharifianjazi et al., 2022). Due to its versatility and numerous properties such as bioactivity (especially biocompatibility and antiinflammation effect), wound healing, bone repair, and hemostasis, chitosan has, indeed, a wide range of applications in dentistry: for example, for dental hygiene (ChitoDent® is a homeopathy-compatible chitosan toothpaste without fluorine) and oral care, as an ingredient in toothpastes (release of fluoride) or in chewing gums with antibacterial effects (e.g., chitosan acts against the bacterial strains responsible for dental caries) or to increase salivary secretion, to prevent diseases (dental caries, periodontitis, erosive lesions, plaque inhibitor), to improve dental alignments, and to prepare biomaterials (adhesives, barrier membranes, bone replacement, tissue regeneration). Toothpastes, mouthwashes, and chewing gums freshen the breath and prevent plaque and caries formulation (Kmiec et al., 2017). Chewing pastes based on chitosan and chitooligosaccharides are proposed to remove food debris ingested during a meal as an alternative to tooth brushing. The saliva releases the substances, thus activating their anticavity role (Hayashi et al., 2007). Another example of chitosan application in dental hygiene concerns dental implants. Dental implants are artificial roots designed to replace missing teeth. As with natural tooth, bacterial plaque can accumulate on the base of dental implants and lead to tissue inflammation. These implants also require regular care. Recently, a chitosan fiber-based dental brush (Labrida BioClean ™) with antibacterial and antiinflammatory properties has been developed and brought to market. The advantages cited are effective and gentle cleaning of the implant surface, significant maintenance of peri-implant health, and most importantly, increased patient comfort compared to titanium curettes. In dental surgery and therapy, chitosan is used as a dental adhesive, bone tissue healing agent, cell protective agent, implants, and scaffolds. For example, chitosan preparations are able to repair early caries lesions (Queiroz et al., 2015). Chitosan-based endodontic cements reduce inflammation and promote bone regeneration (Navarro-Suarez et al., 2018). The applications of chitin in dentistry are rare. It is proposed to prepare biomaterials (implants, cements). Carboxymethyl chitin also has an efficient activity against bacterial strains such as *Streptococcus mutants* (Satitsri and Muanprasat, 2020). The synthesis of recent nanochitosan (nanosphere, nanocapsules) and nanocomposites seems to open new developments for dentistry (Kmiec et al., 2017; Aranaz et al., 2018; Navarro-Suarez et al., 2018; Morin-Crini et al., 2019; Chen et al., 2021; Sharifianjazi et al., 2022).

7.5.5 Veterinary medicine

Applications of chitosan in veterinary medicine include drug delivery (antibiotics, antiparasitics, mucosal formulations) and vaccines, incorporation into analgesic and antimicrobial formulations, wound healing (regenerative medicine), bone regeneration, hemostatic products, body care products, and nutritional ingredients (Şenel and McClure, 2004; Şenel, 2011; Underwood and van Eps, 2012; Drewnowska et al., 2013; Gerdts et al., 2013; Tonda-Turo et al., 2016; Morin-Crini et al., 2019; Maldonado-Cabrera et al., 2021). Maldonado-Cabrera et al. (2021) recently reviewed the therapeutic effects of chitosan in veterinary dermatology, concluding that when applied to animal skin lesions, chitosan produces positive healing effects, potentially becoming a safe biomaterial for skin treatments in veterinary practice. There are several medical products on the market. For example, ChitoClear® Animal, in spray or gel form, is offered to treat irritated and damaged skin and chronic wounds and ulcers of pets. It also relieves pain and itching. Other properties of the formulation are also cited as a hemostatic barrier, antibacterial, antifungal, and antiviral effects. ChitoClear® is also effective against mud-fever in horses. Chitosan-based nutritional supplements such as Epakitin™, IPAKITINE®, and Nutri+Gen® are commercially available for use as essential nutrient supplements or in chronic kidney disease in dogs and cats. The Epakitin™ formulation for animal use contains chitosan, calcium carbonate, lactose, and hydrolyzed soy protein. It decreases serum phosphorus levels and reduces urea and creatinine levels. IPAKITINE® is a dietary supplementary food for the support of renal function in chronic renal failure in cats and dogs. Nutri+Gen® is a probiotic supplement for all pets containing chitosan, vitamins, minerals, and essential nutrients. Many body care products are also available as shampoos, ear cleaners, conditioners, and sprays for pets. For example, ChitoCure® is a hypoallergenic, tear-free shampoo for cats and dogs and is recommended for pets with sensitive skin.

7.6 Cosmetics and personal care products

The cosmetic industry was one of the first markets for chitin and chitosan, and this market is still growing, with in particular the emergence of a new very promising domain, that of cosmetoceuticals (Muñoz et al., 2012; Lima et al., 2012; Senevirathne et al., 2012; Chalongsuk and Sribundit, 2013; Jimtaisong and Saewan, 2014; Ahmed and Ikram, 2017; Bonecco et al., 2017; Costa and Santos, 2017; Nechita, 2017; Philibert et al., 2017; Serrano-Castañeda et al., 2018; Aranaz et al., 2018; Rahangdale and Kumar, 2018; Serrano-Castañeda et al., 2018; Casadidio et al., 2019; Crini and Lichtfouse, 2019a; Morin-Crini et al., 2019; El-Hack et al., 2020; Triunfo et al., 2021; van den Broek and Boeriu, 2020). In cosmetics, the cationic character of chitosan, its bacteriostatic, antifungal, antistatic, film-forming, and moisture retention properties are particularly appreciated by formulators. Chitosan is also interesting because it is easy to handle, compatible with other ingredients used in cosmetic formulations, and especially it allows the controlled release of bioactive agents or drugs. Thus, chitosan is used as a multifunctional ingredient and a delivery system in the formulation of cosmetics (hair care, creams, lotions, coloring, and deodorizant products). It has several functions: antioxidant agent (shampoos), moisture absorption-retention agent (makeup, nail polish), antimicrobial agent (mouthwash), and emulsion stabilizer. There are hand sanitizers based on chitosan, without alcohol and also having a moisturizing, humectant, and even protective role. Chitin is also an effective moisturizing agent and film-forming tensor as it prevents dehydration by providing water. Like chitosan, it is well tolerated by the skin (the biopolymers cannot penetrate the skin due to their high MW). Chitosan and chitin also have chelating properties with respect to metals which are responsible for many contact allergies.

Numerous chitosan- and chitin-based products for cosmetic use are commercially available: Curasan™, Hydamer™, Zenvivo™, Ritachitosan®, etc. Curasan™ is used in shampoos, hair treatment emulsions, lotions, and skin care products. Chitosan, forming a nonocclusive protective film, is claimed to improve the quality of dry, fine, and structurally damaged hair. The chitosan film increases the tensile strength and volume of the hair, making it easier to style. Curasan™ also smoothens rough and cracked skin. Hydamer™ also acts as a film-forming agent and finds applications in the formulation of skin care products as an antioxidant, antimicrobial, and deodorizing agent. Zenvivo™ contains cationic, high molecular weight, vegetal-based chitosan. Due to its antimicrobial efficacy, odor control, moisturizing and improved skin feel, this formulation finds applications in skin care products, shampoo, hair care products (coloring products), conditioners, styling foams, deodorants, oral care (toothpaste, mouthwash, chewing gum), mascara, lipsticks, and nail makeup. Ritachitosan® is used in deodorants and antiperspirants due to the antimicrobial and moisturizing properties of the chitosan in the product. All these commercial formulations highlight the polycationic character of chitosan, the fact that it integrates perfectly with the other ingredients of the formulations in products as diverse as emulsions, fluids, deodorants, sprays and gums, and especially its ability to form films owing to its long macromolecular chains.

Despite the health and economic crisis, the beauty and cosmetics sector is booming, especially the cosmeceuticals market (Aranaz et al., 2018; Morin-Crini et al., 2019; Crini and Lichtfouse, 2019a; Triunfo et al., 2021). Cosmeceuticals are cosmetic products, often in the form of creams, lotions, and ointments, with pharmaceutical/medicinal benefits intented for external application to the human body to produce targeted results. However, the term cosmeceuticals do not have a consensus definition, and these products have not yet been legally recognized. In their conception, cosmeceuticals are similar to nutraceuticals, in that they contain both cosmetics (instead of food in the case of nutraceuticals) and drugs that are considered active ingredients with various functions such as protection, whitening, tanning, antiaging (antiwrinkle), deodorization, antiperspirant effect, and nail and hair care. The formulations contain essential oils, vitamins, enzymes, chemicals (phytochemicals), and other ingredients such as chitin, chitosan, glycogen, alginate, and carrageenan. Chitin is used for its abrasion swelling functions and as a vehicle for controlled delivery, skin protection, and antiaging. Chitosan is used for the same reasons (vehicle for controlled delivery, skin protection), but mainly for its polycationicity, its ability to form films (e.g., in hair setting and repair), and its bioactive properties such as antimicrobial and antioxidant properties. For example, Chitoseen™-K is proposed not only for its cosmetic properties (antimicrobial and moisturizing agent) but also pharmaceutical/medicinal benefits. Chitosan retains moisture in low humidity and maintains the hair's style in high humidity and especially allows the repair of damaged hair. The use of such products promises smooth, soft, strong hair that is resistant to atmospheric aggressions (pollution, UV rays). Another example is the cosmetoceutical product Silver Chitoderm®, in the form of a hydrogel, which is offered as a skin moisturizer and can also reduce inflammation and promote skin cell regeneration, while being gentle and nonaggressive to the skin. It has even been reported to help improve rosacea and eczema.

As for neutraceuticals, the borderline between cosmetic and pharmaceutical products is doubtful. It is the concept of the "borderline product" that has no real status. Further scientific studies are needed to validate the beneficial effects of cosmetoceuticals. Nevertheless, given the strong consumer demand for natural products containing safe ingredients such as chitin and chitosan biopolymers, the cosmetoceutical market will continue to grow strongly in the coming years.

7.7 Agriculture

The agricultural industry and the oenology sector have evolved enormously over the last 50 years not only in mechanization and work techniques but also in the implementation of new agro-environmental practices with the objective of reducing pesticides and energy consumption and the use of ecological plant protection and improvement products. Chitosan and its derivatives are part of this approach. One of the objectives is to replace conventional pesticides with useful substances, often called "biopesticides," to improve crop yields without causing serious environmental and health problems, or at least to decrease the quantities of pesticides used. In addition, it is known that microorganisms harmful to plants are becoming increasingly resistant to the traditional chemical pesticides used today, so what worked for farmers a few decades ago no longer has the same impact today, hence the need for innovation. Restrictions on the use of plant growth regulators also require the use of effective, nontoxic, biodegradable, and biocompatible products, properties that the biopolymer chitosan possesses. The many intrinsic biological properties of chitosan, especially its antifungal activity, also play a key role in its application in agriculture. It is also important to note that in both agriculture and medical applications, low molecular weight chitosans with controlled homogeneity and chitooligosaccharides are more effective than high molecular weight chitosans.

In agriculture, chitosan is used in three main ways (Goosen, 1997; Rabea et al., 2003; El Hadrami et al., 2010; Yin and Du, 2011; Sharp, 2013):

(1) seed pretreatment in order to increase seed germination and its immunity to stress and diseases to strengthen seedlings;

(2) soil treatment to improve the soil environment: increase the growth of plant-beneficial rhizobacteria, stimulate plant root growth, and prevent pathogens and harmful nematodes; in general, chitinous products provide nitrogen when they decompose, contributing to soil and plant enrichment;

(3) foliar spraying in the field: to promote stem, leaf, and fruit growth, to improve plant photosynthesis, and to enhance crop immunity to stress and disease.

The many results described in the literature are astonishing and suggest an important development in agriculture (Muñoz-Bonilla et al., 2014; Katiyar et al., 2014; Xing et al., 2015; Badawy and Rabea, 2016, 2017; Bautista-Baños et al., 2016; Hadwiger, 2017; Ippólito et al., 2017; Orzali et al., 2017; Divya and Jisha, 2018; Grande-Tovar et al., 2018; Sharif et al., 2018; Betchem et al., 2019; Crini and Lichtfouse, 2019a; Morin-Crini et al., 2019; Yuan et al., 2019; Bandara et al., 2020; Maluin and Hussein, 2020; Malerba and Cerana, 2019, 2020; Mujtaba et al., 2020; Qu and Luo, 2020; Shamshina et al., 2020; Singh et al., 2021a; Dave et al., 2021; Yu et al., 2021). For example, chitosan is thus used as a seed soaking agent (soybean, cotton, tomato, cucumber, wheat, and rice), root application agent (potato, soybean lettuce, and spinach), spray agent (peanut, soybean, cabbage, rice, maize, and cotton), and supplement (rice, wheat, peanut), with these activities playing an important role in plant disease control and stress resistance. Chitosan positively affects seed germination and seedling growth of wheat (in maize and cucumber, there is also an increase in seed germination), improves the viability and purity of long-stored seeds (e.g., fenugreek seeds), induces the expression of pathogenesis-related proteins thus enhancing resistance against pathogens in tomato, or enhances the uptake of nutrients, photosynthesis, and growth of treated coffee plants. Chitosan, as a powerful elicitor, stimulates the production of chitinolytic enzymes in plants, which help defend themselves against pathogen aggressors. Indeed, the presence of chitosan activates many defense responses in plants: its use on a plant makes it believe that it is under attack by parasites. Because of this perceived infestation, the plant's immune system goes into protective mode. It can instantly induce defense reactions against various fungus, bacteria, and viruses. It can prevent the onset of disease and prevent the microorganism from spreading to new leaves when the microorganism appears. Studies show that this natural defense mechanism not only stimu lates the plant's immune system but also keeps pathogens and pests at bay. However, the exact mechanism is still not fully understood (Yin and Du, 2011;

Muñoz-Bonilla et al., 2014; Dave et al., 2021). In agriculture, chitin plays a different role from chitosan. Chitin, as a nutrient substrate, promotes the growth of chitinolytic microorganisms, a property that is exploited to improve composts (De Jin et al., 2005) and as a method to identify these microorganisms (Gomez-Ramirez et al., 2004). Malerba and Cerana (2019) also reported that chitin contributes to the protection of seeds, and its addition to the soil improves microbial communities. It can be directly used as a fertilizer to enhance crop growth or after postharvest to protect tomatoes and apples for example. Oligochitins are proposed as biostimulators and elicitors of plant defenses, for example, rice (induction of phytoalexin), soybean and parsley (synthesis of callose), rape (increase chlorophyll), and strawberry (increase yields). Chitin derivatives are proposed for reducing the use of chemical fungicides on tobacco plantations. Shamshina et al. (2020) reviewed the agricultural uses of chitin and its beneficial effects as nutrient substrate, fertilizer, soil conditioning agent, and an ingredient to improve microbial communities in the soil or in composts, protection of seeds, plant disease control agent, ripening retardant, biopesticide, fruit coating, and protective agent after postharvest. Prathibhani et al. (2022) recently discussed the beneficial effects of chitosan coating of kiwifruit after harvest. The results showed improved preservation of the fruit during storage and reduced moisture loss and ethylene production, while maintaining its nutritional quality. However, the mechanisms of action are not known.

There are several products on the market (e.g., DORA®, HYSHIELD™, Chitoseen™-H, Chitosan 6 Rumexo, etc.) used in seed, leaf, fruit, and vegetable coatings, for spraying, and as biocides (bactericidal or bacteriostatic agent), biopesticides, elicitors, and fertilizers. The DORA® formulation, containing a chitosan oligosaccharide, is marketed for its capacity to improve seed germination, increase immunity ability to diseases and stress, and its ability to induce a general regulation of endogenous plant hormones. The HYSHIELD™ product helps stimulate plant defense mechanisms during pest attacks and also keeps pathogens and pests at bay. Chitoseen™-H is used as an antimicrobial and moisturizing agent in agriculture. In China, chitosan is registered as a "plant inducer" by the Ministry of Agriculture and known as a "sugar chain plant vaccine." It is used in tomatoes, eggplant, peppers, coffee, tobacco, fruit trees, and cannabis against plant diseases, pests and insects, plant growth promotion, seed-coating, and postharvest (e.g., chitosan can extend the shelf life of bananas). Chitosan can, indeed, manage postharvest diseases of horticultural crops.

The advantages of the use of chitosan are not only an effective protection against broad spectrum pathogens owing to an effect on the immune system of the plant but also its ease of use (no effect on the farmer), the increase of the yields, and thus of the productivity owing to effects on the nutrition of the plant (increased mineral uptake) and the stimulation of the growth. Other benefits cited are the improvement of the plant's water retention capacity, a better resistance to cold or drought, and the improvement of growth conditions in different environments. Of course, there are problems in the applications of chitosan in agriculture. It is difficult to obtain batches of commercial chitosan with the same characteristics, especially the DA (or DD) and MW parameters, the antimicrobial activity of chitosan depending on these parameters. However, the data in the literature are sometimes contradictory (Badawy and Rabea, 2017; Ardean et al., 2021). Another problem for the wide application of chitosan in agriculture is its insolubility in water. Recently, to overcome this problem, chitosan derivatives and oligomers produced by enzymatic and chemical modifications and nanotechnology have been proposed (Divya and Jisha, 2018). These materials are expected to be promising alternatives in agriculture.

7.8 Aquaculture

In aquaculture, chitosan can be used as functional food, nutritional supplements (probiotics, prebiotics), carrier of bioactive compounds (proteins, pigments), drug delivery, to design vaccine delivery systems, encapsulate pathogens (bacteria) or nucleic acids, or to remove pollutants from water. For example, as probiotics, chitosan improves feed conversion, growth rates, weight gain, immune system, and disease resistance in fish due to its immune-stimulating, antimicrobial, and antioxidant properties (Chung et al., 2005; Chung, 2006; Borgogna et al., 2011; Cerezuela et al., 2011; Alishahi and Aïder, 2012; Harikrishnan et al., 2012; Niu et al., 2013; Zaki et al., 2015; Lian et al., 2016; Bernardi et al., 2018; Vinay et al., 2018; Crini and Lichtfouse, 2019a; Morin-Crini et al., 2019; Yuan et al., 2019; El-Naggar et al., 2022; Mahari et al., 2022). Chitosan nanostructures are of particular interest in aquaculture. They act as natural, nontoxic, and more effective feed additives than granular forms, improve fish growth performance, enhance immune function, and inhibit gut microbial pathogens. Chitosan nanostructures can also be used as antioxidants and drug carriers. Mahari et al. (2022) recently reported that due to their larger surface area, porosity, better bioavailabilty, and deeper penetration into target sites, chitosan nanostructures are more effective than the traditional form of chitosan.

7.9 Textile and paper industries

7.9.1 Textile industry

In textile industry, chitosan and its derivatives are mainly used as active compounds, for example, as antimicrobial finishing of functional textiles, including medical textiles (Giri Dev et al., 2005; Enescu, 2008; Ummu Habeeba et al., 2007; Crini et al., 2009; Francesko et al., 2010; Islam et al., 2013; Şahan and Demir, 2014; Hamed et al., 2016; Voncina et al., 2016; Gutiérrez, 2017; Roy et al., 2017b; Massella et al., 2019; Morin-Crini et al., 2019). However, its mode of action is yet not fully understood (Islam et al., 2013). The other characteristics of chitosan appreciated by industrialists are its cost-effectiveness, nontoxicity, biocompatibility, biodegradability, antistatic activity, chelating property, deodorizing property, ability to form films, chemical reactivity, ability to improve the strength and rigidity of a tissue and dyeing, thickening property, and also its wound healing activity, the latter being of interest in the biomedical field (surgical threads, sanitary fibrous products).

7.9.2 Cosmetotextiles

These last two decades have seen the appearance and the development of so-called "intelligent" technical clothing. These are the cosmetotextiles (Han et al., 2020). As for neutraceuticals and cosmetoceuticals, cosmetotextiles brings together two industrial sectors, cosmetics and textiles. In fact, the technology used by manufacturers is that of microencapsulation. These textiles contain microcapsules containing a cosmetic substance intended to be released on different parts of the human body. The targeted functions are, for example, protection of the skin, the release of a perfume, to promote recovery after effort or as an aid to well-being, or the elimination of body odors. Chitosan can play this role of an encapsulating agent.

There are, indeed, many other possibilities for the development of new textile and cosmetic products (recent field of cosmetotextiles) containing chitosan-based capsules/nanoparticles with advanced properties (non-allergenic fibers, antimicrobial textiles, antiodor textiles, perfuming textiles, encapsulation of fragrances, textile preservative and deodorant, self-cleaning, UV blocking, water repellency, and antifouling property)

(Morin-Crini et al., 2019; Crini and Lichtfouse, 2019a; Han et al., 2020; Raza et al., 2020). There are antiodor socks on the market containing chitosan used as gel finishing agent. There are also antibacterial and deodorizing care socks made of chitosan, with long-lasting effects over time (even after 15 washings). Another commercial example is that of slimming leggings, anticellulite, and allowing to slim down without effort or almost. However, for a widespread use of chitin and chitosan, which are very promising in the field of cosmetotextiles, and in the fields of nutraceuticals and cosmetology, it is essential to conduct scientific studies in order to promote the use of these biopolymers on the basis of proven results and not only attributed to fashionable but unproven effects.

7.9.3 Pulp and paper industry

The main use of chitosan in the pulp and paper industry is to improve the wet strength of paper. It is also used as a retention and drainage agent and as a reinforcing additive. As a functional material, it is capable of interacting with cellulose pulp during paper formation and being film-forming to provide strength at break (Ravi Kumar, 2000; Samyn et al., 2018; Song et al., 2018; Morin-Crini et al., 2019). Chitosan can be used for sizing and finishing paper and applying surface coatings to impart antimicrobial properties to paper. Examples of applications include paperboard, biodegradable packaging for wood wrapping, toilet paper, chromatography paper, carbonless copy paper, and photochromic paper. Chitosan, as a chelating and complexing agent, is used in the decontamination of pulp and paper wastewaters, for example, to remove lignin, color, and unwanted contaminants (Crini et al., 2009; Morin-Crini et al., 2019). Chitin is also used in manufacturing as a strengthener and sizing agent (Phillips, 2017).

7.10 Biotechnology and chemical industries

7.10.1 Biotechnology

Applications of chitosan and its derivatives (chitooligosaccharides) in biotechnology include enzyme and cell immobilization, cell recovery, cell-stimulating materials, protein separation, and as a matrix for affinity and gel permeation chromatography. For example, enzymes such as lysozyme, urease, *E. coli* cells, and amylases are immobilized with chitosan by encapsulation and adsorption mechanisms. Chitosan can also be used for electronic devices, construction of biosensors and biodevices, and metabolic analysis of biological fluids (Krajewska, 2005; Wang, 2012; Suginta et al., 2013; Philibert et al., 2017; Grifoll-Romero et al., 2018; Morin-Crini et al., 2019; Verma et al., 2020; Peter et al., 2021; Yang et al., 2021).

7.10.2 Chemistry

The chemical sectors using chitosan are chromatography, catalysis, green chemistry, green solvents (ionic liquids, deep eutectic solvents), solvent separation, electrochemistry (biosensors), membrane technology, including reverse osmosis, filtration and pervaporation, energy stockage, and membranes for lithium batteries (Suginta et al., 2013; Mati-Baouche et al., 2014; Carneiro et al., 2015; Salehi et al., 2016; Thakur and Voicu, 2016; Silva et al., 2017; Osman and Arof, 2017; Argüelles-Monal et al., 2018; Galiano et al., 2018; Marpu and Benton, 2018; Zdanowicz et al., 2018; Crini and Lichtfouse, 2019a; Morin-Crini et al., 2019; Xie and Yuan, 2019; Jaworska et al., 2020; Boominathan and Sivaramakrishna, 2021; Özel and Elibol, 2021; Peter et al., 2021; Khajavian et al., 2022; Prabhu et al., 2022; Wang et al., 2022; Zhang et al., 2022).

In thin layer chromatography, chitosan is proposed for separation of nucleic acids. Metal loading on chitosan is an interesting tool for catalytic applications because this amino-polysaccharide has macromolecules rich in functional groups, which can efficiently bind weakly active metals or metal nanoparticles. The metal-loaded materials are then used for catalytic organic reactions. Chitosan plays a very important role in catalysis, not only as a catalyst support but also as a green solvent or green electrolyte (this is referred to as biocatalytic processes using ionic liquids or deep eutectic solvents; some authors consider deep eutectic solvents as a subclass of ionic liquids without their drawbacks such as instability, flammability, and toxicity) (Phan et al., 2012; Silva et al., 2017; Özel and Elibol, 2021; Khajavian et al., 2022; Wang et al., 2022). Chitosan, indeed, generates green solvents, increasing the surface exchange capabilities and utilization of ionic liquids, contributing to the implementation of green chemistry principles by minimizing the amount of required products and utilization of renewable raw materials. These solvents open up new fields of nonaqueous biocatalysis and biotechnology (enzymology).

Shamshina et al. recently showed that solution processing of chitin in ionic liquids is also a promising strategy for handling biopolymers for the biomedical market (Shamshina, 2019; Shamshina and Berton, 2020; Shamshina et al., 2019; Achinivu et al., 2022). The use of ionic liquids and deep eutectic solvents in the dissolution and extraction processes of polysaccharides in general, and of chitin in particular, for a sustainable valorization of the biomass, could be another promising way. In particular, deep eutectic solvents as new types of green solvents can play an important role in the chitin and chitosan industry by solving the classical extraction difficulties encountered on the industrial scale. However, their application is still in its infancy, and application and industrial research is needed to make the process mature and even commercialized. Indeed, the processes are mechanistically complex (e.g., the mechanism of chitin dissolution by green solvents is not clear), and their performance depends on the experimental conditions and the solvents used. The process recycling of deep eutectic solvents is also complicated, requiring many steps (filtration, centrifugation, rotary vaporization), environmentally unfriendly, and energy-consuming.

Chitin and chitosan are proposed for the design of electrochemical device applications (Suginta et al., 2013). Membranes containing these biopolymers are also used for separation of organic liquid mixtures (Salehi et al., 2016), for water purification (Thakur and Voicu, 2016; Spoială et al., 2021), and particularly for biomedicine and biotechnology (Galiano et al., 2018). For example, albumin-blended chitosan membranes have been used in hemodialysis, artificial skin, and also drug targeting. Chitosan is also proposed for CO_2 removal and corrosion protection of aluminum.

Another area of high potential concerns the use of chitosan-based materials for energy storage device (supercapacitors, lithium-ion batteries) applications (Roy et al., 2021; Peter et al., 2021; Vinodh et al., 2021). Currently, electrodes and electrolytes of supercapacitors are being made of non-renewable, toxic, and hazardous materials. They have also other problems such as chemical degradation, corrosion, volatility, and flammability. Chitosan as a low-cost and biodegradable biopolymer can be a suitable alternative for replacing these substances. Chitosan-based electrodes exhibit high specific capacitance, electrical conductivity, and long cycle stability. Developments are expected in the near future in all these fields.

7.11 Environmental applications

Chitin and, in particular, chitosan, as low-cost and natural materials are used for environmental applications, including water and wastewater treatment using adsorption- and flocculation-oriented processes and membrane filtration (polymer-assisted ultrafiltration) due to their ease of use, versatility (used in soluble or insoluble forms), outstanding pollutant-binding capacities, and excellent selectivity (Crini and Badot, 2008; Crini et al.,

2009). Chitosan can also be used to reduce odors, in biological denitrification, as antifouling agent, and as a deshydrating agent (sludge treatment, drilling muds).

7.11.1 Wastewater treatment

Numerous recent studies showed that chitosan-based materials as innovative biosorbents represent a promising alternative to replace conventional adsorbents such as activated carbon and ion-exchange resins to remove dyes, metals, and pesticides present in industrial effluents (Vakili et al., 2014; Liu and Bai, 2014; Boamah et al., 2015; Yong et al., 2015; Azarova et al., 2016; Barbusinski et al., 2016; Kos, 2016; Ahmad et al., 2017; Kyzas et al., 2017; Nechita, 2017; Sudha et al., 2017; Alaba et al., 2018; de Andrade et al., 2018; Desbrières and Guibal, 2018; El Halah et al., 2018; Pakdel and Peighambardoust, 2018; Van Tran et al., 2018; Wei et al., 2018; Crini and Lichtfouse, 2019a; Crini et al., 2019; Sarode et al., 2019; Abhinaya et al., 2021; Begum et al., 2021; da Silva Alves et al., 2021; Dave et al., 2021; Karimi-Maleh et al., 2021; Pal et al., 2021; Peter et al., 2021; Sadiq et al., 2021; Saheed et al., 2021; Sheth et al., 2021; Sirajudheen et al., 2021). Chitosan can also be used to recover precious metals and reduce odors (Kong et al., 2022). Anastopoulos et al. (2017), Boulaiche et al. (2019), and Sarode et al. (2019) reported that chitin is also able to remove toxic metals, but studies are less numerous. As an innovative biocoagulating and bioflocculating agent, chitosan can replace metallic salts and synthetic polyelectrolytes used in water treatment for the removal of both particulate and dissolved substances and for the clarification of drinking water, pools, and spas. Chitosan is particularly efficient for the flocculation of cardboard-mill secondary biological wastewater, but the actual applications in industry remain rather rare, as concurrent flocculating agents are cheaper (Renault et al., 2009). Indeed, even if chitosan as a bioflocculant shows better properties, the conventionally cheaper metallic salts and polyacrylamides are sufficient to fulfill current regulatory frameworks (Crini and Lichtfouse, 2019a; Crini et al., 2019). Nevertheless, industrial-scale applications are beginning to develop with the arrival on the market of environmentally friendly products that compete with conventional flocculating and adsorbent agents. These products offer better performance in terms of pollutant removal with values well below the current regulatory frameworks. There are several products on the market to treat wastewaters, stormwater run-off, dewatering water, and groundwater: for example, Fennofloc™ coagulants for water clarification, ChitoVan™, a combination of chitosan and lactic acid, for biofiltration (chitosan-enhanced sand filtration), and coagulation/flocculation of suspended solids (fine suspended sediment) or turbidity reduction.

7.11.2 Sludge dewatering

Chitosan can be used as a nontoxic and environmentally friendly dewatering and flocculating agent (sludge treatment, drilling muds) to reduce the moisture content of the sludge, which also reduces its volume and facilitates its subsequent disposal (Kaseamchochoung et al., 2006; Zemmouri et al., 2014; Lin et al., 2019; Shi et al., 2019; Zhang et al., 2019b). Other advantages are cited: chitosan is easy to use and without endangering the handler and the treated sludge (e.g., from municipal wastewater) can be reused on agricultural land (chitosan is "easily" to biodegrade, even if there is a doubt on this point; Zhang et al., 2019b). In general, a similar efficiency of biological sludge conditioning is obtained with chitosan compared to synthetic polymers but without their drawbacks such as their potential toxicity and biodegradability, pointed out by many authors (Kaseamchochoung et al., 2006; Renault et al., 2009; Zemmouri et al., 2014; Lin et al., 2019; Shi et al., 2019; Zhang et al., 2019b). The performances obtained in a sludge dewatering process involving chitosan are mainly explained by its long macromolecular chains which carry numerous amine functions and their cationic character, which favor the flocculation phenomenon and then the filtration technique (Renault et al., 2009). The process is more ecological than that of traditional coagulants and flocculants because chitosan is nontoxic, biocompatible, and biodegradable, and there is no risk of releasing toxic compounds. However, chitosan-based sludge dewatering processes are relatively expensive because the doses of the biopolymer to be used must be increased in order to achieve similar or even better performance than traditional formulations. In addition, the low solubility of chitosan in water remains an important drawback to overcome and, depending on the type of sludge, the results may also depend on the characteristics (DD, MW) of the chitosans used (Kaseamchochoung et al., 2006; Renault et al., 2009).

7.11.3 Polymer-assisted ultrafiltration

The polymer-assisted ultrafiltration (PAUF) process combines the selectivity of a complexing or chelating agent with the filtration ability of a membrane acting in synergy. The principle is based on a sequence of steps with complexation of metals by a polymer such as chitosan and a step involving the rejection of the complex formed by means of an ultrafiltration membrane. Complexation-ultrafiltration using chitosan is able to extract, concentrate, and separate metals present in solution (Crini et al., 2017). Due to its low-cost, nontoxicity, high binding capacity for many metals, and recyclability, chitosan is a promising alternative to synthetic polyelectrolytes, such as poly(ethyleneimine) polysulphones, poly(vinylalcohol), and poly(acrylic acid), currently in use in PAUF. However, the processes are essentially at the laboratory stage, despite undeniable progress published in the literature. It now remains to transfer this research to an industrial stage (Sarode et al., 2019).

7.12 Other applications

Due to their strong mechanical properties, chitin nanofibers are proposed as a reinforcement agent to improve the mechanical properties of composite materials or reduce the thermal expansion of resins. Chitin nanocrystals as new surfactants (Pickering stabilizers/emulsions) are attracting increasing attention in coating applications (Muñoz-Núñez et al., 2022). These stabilizers offer improved long-term stability, irreversible adsorption at interfaces, and prevent droplet coalescence. Chitin nanocrystals have other advantages over conventional surfactants, such as biocompatibility, biodegradability, and high chemical reactivity, which broadens their potential uses in the fields of cosmetics, healthcare, and composite film formation. Chitin nanofiber sheets are also proposed as a membrane for chiral separation (Ifuku and Saimoto, 2012). In the field of photography, chitosan's optical characteristics, its ability to form films, its reactions with silver complexes, and its ability as a fixing agent are of particular interest (Dutta et al., 2004). In both the photographic and paper industries, chitosan is also used to provide a better surface and resist moisture. Chitosan-coated sheets enhance morphological, mechanical, optical, and aging properties and provide better print quality. Chitosan-based paint formulations have been proposed as environmentally friendly alternatives to metallic (copper, tin) antifouling paints (Heuser et al., 2009; Pelletier et al., 2009; Banerjee et al., 2011; Heuser and Cárdenas, 2014). Also, for ecological reasons, chitosan is proposed for wood protection as a biofungicide.

One area currently being explored is the use of chitosan in the petroleum industry and associated processes, for example, in oil exploration, transportation, and refining processes (formulation of drilling fluids, borehole treatment fluid for water control), as a constituent in petroleum products (fuels, catalysts, composites, lubricants, bitumen), for wastewater treatment in hydrocarbon industry, and for oil spillage treatment (Fig. 7.11). Potential applications for chitosan include the following: additive in biolubricants, thickener and modifier in greases, emulsifier in bitumen emulsions, catalyst in biodiesel production, constituent in composite catalysts, viscosity reducing agent in pipeline transportation, and cleaner for oil spills and refinery wastewater treatment. Chitosan is also a suitable raw material for upstream chemicals due to its sustainable nature

Fig. 7.11 Possible applications of chitosan in the petroleum industry.

(Negi et al., 2021). For the treatment of wastewater from the hydrocarbon industry, it is used as a coagulant, flocculant or adsorbent, or for the immobilization of bacteria. Its use as a catalyst for biofuel production seems to be a very promising way despite some limitations such as the low crystallinity, the difficulty of handling, and the design of the catalyst.

Other application areas include the cement industry (chitosan is used as a water proofing and water repellent), production of bioplastics and papers for the tobacco industry, nanoimprint lithography, semiconductors, luminescent nanoparticles, photonics (as a stabilizing agent or for the preparation of biomimetic photonic materials), imaging applications (bioimaging and cancer research, radiation therapy, cellular imaging, biosensing, detection), and quantum and carbon dots (Cheba, 2011; Dutta, 2016; Galiano et al., 2018; Marpu and Benton, 2018; Chivere et al., 2020; Dave et al., 2021; Lima et al., 2021; Lizundia et al., 2021).

7.13 Challenges and future trends

The chitin and chitosan market is growing rapidly, with demand for chitin twice as high as its industrial production, mainly concentrated in Asia, which is also the main consumer of these biopolymers. Production is, indeed, limited, but there is no doubt that this sector will grow, as world demand is increasingly strong, particularly in the environmental, medical, and cosmetic fields (Crini and Lichtfouse, 2019a,b). However, despite the numerous chitin- and chitosan-based products (more than 2000) and applications on the market and the impressive amount of published scientific results, there are still several questions and challenges for the scientific community and the industrial world (Bastiaens et al., 2019; Jacob et al., 2020; Kumari and Kishor, 2020; Thomas et al., 2020; Ashok et al., 2021; Berrada, 2021; Fenice and Gorrasi, 2021; Nayak et al., 2021; Wertz and Perez, 2021; Cassie, 2022; Savvaidis, 2022; Kumar and Madihally, 2022). These questions and challenges and current research directions and future trends can be summarized as follows:

– The first important point that is still relevant today is obtaining chitin. Indeed, the major problem encountered in the chitin chemistry is its extraction (isolation) and purification to obtain a product with characteristics similar to the original biopolymer, that is, macromolecular mass (MW) and acetylation degree (DA). These two parameters are fundamental to the chemistry and biology of chitin and its industrial use. The characteristics of chitin also influence that of chitosan (MW,

DD), its chemistry, and its future chemical, biological, and technological properties. The chemistry of these two biopolymers is intimately linked. This is the reason why the preparation of these biopolymers and the control of the experimental conditions always arouse a great interest in the scientific literature (Table 6.3). Although biopolymers with a homogenous and controlled composition and structure are marketed, their production remains a challenge.

- From a fundamental point of view, another important parameter of interest to the scientific community is a better understanding of the distribution of *N*-acetylated units along the long macromolecular chain. Indeed, the distribution and the more or less regular presence of these units can confer to the biopolymers more or less hydrophobic characters, and thus variable gelling and viscosifying properties (auto-associative properties), and modify their solubility. A control of these properties could be useful for targeted applications.
- Another important parameter is the physical structure of chitins and their polymorphism. Chitins exist in three forms of allomorphs due to the arrangement of the macromolecular chains. Indeed, chitins found in nature, for example, animal kingdom, yeast, bacteria and plant kingdom differ not only by their MW and their DA but also in the number of chains unit per cell, the size of the unit, the degree of hydratation, and especially their more or less crystalline state. Thus, it is technically challenging to characterize the structure of these biopolymers in their cellular contexts. For example, a challenge is to study the differences between the polymorphic forms of chitin and chitosan and to confirm the existence of a structural distinction between these forms, from which they are derived. Computational methods (molecular dynamics) and NMR spectroscopy coupled with other techniques (X-ray power diffraction, FT-IR, scanning electron microscopy) can be used to obtain these structural informations. Another interesting point of interest for the scientific community concerns the existence of a dependence between the structure of chitin and the type of fungus.
- The last important parameter is the solubility of chitin. It is only soluble in dimethylformamide-lithium chloride, dimethylacetamide-lithium chloride, hexafluoroisoacetone sesquihydrate, hexafluoroisopropanol, $2H_2O$-saturated methanol, and other organic salts (LiSCN, alkali metal/urea system). However, the associated dissolution mechanisms are not clear. Chitin can also be dissolved in high concentrations of hydrochloric, sulfuric, and phosphoric acid solutions, but these

processes induce both degradation and depolymerization of the chains, with a reduction in molecular weight and an increase in the degree of deacetylation, resulting in poor performance of the prepared chitin. So, finding more environmentally friendly and efficient solvents remains a difficult challenge (Hong et al., 2018; Kadokawa, 2019; Pakizeh et al., 2021). These two aspects, solubility and polymorphism, continue to be of interest to researchers, as shown by the number of articles published each year (Table 6.3).

- Although biological processes have several advantages over conventional chemical processes, mainly due to their ecological properties, and the chemical process is still widely used for the production of chitin on a commercial scale. In addition, the chemical extraction of commercial chitin is always carried out from crustacean shell waste. However, there is debate about the intensification of the fishery in general. The availability of crustacean exoskeletons from the fishing industry is also limited by geography and is highly seasonal (see references in Table 6.3). This availability of waste and the aspects of sustainability and ecological impact are also being discussed. Should alternative sources of supply be promoted? The question is being asked.
- The MW and DA (or DD) values of chitosan biopolymers are highly dependent not only on the conditions for deacetylating the chitin but also on the process used to extract and purify chitin. The most important impact is visible on the MW values obtained. In general, the macromolecular mass of chitins varies from 800,000 to 10^6 g/mol, and their DA are above 75%. Commercial chitosans have masses close to 200,000 g/mol and DA below 25%. In all cases, the molar mass values of chitosans are largely low because during the classical processes for the preparation of chitosans from chitin, more or less pronounced depolymerization reactions of the polymer take place. However, on these parameters, there is no standardized supply chain as each company uses its own production protocol. So, there are different commercial batches of chitin and chitosan with their own characteristics (DD, MW, purity) that can vary and also different prices. Products from different suppliers differ, and even for the same supplier, there is no guarantee of finding a batch with the same characteristics from 1 year to the next. It is, therefore, difficult for the same product to have the same biological properties.
- The purity of commercial chitosan is not only a key issue in several application areas but it also is a matter of debate. There are no general rules or consensus on the type of purity to be used. In the

pharmaceutical and medical fields, chitosan is used as an excipient, drug carrier, or therapeutic agent with high levels of purity but which may vary depending on the intended application. However, it is approved, for example by the US Food and Drug Administration, as a biomaterial but not for pharmaceutical use, mainly because of concerns about its purity and also its immunogenicity. There are also warnings and (medical) contraindications for people allergic to shellfish (as in the case of dietary supplements which may use biopolymers of different purity). This is why more and more studies are recommending the use of other sources of chitin that allow higher purities to be obtained without this allergenic risk.

- For the viability of the chitin and chitosan industry, the economic criteria is fundamental. Different manufacturers market different qualities/grades of chitin and chitosan, whose price depends on the degree of purity, which in turn depends on the intended applications. However, the prices remains high. Indeed, although products and formulations containing chitin, chitosan, and their derivatives present advantages, ecological (renewable resources, less impact on the environment), sanitary (healthy products), and societal (local employment) to which the public is more and more aware; their cost remains higher than that of petroleum-based polymers with similar properties and performances. The high prices of some chitins and chitosans are due to the difficulties encountered during the extraction and purification of biopolymers, regardless of their origin and the reliability of available sources which remains a problem. This unfortunately has an impact on its large-scale applications. Chitin is difficult to purify without deeply altering its structure due to the use of the aggressive chemical method. The difficulty of controlling product quality (DD, MW, purity) is often cited as the main drawback of chemical treatments. The extraction of chitin from insects has been proposed as an alternative to overcome these difficulties. However, chitin purification from insects is also known to be very difficult and an area of focus for researchers. In addition, even if local initiatives exist in Asia, Europe, and North America, the price of this type of chitin/chitosan is also high. Other alternatives such as the use of biotechnology (fermentation), green chemistry (green solvents, deep eutectic solvents, and ionic liquids) and microwave technology are yet to be proven (Hong et al., 2018; Kadokawa, 2019; Ma et al., 2022). There is a need to improve the extraction and purification of chitin to standardize the production and to decrease the prices.

- From an environmental point of view, chemical methods are also problematic as they have several other disadvantages, such as the use of hazardous chemicals, severe reaction conditions, high water and energy consumption, and the equally high production of by-products and waste. Chitin production using fermentation processes has been proposed as a green and environmentally friendly alternative (Table 6.3). Nevertheless, biotechnology treatments give lower yield and product quality (although on this aspect, there are contradictory results), and are still at an early stage of development. They have also a higher cost than chemical processes. Nevertheless, advances are expected in this area.
- Chitosans are still little used in Europe due to lack of supply and local production. In fact, all production comes from Asia (Japan, China), and the raw material used is essentially crustaceans, mainly shrimp. The collection of shrimp shells, which are destined for frozen products, is carried out in several places without any control of traceability (for example on the use of antibiotics or other substances in shrimp farms). In order to guarantee local production and better product traceability, initiatives to set up industrial sites are beginning to emerge in Europe, particularly in France, Germany, Norway, and Iceland to produce high-quality chitin and chitosans intended mainly for the pharmaceutical, medical, and cosmetic sectors. The raw materials used are not only shellfish waste but also fungal biomass and insects (fly larvae). The European market should then develop further.
- As already mentioned, the production of chitin and chitosan from fungi and insects not yet been scaled up to the industrial scale, even in Asia (Ma et al., 2022). The main challenge is the optimization and scale-up of its purification process and the development of suitable procedures to improve characterization (Feng et al., 2020) and properties. For fungal chitin and for chitin from insect biomass, information is lacking on the quantitative evaluation of the efficiency of extraction, purification, deacetylation, and bleaching, and on the degree of purification of the products.
- The use of deep eutectic solvents in the dissolution and extraction processes of chitin for a sustainable valorization of the biomass could be another promising way. Two of the main objectives of green chemistry are to minimize the use of hazardous substances in chemical production processes and to replace conventional solvents with renewable sources. As new generation of green solvents, deep eutectic solvents are the most promising alternatives for this purpose due to their nontoxicity, stability,

easy of use, renewable, reusable, and biodegradable and biocompatible character. Deep eutectic solvents can play an important role in the chitin and chitosan industry by solving the classical extraction difficulties encountered on the industrial scale (Kadokawa, 2019; Özel and Elibol, 2021; Khajavian et al., 2022; Wang et al., 2022). Many laboratory-scale results have been published, showing, for example, that chitin extracted by a process using deep eutectic solvents has high purity and crystallinity compared to conventional chemical methods. However, the results depend on the type of green solvents used. Another example concerns the synthesis of chitosan films for the food packaging industry prepared by this technology, which shows improved properties in terms of flexibility and resistance to breakage compared to films prepared by traditional solvents. The use of green solvents for the dissolution, extraction, preparation, and modification of chitin-related materials has made great progress in recent years, and the many published results are encouraging and promising. However, the strategy of using deep eutectic solvents to extract chitin is still in its infancy, and the challenge now is to convince the industry to use these green solvents.

- In general, to demonstrate the interest of using chitin and chitosan biopolymers and their derivatives in a given field of application, several biological properties are often associated and put forward. For example, in the medical field, the beneficial effects of these substances cited are as follows: antimicrobial activity, antioxidant activity, anticoagulant, antiinflammatory, antiobesity, antidiabetic, antihyperlipidemia, antihypertensive, anticarcinogenic, and immune-modularity. However, whatever the field of application, the main biological activity of chitin and chitosan biopolymers most often highlighted in the numerous published studies is their antimicrobial activity (Verlee et al., 2017). This biological activity, like the others, is variable. Indeed, it is associated with the physicochemical characteristics of the raw and modified biopolymer used. The low solubility in water and the lack of defined molecular weight and purity are other major problems. There are also other important parameters that modulate the antimicrobial activity, for example, in medical applications, experimental conditions (pH, dissolved oxygen, time of contact) used, sensitivity to chitosan (target microbe, microbial physiological state), and behavior (microbial agglutination, sedimentation). Despite much successful work (Table 6.3), it is often difficult to compare the results of different studies using biopolymers from different sources and with different characteristics and to clearly define mode of actions. There is still little literature containing a comprehensive review of the comparison between all these data.

- Unlike chitosan, chitin has a rather limited range of exploitable applications, but a great deal of research is being conducted to extend its applications. Indeed, despite solubility problems and a lower versatility, chitin allows the preparation not only of partially hydrolyzed products (chitooligosaccharides) and carboxymethylchitin but also of other materials that can find applications in various fields (Yang et al., 2019). Significant examples are chitin nanofibers for tissue engineering, wound dressing, anticancer therapy, drug delivery, obesity management and cosmetic applications; chitin nanoparticles used in multiple applications due to their biocompatible nature, biodegradable and nontoxic nature; chitin nanocomposites/nanobiocomposites as multiphase materials; nanocrystals for the development of bioactive films of for biomedical applications; and chitin nanoscaffolds/nanowhishkers used as biomedical composites for drug/gene delivery or as nanoscaffolds in tissue engineering (Muñoz-Núñez et al., 2022).
- There is no doubt that the applications of chitin and chitosan and their derivatives in agriculture, as effective, safe, and environmentally friendly tools to increase seed germination, to improve crop productivity and conservation of agronomic products, and to induce defense mechanism, seem very promising (Table 6.3). For example, chitosan induces a defense mechanism in tomatoes, cucumbers, and peppers, effectively controlling fungal diseases (e.g., controlling pepper anthracnose against *Colletotrichum capsici* and suppressing wilt disease against *Fusarium oxysporum*) and nematodes, and even insects. However, most studies are still conducted in the laboratory or greenhouse and field trials are needed. In addition, further research is needed to explain the mechanisms involving chitin and chitosan and their interactions with plants and pathogens and to assess that their derivatives are not toxic to living organisms, while also addressing their fate in soil, water, and air.
- The wine sector is certainly a growing industry, but it is torn between tradition and environmental challenges. Vines are susceptible to fungal diseases, in particular downy mildew. The fight against the latter requires fungicide treatments, the use of which must be reduced for environmental and health reasons. This is why this industrial sector is strongly interested in the use of chitosan. However, in the viticultural field, chitosan is known to be very effective against downy mildew in greenhouses but much less so in the vineyard. Research is conducted

to understand the reasons for this loss of efficiency and to find formulations to remedy it. Another challenge concerns spoilage yeasts in red wines. However, their elimination or limitation is not an easy task for winemakers. Chitosan could be a solution although its effectiveness on the spoilage yeast *Brettanomyces bruxellensis* is yet uncertain (Petrova et al., 2016).

- For the beverage industry, the challenges are to improve processes (clarification) and promote the use of natural resources such as biopolymers in beverages, for example, to increase their quality and shelf life, to offer healthy and functional products to meet consumer demand, and to develop active packaging.
- Chitosan is a highly sought after in the food industry, and the market is growing strongly, especially with applications as a food supplement and functional food (neutraceuticals). The challenges for this sector are mainly in terms of food safety, public health, and new regulatory standards with specific points in terms of quality, chemical and microbiological contamination, the traceability and sustainability of food and beverages, and the issue of single-use packaging. The use of chitosan can help answer many of these questions positively.
- In recent years, the concept of "borderline products" has emerged, refering to neutraceuticals, cosmetoceuticals, and cosmetotextiles containing chitin and chitosan with health claims related to the foods, cosmetics and textiles in these formulations and products. However, this concept has no real status and no legal framework. These products are certainly at the forefront of food, cosmetic, textile, or medical innovation, but regulations must evolve in order to market these products not only on fashionable benefits but also on real scientific data.
- The application of chitosan in the biomedical sector remain limited, mainly due to the extreme difficulty to access sufficient purity and source reliability of the biopolymer. In addition, most of the materials proposed for biomedical applications are at an experimental stage. Nevertheless, the progress made in the manufacture of emergency hemostatic dressings, cardiovascular prostheses, orthopedic prostheses, and nerve regeneration treatments suggests that this industrial sector will develop strongly in the coming years (Table 6.3).
- The use of genetic materials is limited due to rapid degradation by nuclease, large size, poor cellular uptake, high anionic charge density, and also nonspecificity. To overcome these problems, cationic chitosan as nonviral vectors was proposed in gene therapy. This is a promising

delivery biomaterial due to its facility to form complex biodegradability and biocompatibility, although its transfection efficiency and cell specificity are low (Badawy and Rabea, 2017). Most of the results were obtained from experiments in vitro, and further research is needed in vivo.

- The application in the packaging sector has not yet reached the stage of industrialization because of several obstacles: the price of chitosan (still high) and biopolymer-based materials (the sector is essentially looking for low-cost products for packaging), the laboratory results are convincing, but the cases of industrial transfer are very rare, the lack of data on a harmfulness in case of ingestion, the lack of industrial and political will and the lack of certifications for food uses, even if the regulations are starting to evolve (reinforcement of sanitary norms, tendency to eco-design, and willingness to decrease the impact on the environment). In addition, the thermal, mechanical, and water barrier properties of chitosan-based films and coatings are still problematic and need to be improved in order to be transferred to the industrial scale at low cost to compete with traditional petroleum-based plastics. Blends, composites, reinforcements, and bilayers with other polymers have recently proposed to improve the performance of chitosan-based films, pointing to further developments.
- Nanotechnology is an interesting approach in the formulation of active ingredients for pharmaceutical, medical, agricultural, or packaging applications and as functional materials for environmental applications (Bose et al., 2021). For example, nanochitosans are recent and promising materials in agriculture for the controlled release of agrochemicals such as pesticides, nutrients, fertilizers, and plant hormones (Chouhan and Mandal, 2021). This type of material is, nevertheless, expensive to customize into stable and desired nanoforms. The results showed that nanochitosans enhanced antipathogenic and plant growth promoting activity in comparison to chitosan. However, their bioactivity is yet poorly known and their potential toxicity and ecotoxicity. Another important approach is genetic engineering using nanochitosan for sustainable increase of crop productivity (biotech agriculture) and biosynthesis of useful molecules (biomedicine). Nanochitosans also target antiaging beauty products and the manufacture of creams, lipsticks and lotions, or the manufacture of oral care products. Outlets are also expected in hair care. Chitosan-based nanoparticles containing imaging agents are also proposed for radiopharmaceutical applications and magnetic

resonance imaging: these derivatives seem promising. Nanoparticles and nanochitosans are proposed for the treatment of wastewaters to remove total suspended solids, biochemical oxygen demand, chemical oxygen demand, and pollutants such as metals, dyes, and emerging contaminants. In the field of water, the encapsulation of enzymes by chitinous matrices offers new perspectives for the elimination of pollutants. Similarly, the incorporation of chitosan in the composition of filtration membranes is another promising mode of action. Many research studies on nanochitines are also conducted in the agricultural and biomedical fields (Table 6.3).

- The use of chitosan in petroleum industries still remains a major challenge mainly due to its insolubility in most organic/inorganic solvents. Another serious obstacle is directly related to its natural origin and the fact that it is not a single polymer with a defined structure, but a family of macomolecules with differences in their composition, size, and monomer distribution. These properties have a fundamental effect on the technological performance of the polymer.
- To combat the corona virus pandemic, researchers and companies around the world are working to develop vaccines and drugs. As recently published, chitosan and chitin have potential in vaccine development, aerosol production, and functionalization of protective equipment such as respirators (Ejeromedoghene et al., 2020; Tatlow et al., 2020; Meng et al., 2021; Safarzadeh et al., 2021). The two biopolymers have direct antiviral activities and exert immune-boosting effects. Chitosan could be used as a drug delivery vehicle for the treatment of COVID-19 infection (Safarzadeh et al., 2021). The aerosol formulation of chitosan nanoparticles (Novochizol™) has been proposed to treat COVID-19 patients. These nanoparticles are fully biocompatible, strongly adhere to lung epithelial tissues, and provide sustained release without systemic distribution. Chitin also has shown promising results against viral infections. Developments are expected in the coming months (Jaber et al., 2022; Mohammed and Shaaban, 2022; Zhuo et al., 2022).

Conclusions

This book offers a detailed history of chitin and chitosan and highlights the main historical landmarks on the state of their scientific knowledge since the discovery of chitin 210 years ago. This historical review does not pretend to be exhaustive, but it aims to show the progressive evolution of knowledge on these two biopolymers in a historical context illustrated from selected works.

For this purpose, I have divided the history of chitin into five quite distinct periods, each period being illustrated by relevant references that I have chosen to highlight. The first period, from 1799 to 1894, covers the discovery of chitin in fungi by Braconnot in 1811 who called it *fongine* (fungine). Braconnot also proved that *fongine* was a nitrogenous substance and pointed out that its nitrogen content was less than that of protein. However, the first mention of the presence of calcified chitin in invertebrates is due to Hatchett in 1799 who studied the decalcification of shells of crabs, lobsters, prawns, and crayfish by mineral acids. Nevertheless, there is no indication that he knew what he had prepared as a substance, neither in his memoirs nor in the bibliography of the time. The name *chitine* (chitin) was introduced by Odier in 1823 for a similar substance found in insects. Odier proved that *chitine* was the basic material of the exoskeletons of all insects and arachnids and suggested the presence of chitin and its wide distribution on the cell walls of fungi, exoskeletons of crustaceans, radulas of mollusks, and beaks of cephalopods and other marine invertebrates. However, Odier thought that the frameworks of insects and of plants comprised the same substance, cellulose. In addition, he erroneously claimed that insect chitin did not contain nitrogen. This assertion was contradicted 1 year later, in 1824, by the work of Children who revealed the nitrogenous nature of chitin. In 1843, Lassaigne proposed to distinguish chitin produced by arthropods from cellulose produced by plants, as two different substances. The history of the main derivative of chitin, first called *chitine modifiée* (modified chitin) and then chitosan, began in 1859 with the work of Rouget who showed the soluble character of this compound in acidic aqueous solution. Later, in 1894, the name chitosan was introduced by Hoppe-Seyler. The interconversion of chitin and chitosan depending on the conditions used in the hydrolysis reaction was also demonstrated for the first time by Hoppe-Seyler. From 1894 to 1930 came a period of confusion due to the controversy about whether or not chitin was identical with cellulose, although as early as 1843 both Lassaigne and Payen had noted that chitin contained nitrogen. The second important derivative of chitin, namely, glucosamine was discovered in 1876 by Ledderhose who also identified glucosamine as the basic unit of chitin. This was demonstrated 10 years later with the work of Tiemann, Schmiedeberg, Winterstein, and Gilson. In 1894, Fischer and Tiemann suggested for the first time that glucosamin was a dextrorotatory amino sugar to which they recognized the constitution of a α-amino-glucose. One year later, Araki was the first to demonstrate that the acid hydrolysis of chitin was carried out in two successive steps: chitin giving chitosan and then glucosamine and that in the chitin structure, an acetyl group was attached to the amine. The same year, Gilson demonstrated the chemical similarity of fungal and animal chitin. In 1899, Lobry de Bruyn also demonstrated that glucosamine was the basic unit of chitosan. The first structure of the acetylglucosamine obtained during the hydrolysis of chitin and chitosan was proposed by Fränkel in 1901. Two years later, Fischer demonstrated that glucosamine was an α-amino-derivative with D-glucose configuration. This was demonstrated by Irvine at the beginning of the 1910s. In 1929, Karrer and Hofmann described the first enzyme that efficiently degraded chitin, obtained from the crude extracts of a common snail, *Helix pomatia*. The third period, from 1930 to 1950, was marked by the elucidation of the structure of chitin, chitosan, and glucosamine by numerous patents, by the first practical applications, and by significant advances on the chemistry and biochemistry of chitin. During this period, (i) the β-glycosidic linkage was confirmed by Bergmann and Zechmeister; (ii) X-ray analysis became the most reliable physical method for differentiating chitin from cellulose; (iii) the work of Meyer indicated that chitin closely resembled cellulose in many of its properties and in structure, and both consisted of long primary valence chains of glucose residues; and (iv) unequivocal evidence for the structure and configuration of glucosamine was finally demonstrated by Haworth. From 1950 to 1970 came a short period of doubt although numerous fundamental studies and results on chitin and chitosan have been published, which are still valid today. Finally, the period of real utilization has been in progress since 1970 and has seen chitin and its main derivative, chitosan, find numerous industrial applications. Indeed, the early 1970s marked a new stage in their development with the first applications in the food, cosmetics, and pharmaceutical industries, these three markets still being the main outlets for chitosan.

Chitin is the second most abundant substance in nature after cellulose. The main source of production is shellfish waste from the fishing industry.

Chitin is renewable, biodegradable, and biocompatible. More than three-quarters of the world's chitin production is used to produce chitosan, D-glucosamine, and oligosaccharides, with chitosan being the most important commercial product due to its high percentage of nitrogen, cationic nature, attractive polyelectrolyte properties at acidic pH, and numerous biological properties that make it useful in a variety of industrial applications in the form of solutions, particles, sponges, foams, films, membranes, and fibers. All industrial sectors are concerned by the use of chitosan, those concerning chitin being less numerous because of its strong stability, its insolubility in water, and a less good technological versatility. Today, there are more than 2000 applications of chitin, chitosan, and their numerous derivatives, with practical applications in all industrial fields and in our daily lives. The industrial fields are food industry, beverage industry, cosmetology, hygiene and personal care, pharmacy, medicine (wound healing), agriculture, textile and paper industries, chemistry, biotechnology, photography, and wastewater treatment. In recent years, chitosan has also received much attention in chromatography, catalysis, dentistry, dermatology, ophthalmology, veterinary medicine, biomedicine, and packaging industry. Nutraceuticals and cosmeceuticals are actually growing markets. The chitin and chitosan market is also expected to grow rapidly due to its increasing consumption in the beverage industry, water treatment, and medicine. In the latter field, apart from wound healing and tissue engineering, medical applications on the market are still very limited but promising. Indeed, numerous studies on the use of chitosan in bio-imaging, genetic engineering, therapeutic engineering, cancerology (radiation therapy, chemotherapy), and virology suggest future developments in these fields. Agriculture should also be a next market in the development of chitin and chitosan applications. Other new developments on the use of these biopolymers in sectors such as catalysis, green chemistry, sensor applications, packaging materials, and cosmetotextiles are also expected in the near future.

Despite a considerable amount of academic work and an equally large number of industrial products on the market, chitin and chitosan continue to attract the interest of the scientific community and industry and offer new horizons. Due to their numerous intrinsic chemical and biological properties and a wide range of possible modifications and forms, these biopolymers are still considered "novel" materials with high potential and innovative formulations, and products containing chitin and chitosan and their derivatives continue to be reported. Chitin and chitosan never cease to amaze scientists and industrialists, and there is no doubt that other discoveries and applications are still to come, as for example in the field of virology with the studies on COVID-19. Indeed, recent studies on the use of chitosan as a protective agent against COVID-19 show that these natural substances have not yet delivered all their secrets!

References

Abderhalden, E., Heyns, K., 1933. Nachweis vom chitin in flügelresten von Coleopteren des oberen mittleocäns. Biochem. Z. 259, 320–321.
Abhinaya, M., Parthiban, R., Senthil Kumar, P., Vo, D.V.N., 2021. A review on cleaner strategies for extraction of chitosan and its application in toxic pollutant removal. Environ. Res. 196, 110996.
Abidin, N.A.Z., Kormin, F., Abidin, N.A.Z., Anuar, N.A.F.M., Bakar, M.F.A., 2020. The potential of insects as alternative sources of chitin: an overview on the chemical method of extraction from various sources. Int. J. Mol. Sci. 21, 4978.
Abo Elsoud, M.M., El Kady, E.M., 2019. Current trends in fungal biosynthesis of chitin and chitosan. Bull. Natl. Res. Cent. 43, 59.
Achinivu, E.C., Shamshina, J.L., Rogers, R.D., 2022. Chitin extracted from various biomass sources: it's not the same. Fluid Phase Equilib. 552, 113286.
Acosta, N., Jiménez, C., Borau, V., Heras, A., 1993. Extraction and characterization of chitin from crustaceans. Biomass Bioenergy 5, 145–153.
Adiletta, G., Di Matteo, M., Petriccione, M., 2021. Multifunctional role of chitosan edible coatings on antioxidant systems in fruit crops: a review. Int. J. Mol. Sci. 22, 2633.
Ahmad, M., Manzoor, K., Ikram, S., 2017. Versatile nature of hetero-chitosan based derivatives as biodegradable adsorbent for heavy metal ions: a review. Int. J. Biol. Macromol. 105, 190–203.
Ahmed, S., Ikram, S. (Eds.), 2017. Chitosan—Derivatives, Composites and Applications. Scrivener Publishing LLC. Wiley, Beverly, ISBN: 978-1-119-36350-7. 516 p.
Ahmed, S., Annu, A.A., Sheikh, J., 2018. A review on chitosan centred scaffolds and their applications in tissue engineering. Int. J. Biol. Macromol. 116, 849–862.
Akbar, A., Shakeel, A., 2018. A review on chitosan and its nanocomposites in drug delivery. Int. J. Biol. Macromol. 109, 273–286.
Akbari-Alavijeh, S., Shaddel, R., Jafari, S.M., 2020. Encapsulation of food bioactives and nutraceuticals by various chitosan-based nanocarriers. Food Hydrocoll. 105, 105774.
Alaba, P.A., Oladoja, N.A., Sani, Y.M., Ayodele, O.B., Mohammed, I.Y., Olupinla, S.F., Daud, W.M.W., 2018. Insight into wastewater decontamination using polymeric adsorbents. J. Environ. Chem. Eng. 6, 1651–1672.
Al Aboud, A.K., 2018. Eponyms in dermatology literature linked to fibromatoses. Our Dermatol. Online 9, 108–109.
Alexandrov, V.Y., 1934. The permeability of the chitin of some dipterous larvae and a method of investigating it. J. Biol. Mosc. 3, 490–507.
Alexandrov, V.Y., 1935. Permeability of chitin in some dipterous larvae and the method of its study. Acta Zool. 16, 1–19.
Alishahi, A., Aïder, M., 2012. Applications of chitosan in the seafood industry and aquaculture: a review. Food Bioprocess Technol. 5, 817–830.
Aljohani, W., Ullah, M.W., Zhang, X.L., Yang, G., 2018. Bioprinting and its applications in tissue engineering and regenerative medicine. Int. J. Biol. Macromol. 107, 261–275.
Alonso, M.J., Sánchez, A., 2003. The potential of chitosan in ocular drug delivery. J. Pharm. Pharmacol. 55, 1451–1463.
Alsberg, C.L., Hedblom, C.A., 1909. Soluble chitin from *Limulus polyphemus* and its peculiar osmotic behavior. J. Biol. Chem. 6, 483–497.
Amber Jennings, J., Bumgardner, J.D. (Eds.), 2017. Chitosan Based Biomaterials. Fundamentals. Woodhead Publishing Series in Biomaterials, Number 122, vol. 1. Elsevier, Amsterdam, ISBN: 978-0-08-100230-8. 342 p.

Ambronn, H., 1890. Cellulose-reaction bei arthropoden und mollusken. Mittheil. Zool. Stat. Neapel IX, 475–478.
Amirthalingam, S., Rajendran, A.K., Mani, P., Rangasamy, J., 2021. Perspectives and challenges of using chitosan in various biological applications. Adv. Polym. Sci. 287, 1–22.
Anastopoulos, I., Bhatnagar, A., Bikiaris, D.N., Kyzas, G.Z., 2017. Chitin adsorbents for toxic metals: a review. Int. J. Mol. Sci. 18, 114.
Anderson, B.G., Brown, L.A., 1930. Chitin secretion in *Daphnia magna*. Physiol. Zool. 3, 485–493.
Anderson, T.F., Richards Jr., A.G., 1942. An electron microscope study of some structural colours of insects. J. Appl. Phys. 13, 748–758.
Anitha, A., Sowmya, S., Sudheesh Kumar, P.T., Deepthi, S., Chennazhi, K.P., Ehrlich, H., Tsurkan, M., Jayakumar, R., 2014. Chitosan—a versatile semi-synthetic polymer in biomedical applications. Prog. Polym. Sci. 39, 1644–1667.
Araki, T., 1895. Ueber das chitosan. Z. Physiol. Chem. 20, 498–510.
Aranaz, I., Acosta, N., Civera, C., Elorza, B., Mingo, J., Castro, C., de los Llanos Gandía, M., Caballero, A.H., 2018. Cosmetics and cosmeceutical applications of chitin, chitosan and their derivatives. Polymers 10, 213.
Araujo, V.H.S., Chaves de Souza, M.P., Carvalho, G.C., Duarte, J.L., Chorilli, M., 2021. Chitosan-based systems aimed at local application for vaginal infections. Carbohydr. Polym. 261, 117919.
Ardean, C., Davidescu, C.M., Nemeş, N.S., Negrea, A., Ciopec, M., Duteanu, N., Negrea, P., Duda-Seiman, D., Musta, V., 2021. Factors influencing the antibacterial activity of chitosan and chitosan modified by functionalization. Int. J. Mol. Sci. 22, 7448.
Argüelles-Monal, W.M., Lizardi-Mendoza, J., Fernandez-Quiroz, D., Recillas-Mota, M.T., Montiel-Herrera, M., 2018. Chitosan derivatives: introducing new functionalities with a controlled molecular architecture for innovative materials. Polymers 10, 342.
Armbrecht, W., 1919. Chitose. Biochem. Z. 95, 108–123.
Arnold, L.B., 1939. Fibrous Product. US Patent 2,142,986.
Aronson, J.M., 1965. The cell wall. In: Ainsworth, G.C., Sussman, A.S. (Eds.), The Fungi. An Advanced Treatise. Volume I: The Fungal Cell. Academic Press, New York, pp. 49–76 (Chapter 3).
Ashok, N., Pradeep, A., Jayakumar, R., 2021. Synthesis-structure relationship of chitosan based hydrogels. Adv. Polym. Sci. 287, 105–1129.
Astrup, T., Galsmar, I., Volkert, M., 1944. Polysaccharide sulfuric acid esters as anticoagulants. Acta Physiol. Scand. 8, 215–226.
Atkins, A., 1853. Memoir of J.G. Children, ESQ. Some Unpublished poetry by his father and himself, John Bowyer Nichols and Sons, Westminster, pp. 1–313.
Attwood, M.M., Zola, H., 1967. The association between chitin and protein in some chitinous tissues. Comp. Biochem. Physiol. 20, 993–998.
Azarova, Y.A., Pestov, A.V., Bratskaya, S.Z., 2016. Application of chitosan and its derivatives for solid-phase extraction of metal and metalloid ions: a mini-review. Cellulose 23, 2273–2289.
Badawy, M.E.I., Rabea, E.I., 2016. Chitosan and its derivatives as active ingredients against plant pests and diseases. In: Chitosan in the Preservation of Agricultural Commodities, pp. 179–216 (Chapter 7).
Badawy, M.E.I., Rabea, E.I., 2017. Chitosan and its modifications as biologically active compounds in different applications. In: Masuell, M., Renard, D. (Eds.), Advances in Physicochemical Properties of Biopolymers. Bentham e-Books, Bentham Science Publishers, Sharjah, pp. 1–108 (Chapter 1).
Badwan, A.A., Rashid, I., Omari, M.M., Daras, F.H., 2015. Chitin and chitosan as direct compression excipients in pharmaceutical applications. Mar. Drugs 13, 1519–1547.

Bandara, S., Du, H., Carson, L., Bradford, D., Kommalapati, R., 2020. Agricultural and biomedical applications of chitosan-based nanomaterials. Nanomaterials 10, 1903.

Banerjee, I., Pangule, R.C., Kane, R.S., 2011. Antifouling coatings: recent developments in the design of surfaces that prevent fouling by proteins, bacteria, and marine organisms. Adv. Mater. 23, 690–718.

Barbusinski, K., Salwiczek, S., Paszewska, A., 2016. The use of chitosan for removing selected pollutants from water and wastewater - Short review. Archit. Civil Eng. Environ. 9, 107–115.

Barker, S.A., Foster, A.B., Stacey, M., Webber, J.M., 1957. Isolation of a homologous series of oligosaccharides from chitin. Chem. Ind. 7, 208–209.

Barker, S.A., Foster, A.B., Stacey, M., Webber, J.M., 1958. Properties of oligosaccharides obtained by controlled fragmentation of chitin. J. Chem. Soc., 2218–2227.

Barsøe, O.C., Selsø, S., 1946. Anticoagulant action of synthetic polysaccharide polysulfuric acids. Acta Pharmacol. Toxicol. 2, 367–370.

Bastiaens, L., Soetemans, L., D'Hondt, E., Elst, K., 2019. Sources of chitin and chitosan and their isolation. In: Van den Broek, L.A.M., Boeriu, C.G. (Eds.), Chitin and Chitosan: Properties and Applications. John Wiley & Sons, Ltd, pp. 1–34 (Chapter 1).

Baumann, E., Kossel, A., 1895. Felix Hoppe-Seyler. Ber. Dtsch. Chem. Ges. 28, 1147–1192.

Baumann, E., Kossel, A., 1895. Zur erinnerung an Felix Hoppe-Seyler. Z. Physiol. Chem. 21, I–LXI.

Baur, A., 1860. Über den bau der chitinsehnen am kiefer der flusskrebse und ihr verhalten beim schalenwechsel. Arch. Anat. Physiol. Wiss. Med., 113–144.

Bautista-Baños, S., Romanazzi, G., Jiménez-Aparicio, A. (Eds.), 2016. Chitosan in the Preservation of Agricultural Commodities. Academic Press, Elsevier Inc., Oxford. 384 p.

Becking, L.B., Chamberlin, J.C., 1925. A note on the refractive index of chitin. Proc. Soc. Exp. Biol. Med. 2, 256.

Beer, F.J., 1977. Le professeur Paul Karrer et la liberté de la science. In: Société Française d'Histoire de la Médicine, Séance du 23 Avril 1977, pp. 221–231.

Begum, S., Yuhana, N.Y., Md Saleh, M., Kamarudin, N.H.N., Sulong, A.B., 2021. Review of chitosan composite as a heavy metal adsorbent: material preparation and properties. Carbohydr. Polym. 259, 117613.

Bell, D.J., 1949. Carbohydrate chemistry. Annu. Rev. Biochem. 18, 87–96.

BeMiller, J.N., 1965. Chitin. In: Whistler, R.L. (Ed.), Methods Carbohydrate Chemistry. vol. V, pp. 103–105.

Benecke, W., 1905. Über *Bacillus chitinovorus*, einen chitin zerstzenden spaltpilz. Bot. Z. 63, 227–242.

Berezina, N., 2016. Production and application of chitin. In: Luque, R., Xu, C.P. (Eds.), Biomaterials. Biological Production of Fuels and Chemicals. Physical Sciences Reviews, De Gruyter, Berlin/Boston, pp. 1–8 (Chapter 3).

Berger, L.R., Reynolds, D.M., 1958. The chitinases system of a strain of *Streptomyces Griseus*. Biochim. Biophys. Acta 29, 522–534.

Bergmann, M., Zervas, L., 1931. Synthesen mit glucosamin. Ber. Dtsch. Chem. Ges. 64B, 975–980.

Bergmann, M., Zervas, L., Silberkweit, E., 1931a. Über die biose des chitins. Naturwissenschaften 19, 20.

Bergmann, M., Zervas, L., Silberkweit, E., 1931b. Über glucosaminsäure und ihre desaminierung. Ber. Dtsch. Chem. Ges. 64, 2428–2436.

Bergmann, M., Zervas, L., Silberkweit, E., 1931c. Über chitin und chitobiose. Ber. Dtsch. Chem. Ges. 64, 2436–2440.

Bergmann, M., Rinke, H., Schleich, H., 1934. Über dipeptide von epimeren glucosaminsäuren und ihr verhalten gegen dipeptidase. Konfiguration des d-glucosamins. Z. Physiol. Chem. 224, 33–39.

Bergström, S., 1935. Über die Wirkungsgruppe des Heparins. Naturwissenschaften 23, 706.

Bergström, S., 1936. Polysaccharide sulphuric acids with heparin action. Z. Physiol. Chem. 238, 163–168.

Berlese, A., 1929. Gli insetti: Embriologia e morfologia. Società Editrice Libraria, Milano. 1004 p.

Bernardi, F., Zadinelo, I.W., Alves, H.J., Meurer, F., dos Santos, L.D., 2018. Chitins and chitosans for the removal of total ammonic of aquaculture effluents. Aquaculture 483, 203–212.

Bernfeld, P., 1963. Biogenesis of Natural Compounds. Pergamon Press, Oxford. 1224 p.

Bernkop-Schnürch, A., Dünnhaupt, S., 2012. Chitosan-based drug delivery. Eur. J. Pharm. Biopharm. 81, 463–469.

Berrada, M., 2021. Chitin and Chitosan. Physicochemical Properties and Industrial Applications. IntechOpen, ISBN: 9781789844252. 290 p.

Berthelot, M., 1859. Sur la transformation en sucre de la chitine et de la tunicine, principes immédiats contenus dans les tissus des animaux invertébrés. Ann. Chim. Phys. 56, 149–156.

Betchem, G., Johnson, N.A.N., Wang, Y., 2019. The application of chitosan in the control of post-harvest-diseases: a review. J. Plant Dis. Prot. 126, 495–507.

Bettany, G.T., 1887. Children John George. In: Stephen, L. (Ed.), Dictionary of National Biography. Smith, Elder & Co., London, pp. 249–250.

Biedermann, W., 1907. Über die struktur des chitins bei insekten und crustaceeen. Anat. Anz. 21, 485–490.

Biedermann, W., 1914. In: Winterstein, W. (Ed.), Handbuch der Vergleichenden Physiologie. vol. III, part 1. Fisher, Jena, pp. 319–1188.

Bierry, H., Gouzon, B., Magnan, C., 1939. N-acétylglucosamine et sucre protéidique. C. R. Soc. Biol. 130, 411–413.

Blackwell, J., 1969. Structure of β-chitin or parallel chain systems of poly-β-(1→4)-N-acetyl-D-glucosamine. Biopolymers 7, 281–298.

Blackwell, J., 1988. Physical methods for the determination of chitin structure and conformation. Methods Enzymol. 161, 435–442.

Blackwell, J., Parker, K.D., Rudall, K.M., 1965. Chitin in pogonophore tubes. J. Mar. Biol. Assoc. U. K. 45, 659–661.

Blackwell, J., Parker, K.D., Rudall, K.M., 1967. Chitin fibres of the diatoms *Thalassiosira fluviatilis* and *Cyclotella cryptica*. J. Mol. Biol. 28, 383–385.

Blackwell, J., Gardner, K.H., Kolpak, F.J., Minke, R., Claffey, W.B., 1980. Refinement of cellulose and chitin structures. In: French, A.D., Gardner, K.C.H. (Eds.), Fiber Diffraction Methods. ACS Symposium Series, vol. 141, pp. 315–334 (Chapter 19).

Blank, F., 1954. On the cell walls of dimorphic fungi causing systemic infections. Can. J. Microbiol. 1, 1–5.

Blumberg, R., Southall, C.L., Van Rensburg, N.J., Volckman, O.B., 1951. South African fish products. XXXII. The rock lobster: a study of chitin production from processing wastes. J. Sci. Food Agric. 2, 571–576.

Blumenthal, H.J., Roseman, S., 1957. Quantitative estimation of chitin in fungi. J. Bacteriol. 74, 222–224.

Boamah, P.O., Huang, Y., Hua, M.Q., Zhang, Q., Wu, J.B., Onumah, J., Sam-Amoah, L. K., Boamah, P.O., 2015. Sorption of heavy metal ions onto carboxylate chitosan derivatives – a mini-review. Ecotoxicol. Environ. Saf. 116, 113–120.

Bonecco, M.B., Martínez Sáenz, M.G., Buffa, L.M., 2017. Chitosan, from residue to industry. In: Masuell, M., Renard, D. (Eds.), Advances in Physicochemical Properties of Biopolymers. Bentham e-Books, Bentham Science Publishers, Sharjah, ISBN: 978-1-68108-545-6, pp. 224–256 (Chapter 4).

Boominathan, T., Sivaramakrishna, A., 2021. Recent advances in the synthesis, properties, and applications of modified chitosan derivatives: challenges and opportunities. Top. Curr. Chem. 379, 19.
Borgogna, M., Bellich, B., Cesaro, A., 2011. Marine polysaccharides in microencapsulation and application to aquaculture: from sea to sea. Mar. Drugs 9, 2572–2604.
Boroumand, H., Badie, F., Mazaheri, S., Seyedi, Z.S., Nahand, J.S., Nejati, M., Baghi, H.B., Abbasi-Kolli, M., Badehnoosh, B., Ghandali, M., Hamblin, M.R., Mirzaei, H., 2021. Chitosan-based nanoparticles against viral infections. Front. Cell. Infect. Microbiol. 11, 643953.
Bose, I., Singh, R., Negi, P., Singh, Y., 2021. Chitin as bio-based nanomaterial in packaging: a review. Mater. Today Proc. 46, 11254–11257.
Boulaiche, W., Hamdi, B., Trari, M., 2019. Removal of heavy metals by chitin: equilibrium, kinetic and thermodynamic studies. Appl Water Sci 9, 39.
Bouligand, Y., 1965. Sur une architecture torsadée répandue dans de nombreuses cuticles d'arthropodes. C. R. Hebd. Séanc. Acad. Sci. Paris 261, 3665–3668.
Bounoure, L., 1911. La sécrétion de la chitine chez les coléoptères carnivores. C. R. Assoc. Fr. Avanc. Sci. Paris 40, 523–526.
Bounoure, L., 1912a. Sur l'influence du régime alimentaire sur la production de la chitine chez les coléoptères. In: Comptes Rendus du Congrès Savantes de Paris et des Départements. Mémoire n° XXIV, pp. 189–193.
Bounoure, L., 1912b. L'influence du régime alimentaire sur la production de la chitine chez les coléoptères. In: Comptes Rendus du Congrès des Sociétés Savantes, Section Sciences 189–193.
Bounoure, L., 1913. L'influence de la taille des insectes sur la production de la chitine, sécrétion de surface. C. R. Acad. Sci. 157, 140–142.
Bounoure, L., 1919. Généralités sur la chitine. In: Houssay, F. (Ed.), Aliments, Chitine et Tube Digestif chez les Coléoptères. Collection de Morphologie Dynamique. Librairie Scientifique A. Hermann et Fils, Paris. 294 p.
Brach, H., 1912. Untersuchungen über den chemischen aufbau des chitins. Biochem. Z. 38, 468–491.
Brach, H., von Fürth, O., 1912. The chemical constitution of chitin. Biochem. Z. 38, 468–491.
Braconnot, H., 1811a. Sur la nature des champignons. In: Klostermann, J. (Ed.), Recueil de Mémoires Concernant la Chimie et les Arts Qui en Dépendent et Spécialement la Pharmacie. 31 Juillet 1811. Annales de Chimie, vol. 79. Librairie des Ecoles Impériales Polytechnique et des Ponts et Chaussées, Paris, pp. 265–304.
Braconnot, H., 1811b. Des recherches analytiques sur la nature des champignons. In: Klostermann, J. (Ed.), Recueil de Mémoires Concernant la Chimie et les Arts Qui en Dépendent et Spécialement la Pharmacie. 31 Octobre 1811. Annales de Chimie, vol. 79. Librairie des Ecoles Impériales Polytechnique et des Ponts et Chaussées, Paris, pp. 272–292.
Braconnot, H., 1811. De la fongine, ou analyse des champignons. J. Phys. Chim. Hist. Nat. Arts LXXIII, 130–135.
Braconnot, H., 1813. Nouvelles recherches analytiques sur la nature des champignons, pour servir de suite à celles qui ont été insérés dans les tomes LXXIX et LXXX des Annales de chimie. Annales de Chimie. In: Klostermann, J. (Ed.), Recueil de Mémoires Concernant la Chimie et les Arts Qui en Dépendent et Spécialement la Pharmacie. 31 Août 1813. Annales de Chimie, vol. 79. Librairie des Ecoles Impériales Polytechnique et des Ponts et Chaussées, Paris, pp. 237–270.
Brasselet, C., Pierre, G., Dubessay, P., Dols-Lafargue, M., Coulon, J., Maupeu, J., Vallet-Courbin, A., de Baynast, H., Doco, T., Michaud, P., Delattre, C., 2019. Modification of chitosan for the generation of functional derivatives. Appl. Sci. 9, 1321.

Breuer, R., 1898. Ueber das freie chitosamin. Ber. Dtsch. Chem. Ges. 31, 2193–2200.
Brimacombe, J.S., 1973. Carbohydrate Chemistry. vol. 6 The Chemical Society, London.
Brimacombe, J.S., Webber, J.M., 1964. Mucopolysaccharides: Chemical Structure, Distribution And Isolation. vol. 6 BBA Library, Elsevier, Amsterdam. 181 p.
Brinchmann, B.C., Bayat, M., Brogger, T., Muttuvelu, D.V., Tjonneland, A., Sigsgaard, T., 2011. A possible role of chitin in the pathogenesis of asthma and allergy. Ann. Agric. Environ. Med. 18, 7–12.
Brock, N., 1957. Infrared spectra of carbohydrates. Adv. Carbohydr. Chem. 12, 13–33.
Brock, W.H., 2013. Hoppe-Seyler, Ernst Felix Immanuel. In: eLs. John Wiley & Sons, Ltd., Chichester. 1 p.
Brongniart, A.D., 1823. Résumé des Travaux de la Société D'histoire Naturelle de Paris. Mémoires de la Société d'Histoire Naturelle de Paris, vol. 1 Baudouin Frères Libraires-Éditeurs, Paris, pp. 15–27.
Broussignac, P., 1968. Haut polymère naturel connu dans l'industrie: le chitosane. Chim. Ind. Genie Chim. 99, 1241–1247.
Brubaker, M.M., 1937. Protection of Materials Subject to Parasitic Attack. US Patent 2,098,942.
Brug, S.L., 1932. Chitinisation of parasites in mosquitoes. Bull. Entomol. Res. 23, 229–231.
Brunswick, H., 1921. Über die mikrochemie der chitosanverbindungen. Biochem. Z. 113, 111–124.
Bunge, G., 1912. The glucosides. In: Text-Book of Organic Chemistry for Medical Students. Cornell University Library, pp. 122–130. Translated by Aders Plimmer RH. Longmans, Green and Co. Lecture IX.
Burkatovskaya, M., Tegos, G.P., Swietlik, E., Demidova, T.N., Castano, A.P., Hamblin, M. R., 2006. Use of chitosan bandage to prevent fatal infections developing from highl contaminated wounds in mice. Biomaterials 27, 4157–4164.
Bütschli, O., 1874. Einiges über das chitin. Arch. Anat. Physiol. Wiss., 362–370.
Campbell, F.L., 1929. The detection and estimation of insect chitin; and the irrelation of "chitinization" to hardness and pigmentation of the cuticula of the American cockroach, *Periplaneta americana* L. Ann. Entomol. Soc. Am. XXII, 401–426.
Cao, S.L., Liu, Y.X., Shi, L.M., Zhu, W.B., Wang, H.L., 2022. *N*-Acetylglucosamine as a platform chemical produced from renewable resources: opportunity, challenge, and future prospects. Green Chem. 24, 493.
Carlström, D., 1957. The crystal structure of α-chitin (poly-N-acetyl-D-glucosamine). J. Biophys. Biochem. Cytol. 3, 669–683.
Carlström, D., 1962. The polysaccharide chain of chitin. Biochim. Biophys. Acta 59, 361–364.
Carneiro, J., Tedim, J., Ferreira, M.G.S., 2015. Chitosan as a smart coating for corrosion protection of aluminum alloy 2024: a review. Prog. Org. Coat. 89, 348–356.
Casadidio, C., Peregrina, D.V., Gigliobianco, M.R., Deng, S., Censi, R., Di Martino, P., 2019. Chitin and chitosans: characteristics, eco-friendly processes, and applications in cosmetic science. Mar. Drugs 17, 369.
Cassie, J., 2022. Handbook of Chitin and Chitosan. Preparation and Properties. States Academic Press, ISBN: 9781639892532. 253 p.
Castle, E.S., 1936. The double refraction of chitin. J. Gen. Physiol. 19, 797–805.
Castro Marín, A., Colangelo, D., Lambri, M., Riponi, C., Chinnici, F., 2020. Relevance and perspectives of the use of chitosan in winemaking: a review. Crit. Rev. Food Sci. Nutr. 61, 3450–3464.
Cazon, P., Vazquez, M., 2019. Mechanical and barrier properties of chitosan combined with other components as food packaging film. Environ. Chem. Lett. 18, 257–267.

Cerezuela, R., Mesequer, J., Angeles Esteban, M., 2011. Current knowledge in symbiotic use for fish aquaculture: a review. J. Aquac. Res. Dev. S1, 008.
Chalongsuk, R., Sribundit, N., 2013. Usage of chitosan in Thai pharmaceutical and cosmetic industries. Silpakorn Univ. Sci. Technol. J. 7, 49–53.
Chatin, J., 1892. Sur l'origine et la formation du revêtement chitineux chez les larves des libellules. C. R. Acad. Sci. 114, 1135–1138.
Cheba, B.A., 2011. Chitin and chitosan: marine biopolymers with unique properties and versatile applications. Glob. J. Biotechnol. Biochem. 6, 149–153.
Chen, I.H., Lee, T.M., Huang, C.L., 2021. Biopolymers hybrid particles used in dentistry. Gels 7, 31.
Chevallier, A., Richard, A., 1827. Dictionnaire des Drogues Simples et Composées. vol. 2 Bécher Jeune, Libraire de l'Académie Royale de Médecine, Bruxelles, p. 64.
Children, J.G., 1824. Memoir on the chemical composition of the corneous parts of insects; by Augustus Odier. Translated from the original French, with some additional remarks and experiments. Bell T, Children JG, Sowerby JDC and Sowerby GB, editors. London. March, 1824, n° I, article XV, Zool. J. 1, 101–115.
Chivere, V.T., Kondiah, P.P.D., Choonara, Y.E., Pillay, V., 2020. Nanotechnology-based biopolymeric oral delivery platforms for advanced cancer treatment. Cancers 12, 522.
Chouhan, D., Mandal, P., 2021. Applications of chitosan and chitosan based metallic nanoparticles in agrosciences—a review. Int. J. Biol. Macromol. 166, 1554–1569.
Chung, Y.C., 2006. Improvement of aquaculture wastewater using chitosan of different degrees of deacetylation. Environ. Technol. 27, 1199–1208.
Chung, Y.C., Li, Y.H., Chen, C.C., 2005. Pollutant removal from aquaculture wastewater using the biopolymer chitosan at different molecular weights. J. Environ. Sci. Health A 40, 1755–1790.
Clark, G.L., 1934. The macromolecule and the micelle as structural units in biological materials with special reference to cellulose. Cold Spring Harb. Symp. Quant. Biol. 2, 28–38.
Clark, G.L., Smith, A.F., 1936. X-ray diffraction studies of chitin, chitosan, and derivatives. J. Phys. Chem. 40, 863–879.
Codex Alimentarius, 2013. Normes Alimentaires Internationales. Organisation Mondiale de la Santé. https://www.fao.org/fao-who-codexalimentarius/codex-texts/guidelines/fr/.
Cohen, E., 1987a. Interference with chitin biosynthesis in insects. In: Wright, J.E., Retnakaran, A. (Eds.), Chitin and Benzophenyl Ureas. Dr W. Junk Publishers, Dordrecht, pp. 43–74 (Chapter 3).
Cohen, E., 1987b. Chitin biochemistry—synthesis and inhibition. Annu. Rev. Entomol. 32, 71–93.
Cohen, E., 2010. Chitin biochemistry. synthesis, hydrolysis and inhibition. In: Casas, J., Simpson, S.J. (Eds.), Advances in Insect Physiology: Insect Integument and Colour. Advances in Insect Physiology, vol. 38, pp. 5–74.
Conrad, J., 1966. Chitin. In: Encyclopedia of Polymer Science and Technology. vol. 3. Interscience, New York, pp. 695–705.
Cosme, F., Vilela, A., 2021. Chitin and chitosan in the alcoholic and non-alcoholic beverage industry: an overview. Appl. Sci. 11, 11427.
Costa, R., Santos, L., 2017. Delivery systems for cosmetics—from manufacturing to the skin of natural antioxidants. Powder Technol. 322, 402–416.
Cox, E.G., Jeffrey, G.A., 1939. Crystal structure of glucosamine hydrobromide. Nature 143, 894–895.
Cox, E.G., Goodwin, T.H., Wagstaff, A.I., 1935. The crystalline structure of the sugars. 2. Methylated sugars and the conformation of the pyranose ring. J. Chem. Soc., 1495–1504.
Crini, G., 2019. Historical review on chitin and chitosan polymers. Environ. Chem. Lett. 17, 1623–1643.

Crini, G., 2021. The contribution of professor Paul Karrer (1889-1971) to dextrins. J. Incl. Phenom. Macrocycl. Chem. 99, 115–167.

Crini, G., Badot, P.M., 2008. Application of chitosan, a natural aminopolysaccharide, for dye removal from aqueous solutions by adsorption processes using batch studies: a review of recent literature. Prog. Polym. Sci. 33, 399–447.

Crini, G., Lichtfouse, E. (Eds.), 2019a. Chitin and Chitosan: Applications in Food, Agriculture, Pharmacy, Medicine and Wastewater Treatment. Sustainable Agriculture Reviews, vol. 36. Springer Nature, Cham, ISBN: 978-3-030-16580-2.

Crini, G., Lichtfouse, E. (Eds.), 2019b. Chitin and Chitosan: History, Fundamentals and Innovations. In: Crini, G., Lichtfouse, E. (Eds.), Sustainable Agriculture Reviews, vol. 35. Springer Nature, Cham, ISBN: 978-3-030-16537-6.

Crini, G., Badot, P.M., Guibal, E. (Eds.), 2009. Chitine et Chitosane—Du Biopolymère à L'application. PUFC, Besançon, France. 303 p.

Crini, G., Morin-Crini, N., Fatin-Rouge, N., Déon, S., Fievet, P., 2017. Metal removal from aqueous media by polymer-assisted ultrafiltration with chitosan. Arab. J. Chem. 10, S3826–S3839.

Crini, G., Torri, G., Lichtfouse, E., Kyzas, G.Z., Wilson, L.D., Morin-Crini, N., 2019. Dye removal by biosorption using cross-linking chitosan-based hydrogels. Environ. Chem. Lett. 17, 1645–1666.

Croisier, F., Jérôme, C., 2013. Chitosan-based biomaterials for tissue engineering. Eur. Polym. J. 49, 780–792.

Cutler, W.O., Peat, S., 1939. Some derivatives of methylated glucosamine. J. Chem. Soc., 274–279.

Cutler, W.O., Haworth, W.N., Peat, S., 1937. Methylation of glucosamine. J. Chem. Soc., 1979–1983.

da Silva Alves, D.C., Healy, B., de Almeida Pinto, L.A., Sant'Anna Cadaval Jr., T.R., Breslin, C.B., 2021. Recent developments in chitosan-based adsorbents for the removal of pollutants from aqueous environments. Molecules 26, 594.

Dahn, H., Cherbuliez, E., Chekhitliez, E., 1969. Professor Dr. Paul Karrer zum 80. Geburtstag. Helv. Chim. Acta 52. 568A–568C.

Darmon, S.E., Rudall, K.M., 1950. Infra-red and X-ray studies of chitin. Discuss. Faraday Soc. 9, 251–260.

Dash, M., Chiellini, F., Ottenbrite, R.M., Chiellini, E., 2011. Chitosan—a versatile semi-synthetic polymer in biomedical applications. Prog. Polym. Sci. 36, 981–1014.

Dave, U., Somanader, E., Baharlouei, P., Pham, L., Rahman, M.A., 2021. Applications of chitin in medical, environmental, and agricultural industries. J. Mar. Sci. Eng. 9, 1173.

Davis, S.P., 2011. In: Davis, S.P. (Ed.), Chitosan: Manufacture, Properties, and Usage. Biotechnology in Agriculture, Industry and Medicine, Nova Science Publishers, Inc., New York, ISBN: 9781617288319. 507 p.

de Alvarenga, E.S., 2011. Characterization and properties of chitosan. In: Elnashar, M. (Ed.), Biotechnology of Biopolymers. InTech, Croatia, pp. 92–108 (Chapter 5).

de Andrade, J.R., Oliveira, M.F., da Silva, M.G.C., Vieira, M.G.A., 2018. Adsorption of pharmaceuticals from water and wastewater using nonconventional low-cost materials: a review. Ind. Eng. Chem. Res. 57, 3103–3127.

de Bary, A., 1887. Comparative Morphology and Biology of the Fungi, Mycetozoa, and Bacteria. The Clarendon Press London, Oxford. 525 p.

De Jin, R., Suh, J.W., Park, R.D., Kim, Y.W., Krishnan, H.B., Kim, K.Y., 2005. Effect of chitin compost and broth on biological control of Meloidogyne incognita on tomato (*Lycopersicon esculentum* Mill.). Nematology 7, 125–132.

de Lima Batista, A.C., de Souza Neto, F.E., de Souza, P.W., 2018. Review of fungal chitosan: past, present and perspectives in Brazil. Polimeros 28, 275–283.

de Saint-Vincent, R., 1825. Dictionnaire Classique D'histoire Naturelle. vol. 8 Rey et Gravier, Libraires-Éditeurs, Paris, pp. 541–542.
Desbrières, J., Guibal, E., 2018. Chitosan for wastewater treatment. Polym. Int. 67, 7–14.
Dennell, R., 1946. A study of an insect cuticle—the larval cuticle of *Sarcophaga falculata* pand (Diptera). Proc. R. Soc. Lond. B Biol. Sci. 133, 348–373.
Dennell, R., 1947a. A study of an insect cuticle—the formation of the puparium of *Sarcophaga falculata* pand (Diptera). Proc. R. Soc. Lond. B Biol. Sci. 134, 79–110.
Dennell, R., 1947b. The occurrence and significance of phenolic hardening in the newly formed cuticle of crustacea-decapoda. Proc. R. Soc. Lond. B Biol. Sci. 134, 485–503.
Dennell, R., 1958. The hardening of insect cuticles. Biol. Rev. 33, 178–196.
Diehl, J.M., 1936. Vegetable chitin. Chem. Weekbl. 33, 36–38.
Diehl, J.M., van Iterson Jr., G., 1935. Die doppelbrechung von chitinsehnen. Kolloid Z. 73, 142–146.
Dima, J.B., Sequeiros, C., Zaritzky, N., 2017. Chitosan from marine crustaceans: production, characterization and applications. In: Shalaby, E.A. (Ed.), Biological Activities and Application of Marine Polysaccharides. InTech, Rijeka, Croatia, pp. 39–56 (Chapter 3).
Dimassi, S., Tabary, N., Chai, F., Blanchemain, N., Martel, B., 2018. Sulfonated and sulfated chitosan for biomedical applications: a review. Carbohydr. Polym. 202, 382–396.
Divya, K., Jisha, M.S., 2018. Chitosan nanoparticles preparation and applications. Environ. Chem. Lett. 16, 101–112.
Doczi, J., Fischman, A., King, J.A., 1953. Direct evidence of the influence of sulfamic acid linkages on the activity of heparin-like anticoagulants. J. Am. Chem. Soc. 75, 1512–1513.
Dodane, V., Vilivalam, V.D., 1998. Pharmaceutical applications of chitosan. Pharm. Sci. Technol. Today 1, 246–253.
Dongre, R.S., 2018. Chitin-Chitosan. Myriad Functionalities in Science and Technology. IntechOpen, London, ISBN: 978-1-78923-407-7.
Donzelot, P., 1953. Henri Braconnot 1780-1855. Figures Pharmaceutiques Françaises. Notes Historiques et Portraits 1803-1953. Masson, Paris, pp. 53–58.
Dotto, G.L., Campana-Filho, S.P., de Almeida Pinto, L.A., 2017. Chitosan Based Materials and its Applications. vol. 3 Bentham Science Publishers, Sharjah, UAE, ISBN: 978-1-68108-486-2.
Dous, Ziegenspeck, 1926. Das chitin der pilze. Arch. Pharm. 264, 751–753.
Drapiez, P.A.J., 1839. Dictionnaire Classique des Sciences Naturelles. vol. 6 Meline, Cans et Compagnie, Librairie, Imprimerie, Fonderie, Bruxelles, p. 68.
Drewnowska, O., Turek, B., Cartanjen, B., Gajewski, Z., 2013. Chitosan—a promising biomaterial in veterinary medicine. Pol. J. Vet. Sci. 16, 843–848.
Duarte, A.R., Silva, S.S., Reis, R.L., 2018. Chitin/chitosan based aerogels: processing and morphology. In: Thomas, S., Pothan, L.A., Mavelil-Sam, R. (Eds.), Biobased Aerogels: Polysaccharide and Protein-Based Materials. Royal Society of Chemistry, pp. 9–24 (Chapter 2).
Dumitriu, S. (Ed.), 2005. Polysaccharides—Structural Diversity and Functional Versatility, second ed. Marcel Dekker, New York. ISBN: 00-8247-5480-8. 1204 p.
Dutta, P.K. (Ed.), 2016. Chitin and Chitosan for Regenerative Medicine. Springer India, ISBN: 9788132234647. 389 p.
Dutta, P.K., Dutta, J., Tripathi, V.S., 2004. Chitin and chitosan: chemistry, properties and applications. J. Sci. Ind. Res. 63, 20–31.
Dutta, P.K., Yadav, S., Mehrotra, G.K., 2021. Modified chitosan films/coatings for active food packaging. Adv. Polym. Sci. 287, 203–232.
Dweltz, N.E., 1960. The structure of chitin. Biochim. Biophys. Acta 44, 416–435.
Dweltz, N.E., 1961. The structure of β-chitin. Biochim. Biophys. Acta 51, 283–294.

Dweltz, N.E., Anand, N., 1961. Nature of the amino sugar in β-chitin. Biochim. Biophys. Acta 50, 357.

Eidmann, H., 1922. Die durchlässigkeit des chitins bei osmotischen vorgängen. Biol. Zentralbl. 42, 429–435.

Ejeromedoghene, O., Oderinde, O., Egejuru, G., Adewuyi, S., 2020. Chitosan-drug encapsulation as a potential candidate for COVID-19 drug delivery systems: a review. J. Turk. Chem. Soc. 7, 851–864.

El Hadrami, A., Adam, L.R., El Hadrami, I., Daayf, F., 2010. Chitosan in plant protection. Mar. Drugs 8, 968–987.

El Halah, A., López-Carrasquero, F., Contreras, J., 2018. Applications of hydrogels in the adsorption of metallic ions. Rev. Cienca Ingenieria 39, 57–70.

El Knidri, H., Laajeb, A., Lahsini, A., 2020. Chitin and chitosan: chemistry, solubility, fiber formation, and their potential applications. In: Gopi, S., Thomas, S., Pius, A. (Eds.), Handbook of Chitin and Chitosan. Volume 1: Preparation and Properties. Elsevier, pp. 35–57 (Chapter 2).

El-Hack, M.E.A., El-Saadony, M.T., Shafi, M.E., Zabermawi, N.M., Arif, M., Batiha, G.E., Khafaga, A.F., El-Hakim, Y.M.A., Al-Sagheer, A.A., 2020. Antimicrobial and antioxidant properties of chitosan and its derivatives and their applications: a review. Int. J. Biol. Macromol. 164, 2726–2744.

Elieh-Ali-Komi, D., Hamblin, M.R., 2016. Chitin and chitosan: production and application of versatile biomedical nanomaterials. Int. J. Adv. Res. 4, 411–427.

El-Naggar, M., Medhat, F., Taha, A., 2022. Applications of chitosan and chitosan nanoparticles in fish aquaculture. Egypt. J. Aquat. Biol. Fish. 26, 23–43.

Elson, L.A., Morgan, W.T.J., 1933. A colourimetric method for the determination of glucosamine and chondrosamine. Biochem. J. 27, 1824–1828.

Enescu, D., 2008. Use of chitosan in surface modification of textile materials. Roum. Biotechnol. Lett. 13, 4037–4048.

Ettre, L.S., 1989. László Zechmeister: a pionner of chromatography. Anal. Chem. 61, 1315A–1322A.

Eugster, C.H., 1972. Das Portrait: Paul Karrer 1889-1971. Chem. Unserer Zeit 6, 147–153.

Farr, W.K., Sisson, W.A., 1934. X-ray diffraction patterns of cellulose particles and interpretations of cellulose diffraction data. Contrib. Boyce Thompson Inst. 6, 315–321.

Felt, O., Nuri, P., Gurny, R., 1998. Chitosan: a unique polysaccharide for drug delivery. Drug Dev. Ind. Pharm. 24, 979–993.

Felt, O., Furrer, P., Mayer, J.M., Plazonnet, B., Buri, P., Gurny, R., 1999. Topical use of chitosan in ophthalmology: tolerance assessment and evaluation of precorneal retention. Int. J. Pharm. 15, 185–193.

Feng, M., Lu, X.M., Hou, D.F., Zhang, S.J., 2020. Solubility, chain characterization, and derivatives of chitin. In: Gopi, S., Thomas, S., Pius, A. (Eds.), Handbook of Chitin and Chitosan. Volume 1: Preparation and Properties. Elsevier, pp. 101–129 (Chapter 4).

Fenice, M., Gorrasi, S., 2021. Advances in chitin and chitosan science. Molecules 26, 1805.

Feofilova, E.P., 1984. Biological functions and the practical use of chitin. Prikl. Biokhim. Mikrobiol. 20, 147–160.

Ferguson, A.N., O'Neill, A.G., 2011. In: Ferguson, A.N., O'Neill, A.G. (Eds.), Focus on Chitosan Research. Nova Science Publishers. 477 p.

Ferris, G.F., Chamberlin, J.C., 1928. On the use of the word "chitinized". Entomol. News 39, 212–215.

Fikentscher, H., 1932. Systematik der cellulose auf grund ihrer viskositat in losung. Cellulosechemie 13, 58–64.

Finney, N.S., Siegel, J.S., 2008. In Memoriam—Albert Hofmann (1906-2008). Chimia 62, 444–447.

Fischer, E., 1884. Verbindungen des phenylhydrazins mit den zuckerarten. Ber. Dtsch. Chem. Ges. 17, 579–584.

Fischer, E., 1912. Syntheses in the purine and sugar group. In: Nobel Lecture. December 12, 1902, 15 p.

Fischer, E., Andreae, E., 1903. Über chitonsäure und chitarsäure. Ber. Dtsch. Chem. Ges. 36, 2587–2592.

Fischer, E., Leuchs, H., 1902. Synthese des serins, der l-glucosaminsäure und anderer oxyaminosäuren. Ber. Dtsch. Chem. Ges. 35, 3787–3805.

Fischer, E., Leuchs, H., 1903. Synthese das d-glucosamins. Ber. Dtsch. Chem. Ges. 36, 24–29.

Fischer, E., Tiemann, F., 1894. Ueber das glucosamin. Ber. Dtsch. Chem. Ges. 27, 138–147.

Forbes, W.T., 1930. What is chitine? Science 72, 397.

Foster, J.W., 1949. Chemical Activities of Fungi. Academic Press Inc., New York. 648 p.

Foster, A.B., Hackman, R.H., 1957. Application of ethylenediaminetetra-acetic acid in the isolation of crustacean chitin. Nature 180, 40–41.

Foster, A.B., Horton, D., 1959. Aspects of the chemistry of aminosugars. Adv. Carbohydr. Chem. 14, 213–281.

Foster, A.B., Stacey, M., 1952. The chemistry of the 2-amino sugars (2-amino-2-deoxy-sugar). Adv. Carbohydr. Chem. 7, 247–288.

Foster, A.B., Stacey, M., 1958. The aminosugars and chitin. In: Ruhland, W. (Ed.), Formation—Storage—Mobilization and transformation of carbohydrates. Encyclopedia of Plant Physiology, vol. 6. Springer Verlag, Berlin, pp. 518–529. Book series 532. Part VI.

Foster, A.B., Webber, J.M., 1961. Chitin. Adv. Carbohydr. Chem. 15, 371–393.

Foulk, C.W., 1934. Paul Karrer. Ohio State University, 1932. J. Educ. 11, 1.

Fraenkel, S., Kelly, A., 1903. In: Masson et Cie (Ed.), Sur la Constitution de la Chitine. Bulletin de la Société Chimique de Paris. vol. XXX. Libraires de l'Académie de Médecine, p. 372. Troisième série.

Fraenkel, G., Rudall, K.M., 1940. A study of the physical and chemical properties of the insect cuticle. Proc. R. Soc. Lond. B129, 1–35.

Fraenkel, G., Rudall, K.M., 1947. The structure of insect cuticles. Proc. R. Soc. Med. B Biol. Sci. B134, 111–143.

Francesko, A., Díaz González, M., Lozano, G.R., Tzanov, T., 2010. Developments in the processing of chitin, chitosan and bacterial cellulose for textile and other applications. In: Advances in Textile Biotechnology. A Volume in Woodhead Publishing Series in Textiles, Elsevier, pp. 288–311 (Chapter 12).

Fränkel, S., 1898. Über die reindarstellung der sogenannten kohlehydrat-gruppe des eiweisses. Monatsh. Chem. Verw. Teile Anderer Wiss. 19, 747–769.

Fränkel, S., Jellinek, C., 1927. Limulus polyphemus. Biochem. Z. 185, 384–388.

Fränkel, S., Kelly, A., 1901a. Beiträge zur constitution des chitins. Das vorkommen von chitin und seine verwertung als systematisch-phylogenetisches merkmal im pflanzenreich. In: Sitzungsberichte der Kaiserlichen Akademie der Wissenschaften Mathematisch Naturwissenchaftliche Klasse Abt IIB. vol. 110, pp. 1147–1156.

Fränkel, S., Kelly, A., 1901b. Beiträge zur constitution des chitins. Monatsh. Chem. Verw. Teile Anderer Wiss. 23, 123–132.

Frémy, E., 1847. Recherches sur les matières gélatineuses des végétaux. C. R. Hebd. Seances Acad. Sci. 24, 1046–1050.

Frémy, E., 1855. Recherches chimiques sur les os. Ann. Chim. Phys. XLIII, 93–96. 3ème Série.

Freudenberg, K., 1967. Emil Fischer and his contribution to carbohydrate chemistry. Adv. Carbohydr. Chem. 21, 1–38.

Freudenberg, K., Eichel, H., 1935. Über spezifische kohlenhydrate der blutgruppen. II. Liebigs Ann. 518, 97–102.

Freudenberg, K., Walch, H., Molter, H., 1942. The separation of sugars, amino sugars, and amino acids. Naturwissenschaften 30, 87.
Frey, R., 1950. Chitin und zellulose in Pilzzell-wänden. Ber. Schweiz. Bot. Ges. 60, 199–230.
Friedman, S., 1970. Metabolism of carbohydrates in insects. In: Florkin, M., Scheer, B.T. (Eds.), Chemical Zoology. vol. 5. Academic Press, New York.
Fruton, J.S., 1990. Felix Hoppe-Seyler and Willy Kühne. In: Contrasts in Scientific Style. Research Groups in the Chemical and Biochemical Sciences. vol. 191. Memoirs of the American Philosophical Society, Philadelphia, ISBN: 0-87169-191-4. (Chapter III), pp. 72-117 and pp. 308-321.
Galiano, F., Briceno, K., Marino, T., Molino, A., Christensen, K.V., Figoli, A., 2018. Advances in biopolymer-based membrane preparation and applications. J. Membr. Sci. 564, 562–586.
Gallo, M., Naviglio, D., Caruso, A.A., Ferrara, L., 2016. Applications of chitosan as a functional food. In: Grumezescu, A. (Ed.), Novel Approaches of Nanotechnology in Food. vol. 1. Academic Press, pp. 425–464 (Chapter 13).
Gamgee, A., 1895. The late professor Hoppe-Seyler. Nature 52, 575–576.
Gao, Y., Wu, Y., 2022. Recent advances of chitosan-based nanoparticles for biomedical and biotechnological applications. Int. J. Biol. Macromol. 203, 379–388.
Gardner, K.H., Blackwell, J., 1975. Refinement of the structure of β-chitin. Biopolymers 14, 1581–1595.
Gassara, F., Antzak, C., Ajila, C.M., Sarma, S.J., Brar, S.K., Verma, M., 2015. Chitin and chitosan as natural flocculants for beer clarification. J. Food Eng. 166, 80–85.
Gerdts, V., Mutwiri, G., Richards, J., van Drunen Littel-van den Hurk, S., Potter, A.A., 2013. Carrier molecules for use in veterinary vaccines. Vaccine 31, 596–602.
Ghannoum, M.A., Long, L., Isham, N., Bulgheroni, A., Setaro, M., Caserini, M., Palmieri, R., Mailland, F., 2015. Ability of hydroxypropyl chitosan nail lacquer to protect against dermatophyte nail infection. Antimicrob. Agents Chemother. 59, 1844–1848.
Ghormade, V., Pathan, E.K., Deshpande, M.V., 2017. Can fungi compete with marine sources for chitosan production? Int. J. Biol. Macromol. 104, 1415–1421.
Giles, C.H., Hassan, A.S.A., Laidlaw, M., Subramanian, R.V.R., 1958. Adsorption at organic surfaces. III. Some observations on the constitution of chitin and on its adsorption of inorganic and organic acids from aqueous solution. J. Soc. Dye. Colour. 74, 647–654.
Gilson, E., 1893. La cristallisation de la cellulose et la composition chimique de la membrane végétale. Cellule 9, 397–441.
Gilson, E., 1894. Recherches chimiques sur la membrane cellulaire des champignons. Bull. Soc. Chim. Fr. 3, 1099–1102.
Gilson, E., 1894. Recherches chimiques sur la membrane cellulaire des champignons. Cellule XI, 7. 1er fascicule.
Gilson, E., 1894. Note sur le corps azote de la membrane cellulaire des champignons. Bull. Soc. Chim. Paris 23, 590.
Gilson, E., 1895. Das chitin und die membranen der pilzzellen. Ber. Dtsch. Chem. Ges. 28, 821–822.
Gilson, E., 1895. De la présence de la chitine dans la membrane cellulaire des champignons. In: Compte Rendus des Séances de l'Académie des Sciences de Paris. 6 Mai 1895.
Giri Dev, V.R., Neelakandan, R., Sudha, S., Shamugasundram, O.L., Nadaraj, R.N., 2005. Chitosan—a polymer with wider applications. Text. Mag. 46 (83), 86.
Gomez-Ramirez, M., Rojas Avelizapa, L.I., Rojas Avelizapa, N.G., Camarillo, C.R., 2004. Colloidal chitin stained with Remazol Brilliant Blue R, a useful substrate to select chitinolytic microorganisms and to evaluate chitinases. J. Microbiol. Methods 56, 213–219.

Gonell, H.W., 1926. Röentenographische studien an chitin. Hoppe-Seylers Z. Physiol. Chem. 152, 18–30.
Gong, P., Wang, J., Liu, B., Ru, G., Feng, J., 2016. Dissolution of chitin in aqueous KOH. Cellulose 23, 1705–1711.
Good, J.M., Gregory, O., Bosworth, N., 1813. Pantologia: A New Cyclopaedia. London, vol. 3. Oxford University.
Goosen, M.F.A., 1997. In: Goosen, M.F.A. (Ed.), Applications of Chitin and Chitosan. CRC Press LLC., Boca Raton. 336 p.
Grande-Tovar, C.D., Chaves-Lopez, C., Serio, A., Rossi, C., Paparella, A., 2018. Chitosan coatings enriched with essential oils: effects on fungi involved in fruit decay and mechanisms of action. Trends Food Sci. Technol. 78, 61–71.
Gréhant, N., 1904a. Charles Rouget, notice nécrologique. Nouv. Arch. Mus. Hist. Nat. 6, I–V.
Gréhant, N., 1904b. Liste des ouvrages et mémoires publiés par Ch. Rouget. Nouv. Arch. Mus. Hist. Nat. 6, VI–XII.
Griffith, W.P., Morris, P.J.T., 2003. Charles Hatchett FRS (1765-1847), chemist and discover of niobium. Notes Rec. R. Soc. Lond. J. Hist. Sci. 57, 299–316.
Griffiths, A.B., 1892. La pupine, nouvelle substance animale. C. R. Acad. Sci. 115, 320–321.
Griffiths, A.B., 1892. La pupine, nouvelle substance animale. Bull. Acad. R. Belg. 24, 592.
Grifoll-Romero, L., Pascual, S., Aragunde, H., Biarnes, X., Planas, A., 2018. Chitin deacetylases: structures, specificities, and biotech applications. Polymers 10, 352.
Guarino, V., Altobelli, R., Ambrosio, L., 2016. Chitosan microgels and nanoparticles via electrofluidodynamic techniques for biomedical applications. Gels 2, 2.
Gunther, A.E., 1978. John George Children FRS (1777-1852) of the British Museum. Mineralogist and Reluctant Keeper of Zoology. Bulletin of the British Museum (Natural History). Historical Series, vol. 6, pp. 75–108.
Gutiérrez, T.J., 2017. Chitosan applications in textile and food industry. In: Ahmed, S., Ikram, S. (Eds.), Chitosan—Derivatives, Composites and Applications. Scrivener Publishing LLC, Wiley, pp. 185–232 (Chapter 8).
Gwynne-Vaughan, H.C.I., Barnes, B., 1927. The Structure and Development of the Fungi. Cambridge University Press, Cambridge.
Hackman, R.H., 1953a. Chemistry of insect cuticle. 1. The water-soluble proteins. Biochem. J. 54, 362–367.
Hackman, R.H., 1953b. Chemistry of insect cuticle. 3. Hardening and darkening of the cuticle. Biochem. J. 54, 371–377.
Hackman, R.H., 1954. Studies on chitin. I. Enzymatic degradation of chitin and chitin esters. Aust. J. Biol. Sci. 7, 168–178.
Hackman, R.H., 1955. Studies on chitin. III. Adsorption of proteins to chitin. Aust. J. Biol. Sci. 8, 530–536.
Hackman, R.H., 1959. Biochemistry of the insect cuticle. In: Levenbook, L. (Ed.), Proceedings of the 4th International Congress Biochemistry Held in Vienna, September 1-6, 1958. vol. 12. Pergamon Press, pp. 48–62.
Hackman, R.H., 1960. Studies on chitin. IV. The occurrence of complexes in which chitin and protein are covalently linked. Aust. J. Biol. Sci. 13, 568–577.
Hackman, R.H., 1962. Studies on chitin. V. The action of mineral acids on chitin. Aust. J. Biol. Sci. 15, 526–537.
Hackman, R.H., 1984. Cuticle biochemistry. In: Bereiter-Hahn, J., Matolsty, A.G., Richards, K.S. (Eds.), Biology of the Integument. vol. 1. Springer, Berlin, pp. 583–610.
Hackman, R.H., 1987. Chitin and the fine structure of cuticles. In: Wright, J.E., Retnakaran, A. (Eds.), Chitin and Benzophenyl Ureas. Dr W. Junk Publishers, Dordrecht, pp. 1–32 (Chapter 1).

Hadwiger, L.A., 2017. Chitosan—molecular forms with potential in agriculture and medicine. J. Drug Des. Res. 4, 1036.
Hagenbach, D., Werthmüller, L., Grof, S., 2013. Mystic Chemist: The Life of Albert Hofmann and his Discovery of LSD (First English edition). Synergetic Press, Santa Fe, NM, ISBN: 978-0-907791-46-1, p. 16.
Hahn, T., Tafi, E., Paul, A., Salvia, R., Falabella, P., Zibek, S., 2020. Current state of chitin purification and chitosan production from insects. J. Chem. Technol. Biotechnol. 95, 2775–2795.
Hajjar, N.P., Casida, J.E., 1979. Structure-activity relationships of a series of benzoylphenyl ureas as toxicants and chitin synthesis inhibitors in *Oncopeltus fasciatus*. Pestic. Biochem. Physiol. 11, 33–45.
Hale, A.J., 1957. The histochemistry of polysaccharides. Int. Rev. Cytol. 6, 193–263.
Haliburton, W.D., 1885. On the occurrence of chitin as a constituent of the cartilages of *Limulus* and *Sepia*. Q. J. Microsc. Sci. 25, 173–181.
Hamed, I., Özogul, F., Regenstein, J.M., 2016. Industrial applications of crustacean by-products (chitin, chitosan, and chitooligosaccharides): a review. Trends Food Sci. Technol. 48, 40–50.
Hamedi, H., Moradi, S., Hudson, S.M., Tonelli, A.E., King, M.W., 2022. Chitosan based bioadhesives for biomedical applications: a review. Carbohydr. Polym. 282, 119100.
Hamlin, M.L., 1911. A tetra-acetyl aminoglucoside. J. Am. Chem. Soc. 33, 766–769.
Han, J.W., Ruiz-Garcia, L., Qian, J.P., Yang, X.T., 2018. Food packaging: a comprehensive review and future trends. Compr. Rev. Food Sci. Food Saf. 17, 860–877.
Han, J.J., Liu, L., Fan, Z.H., Zhang, Z.L., Yang, S.Y., Tang, Y., 2020. Grafting cosmetic active ingredients for the functionalization of Cosmetotextiles. IOP Conf. Ser. Mater. Sci. Eng. 782, 022026.
Harikrishnan, R., Kim, J.S., Balasundaram, C., Heo, M.S., 2012. Immunomodulatory effects of chitin and chitosan enriched diets in *Epinephelus bruneus* against *Vibrio alginolyticus* infection. Aquaculture 326–329, 46–52.
Hase, W., 1916. Über die struktur des chitins bei arthropoden. Arch. Anat. Physiol. Wiss. 5 (6), 295–338.
Hatchett, C., 1799. XVIII. Experiments and observations on shell and bone. Philos. Trans. R. Soc. Lond. 89, 315–334. Bowyer W and Nichols J, Eds.
Hatchett, C., 1800. VI. Experiments and obfervations on shell and bone. J. Nat. Philos. Chem. Arts III, 500–506. Nicholson W, Ed.
Hatchett, C., 1800. II. Experiments and obfervations on shell and bone (concluded). J. Nat. Philos. Chem. Arts III, 529–534. Nicholson W, Ed.
Haworth, W.N., 1946. The structure, function and synthesis of polysaccharides. Proc. R. Soc. Lond. Ser. A 186, 1–19.
Haworth, W.N., Lake, W.H.G., Peat, S., 1939. The configuration of glucosamine (chitosamine). J. Chem. Soc., 271–274.
Hayashi, Y., Ohara, N., Ganno, T., Yamaguchi, K., Ishizaki, T., Nakamura, T., Sato, M., 2007. Chewing chitosan containing gum effectively inhibits the growth of cariogenic bacteria. Arch. Oral Biol. 52, 290–294.
Hayashi, Y., Yanagiguchi, K., Koyama, Z., Ikeda, T., Yamada, S., 2013. Chitosan application in dentistry. In: Kim, S.K. (Ed.), Marine Nutraceuticals. Prospects and perspectives. CRC Press, Taylor & Francis Group, Boca Raton, pp. 233–242 (Chapter 16).
Hayes, M., Carney, B., Slater, J., Brück, W., 2008a. Mining marine shellfish wastes for bioactive molecules: chitin and chitosan. Part B: applications. Biotechnol. J. 3, 878–889.
Hayes, M., Carney, B., Slater, J., Brück, W., 2008b. Mining marine shellfish wastes for bioactive molecules: chitin and chitosan. Part A: extraction methods. Biotechnol. J. 3, 871–877.
Heckert, W.W., 1937. Textile. US Patent 2,099,363.

Helferich, B., 1969. Max Bergmann 1886-1944. Chem. Ber. 102, I–XXVI.

Helgason, T., Gislason, J., McClements, D.J., Kristbergsson, K., Weiss, J., 2009. Influence of molecular character of chitosan on the adsorption of chitosan to oil droplet interfaces in an *in vitro* digestion model. Food Hydrocoll. 23, 2243–2253.

Herzog, R.O., 1924. Über den feinbau der faserstoffe. Naturwissenschaften 12, 955–960.

Herzog, R.O., Gonell, H.W., 1924. Weitere untersuchungen an naturstoffen und biologischen strukturen mittels röntgenstrahlen. Naturwissenschaften 12, 1153–1155.

Heuser, M., Cárdenas, G., 2014. Chitosan-copper paint types as antifouling. J. Chil. Chem. Soc. 59, 2415–2419.

Heuser, M., Rivera, C., Nuñez, C., Cárdenas, G., 2009. Chitosan-copper paint types as antifouling. J. Chil. Chem. Soc. 54, 448–453.

Heyn, A.N.J., 1936a. X ray investigations on the molecular structure of chitin in cell walls (Preliminary note). Proc. K. Akad. Wet. Amst. 39, 132–135.

Heyn, A.N.J., 1936b. Further investigations on the mechanism of cell elongation and the properties of the cell wall. IV. Investigations on the molecular structure of chitin cell wall of sporangiophores of phycoyces and its probable bearing on the phenomenon of spiral growth. Protoplasma 25, 372–396.

Heyn, A.N.J., 1936c. Molecular structure of chitin in plant cell-walls. Nature 137, 277–278.

Hirano, S., 1989. Production and application of chitin and chitosan in Japan. In: Skjåk-Braek, G., Anthonsen, T., Sandford, P.A. (Eds.), Chitin and Chitosan. Sources, Chemistry, Biochemistry, Physical Properties and Applications. Proceedings From the 4th International Conference on Chitin and Chitosan Held in Trondheim, Norway, August 22-24 1988. Elsevier Applied Science, New York, pp. 37–43.

Hirano, S., 1996. Chitin biotechnology applications. Biotechnol. Annu. Rev. 2, 237–258.

Hirano, S., Inui, H., Kosaki, H., Uno, T., Toda, T., 1994. Chitin and chitosan: ecologically bioactive polymers in biotechnology and bioactive polymers. In: Gebelein, C., Carraher, C. (Eds.), Biotechnology and Bioactive Polymers. Plenum Press, New York, pp. 43–54.

Hirst, E.L., 1953. James Colquhoun Irvine 1877-1952. Adv. Carbohydr. Chem. 8, XI–XVII.

Hofmann, A., 1929. Über den Enzymatischen Abbau des Chitins und Chitosans (Dissertation). Universität Zürich. 50 p.

Holmgren, N., 1902a. Über das verhalten des chitins und epithels zu den unterliegenden gewebearten bei insekten. Anat. Anz. 20, 480–488.

Holmgren, N., 1902b. Über die morphologische bedeutung des chitins bei insekten. Anat. Anz. 21, 372–378.

Holmgren, N., 1910. Über die muskelinsertion an das chitin bei den arthropoden. Anat. Anz. 36, 116–122.

Hong, S., Yuan, Y., Yang, Q.R., Zhu, P., Lian, H.L., 2018. Versatile acid base sustainable solvent for fast extraction of various molecular weight chitin from lobster shell. Carbohydr. Polym. 201, 211–217.

Honke, L., Scheer, B.T., 1970. Carbohydrate metabolism in Crustacea. In: Florkin, M., Scheer, B. (Eds.), Chemical Zoology. vol. 5. Academic Press, New York.

Hopff, H., 1953. In memoriam Kurt H. Meyer 1883-1952. Stärke 5, 53–55.

Hopkins, E.W., 1929. Microchemical tests on the cell walls of certain fungi. Cellulose and chitin. Trans. Wis. Acad. Sci. Arts Lett. XXIV, 187–196.

Hoppe-Seyler, F., 1881. Ueber die veränderungen des blutes bei verbrennungen der haut. Z. Physiol. Chem. 5, 1–9.

Hoppe-Seyler, F., 1894. Ueber chitin und cellulose. Ber. Dtsch. Chem. Ges. 27, 3329–3331.

Hoppe-Seyler, F., 1895. Ueber umwandlungen des chitins. Ber. Dtsch. Chem. Ges. 28, 82.

Horowitz, S.T., Roseman, S., Blumenthal, H.J., 1957. The preparation of glucosamine oligosaccharides. 1. Separation. J. Am. Chem. Soc. 79, 5046–5049.

Horst, M.N., Walker, A.N., Klar, E., 1993. The pathway of crustacean chitin synthesis. In: Horst, M.N., Freeman, J.A. (Eds.), The Crustacean Integument—Morphology and Biochemistry. CRC Press, Boca Raton, pp. 113–149 (Chapter 4).

Hossin, M.A., Al Shaqsi, N.H.K., Al Touby, S.S.J., Al Sibani, M.A., 2021. A review of polymeric chitin extraction, characterization, and applications. Arab. J. Geosci. 14, 1870.

Houwink, A.L., Kreger, D.R., 1953. The cell wall of yeasts. Electron microscope and X-ray diffraction study. Antonie Van Leeuwenhoek 19, 1–24.

Hudson, S.M., Smith, C., 1998. Polysaccharides: chitin and chitosan. Chemistry and technology of their use as structural materials. In: Kaplan, D.L. (Ed.), Biopolymers From Renewable Resources. Springer-Verlag, Berlin, pp. 96–118 (Chapter 4).

Huq, T., Khan, A., Brown, D., Dhayagude, N., He, Z., Ni, Y., 2022. Sources, production and commercial applications of fungal chitosan: a review. J. Bioresour. Bioprod. 14, 50.

Iber, B.T., Kasan, N.A., Torsabo, D., Omuwa, J.W., 2022. A review of various sources of chitin and chitosan in nature. J. Renew. Mater. 10, 1097–1123.

Ifuku, S., Saimoto, H., 2012. Chitin nanofibers: preparations, modifications, and applications. Nanoscale 4, 3308–3318.

Ilkewitsch, K., 1908. Fungal chitin named mycetin. Bull. Acad. Imp. Sci. St. Pétersbg., 571–588.

Illum, L., 1998. Chitosan and its use as a pharmaceutical excipient. Pharm. Res. 15, 1326–1331.

Illum, L., Davis, S.B.S., 2004. Chitosan as delivery system for the transmucosal administration of drugs. In: Dumitriu, S. (Ed.), Polysaccharides. Structural Diversity and Functional Versatility. Marcel Dekker, New York, pp. 643–659 (Chapter 27).

Imms, A.D., 1924. A General Textbook of Entomology. Methuen & Co., Publisher, London.

Ippólito, S.D., Mendieta, J.R., Terrile, M.C., Tonón, C.V., Mansilla, A.Y., Colman, S., Albertengo, L., Rodríguez, M.S., Casalongué, C.A., 2017. Chitosan as source for pesticides formulations. In: Shalaby, E. (Ed.), Biological Activities and Applications of Marine Polysaccharides. InTech Publishers, pp. 3–15 (Chapter 1).

Irvine, J.C., 1909. A polarimetric method of identifying chitin. J. Chem. Soc. 95, 564–570.

Irvine, J.C., Hynd, A., 1912. The conversion of d-glucosamine into d-glucose. J. Chem. Soc. 101, 1128–1146.

Irvine, J.C., Hynd, A., 1914. The conversion of d-glucosamine into d-mannose. J. Chem. Soc. 105, 698–710.

Irvine, J.C., McNicoll, D., Hynd, A., 1911. New derivatives of d-glucosamine. J. Chem. Soc. 99, 250–261.

Islam, S.U., Shahid, M., Mohammad, F., 2013. Green chemistry approaches to develop antimicrobial textiles based on sustainable biopolymers—a review. Ind. Eng. Chem. Res. 52, 5245–5260.

Ito, H., 1924. Contribution histologique et physiologique à l'étude des annexes des organes génitaux des orthopèdes (tubes glandulaires ou glandes annexes, spermathèques, vésicules séminales, glandes prostatiques). Arch. Anat. Microsc. 20, 343–460.

Jaber, N., Al-Ramawi, M., Al-Akayleh, F., Al-Muhtaseb, N., Al-Adham, I.S.I., Collier, P.J., 2022. A review of the antiviral activity of chitosan, including patented applications and its potential use against COVID-19. J. Appl. Microbiol. 132, 41–58.

Jacob, J., Loganathan, S., Thomas, S., 2020. Chitin- and Chitosan-Based Biocomposites for Food Packaging Applications. CRC Press, Boca Raton, ISBN: 978-0-367-28090-1.

Jana, S., Jana, S., 2019. Functional Chitosan. Drug Delivery and Biomedical Applications. Springer Nature Singapore Pte. Ltd., ISBN: 978-981-15-0262-0.

Jangid, N.K., Hada, D., Rathore, K., 2019. Chitosan as an emerging object for biological and biomedical applications. J. Polym. Eng. 39, 689–703.

Jaworska, M., Kozlecki, T., Gorak, A., 2012. Review of the application of ionic liquids as solvents for chitin. J. Polym. Eng. 32, 67–69.

Jaworska, M.M., Antos, D., Gorak, A., 2020. Review on the application of chitin and chitosan in chromatography. React. Funct. Polym. 152, 104606.

Jayakumar, R., Prabaharan, M., Muzzarelli, R.A.A., 2011. In: Jayakumar, R., Prabaharan, M., Muzzarelli, R.A.A. (Eds.), Chitosan for Biomaterials I. Springer, Heidelberg, ISBN: 978-3-642-23113-1. 236 p.

Je, J.Y., Kim, S.K., 2012a. Chitosan as potential marine nutraceutical. Adv. Food Nutr. Res. 65, 121–135.

Je, J.Y., Kim, S.K., 2012b. Chitooligosaccharides as potential nutraceuticals: production and bioactivities. Adv. Food Nutr. Res. 65, 321–336.

Jeanloz, R., 1950. Structure de la chitine et de l'acide hyaluronique et oxydation des sucres aminés par l'ion périodate. Experentia 6, 52–53.

Jeanloz, R.W., 1956. Kurt Heinrich Meyer: 1883-1952. Adv. Carbohydr. Chem. 11, XIII–XVIII.

Jeanloz, R., Forchielli, E., 1950. Recherches sur l'acide hyaluronique et les substances apparentées III. La détermination de la structure de la chitine par oxydation avec l'ion periodate. Helv. Chim. Acta 33, 1690–1697.

Jeanloz, R., Forchielli, E., 1951. Studies on hyaluronic acid and related substances. 2. Periodate oxidation of glucosamine and derivatives. J. Biol. Chem. 188, 361–369.

Jeuniaux, C., 1950. Production d'une exochitinase par des bactéries chitinolytiques isolées du contenu intestinal de l'escargot. Arch. Int. Physiol. 58, 352–353.

Jeuniaux, C., 1963. Chitine et Chitinolyse, un Chapitre de la Biologie Moléculaire. Masson, Paris.

Jeuniaux, C., 1971. Chitinous structures. In: Florkin, M., Stotz, E.H. (Eds.), Comprehensive Biochemistry. vol. 26C. Elsevier, Amsterdam, pp. 595–632.

Jeuniaux, C., 1982. La chitine dans le règne animal. Bull. Soc. Zool. Fr. 107, 363–386.

Jiménez-Ocampo, R., Valencia-Salazar, S., Elisa Pinzón-Díaz, C., Herrera-Torres, E., Aguilar-Pérez, C.F., Arango, J., Ku-Vera, J.C., 2019. The role of chitosan as a possible agent for enteric methane mitigation in ruminants. Animals 9, 942.

Jimtaisong, A., Saewan, N., 2014. Utilization of carboxymethyl chitosan in cosmetics. Int. J. Cosmet. Sci. 36, 12–21.

Jo, G.H., Park, R.D., Jung, W.J., 2011. Enzymatic production of chitin from crustacean shell waste. In: Kim, S.K. (Ed.), Chitin, Chitosan, Oligosaccharides and their Derivatives: Biological Activities and Applications. CRC Press, Taylor & Francis Group LLC, Boca Raton, pp. 37–45 (Chapter 4).

Johnson, W.B., 1803. History of the Progress and Present State of Animal Chemistry. vol. 1 New York Public Library. Printed for Johnson J, St Paul's Churchyard. 411 p.

Joseph, S.M., Krishnamoorthy, S., Paranthaman, R., Moses, J.A., Anandharamakrishnan, C., 2021. A review on source-specific chemistry, functionality, and applications of chitin and chitosan. Carbohydr. Polym. Technol. Appl. 2, 100036.

Kaczmarek, M.B., Struszczyk-Swita, K., Li, X.K., Szczesna-Antczak, M., Daroch, M., 2019. Enzymatic modifications of chitin, chitosan and chitooligosaccharides. Front. Bioeng. Biotechnol. 7, 243.

Kadokawa, J.I., 2019. Dissolution, derivatization, and functionalization of chitin in ionic liquid. Int. J. Biol. Macromol. 123, 732–737.

Kaminsky, B., 2018. John George Children (1777-1852). Arizona State University. School of Life Sciences. Center for Biology and Socieyt.

Karadeniz, F., Kim, S.K., 2014. Antidiabetic applications of chitosan and its derivatives. In: Kim, S.K. (Ed.), Chitin and Chitosan Derivatives. Advances in Drug Discovery and Developments. CRC Press, Taylor & Francis Group LLC, Boca Raton, pp. 191–200 (Part II, Chapter 10).

Karimi-Maleh, H., Ayati, A., Davoodi, R., Tanhaei, B., Karimi, F., Malekmohammadi, S., Orooji, Y., Fu, L., Sillanpää, M., 2021. Recent advances in using of chitosan-based adsorbents for removal of pharmaceutical contaminants: a review. J. Clean. Prod. 291, 125880.

Karrer, P., 1930. Der enzymatische abbau von nativer und umgefällter zellulose, von kunstseiden und von chitin. Kolloid Z. 52, 304–319.

Karrer, P., Hofmann, A., 1929. Polysaccharide XXXIX. Über den enzymatischen abbau von chitin und chitosan I. Helv. Chim. Acta 12, 616–637.

Karrer, P., Mayer, J., 1937. Ein neuer abbau der glucosaminsäure. Die Konfiguration der glucosamin- und chondrosalinsäure. Helv. Chim. Acta 20, 407–417.

Karrer, P., Smirnoff, A.P., 1922. Polysaccharide XVII. Beitrag zur kenntnis des chitins. Helv. Chim. Acta 5, 832–852.

Karrer, P., von François, G., 1929. Polysaccharide XXXX. Über den enzymatischen abbau von chitin II. Helv. Chim. Acta 12, 986–988.

Karrer, P., White, S.M., 1930. Polysaccharide XLIV. Weitere beiträge zur kenntnis des chitins. Helv. Chim. Acta 13, 1105–1113.

Karrer, P., Schnider, O., Smirnoff, A.P., 1924. Polysaccharide XXIX. Zur kenntnis des chitins II und konfiguration des glucosamins. Helv. Chim. Acta 7, 1039–1045.

Karrer, P., Koenig, H., Usteri, E., 1943. Zur kenntnis blutgerinnungshemmender polysaccharid-poly-schwefelsäure-ester und ähnilicher verbindungen. Helv. Chim. Acta 26, 1296–1315.

Kaseamchochoung, C., Lertsutthiwong, P., Phalakornkule, C., 2006. Influence of chitosan characteristics and environmental conditions on flocculation of anaerobic sludge. Water Environ. Res. 78, 2210–2216.

Kasiri, M.B., 2018. Application of chitosan derivatives as promising adsorbents for treatment of textile wastewater. In: The Impact and Prospects of Green Chemistry for Textile Technology. The Textile Institute Book Series, Woodhead Publishing, pp. 417–469 (Chapter 14).

Katiyar, D., Hemantaranjan, A., Singh, B., Bhanu, N., 2014. A future perspective in crop protection: chitosan and its oligosaccharides. Adv. Plants Agric. Res. 1, 00006.

Kato, Y., Onishi, H., Machida, Y., 2003. Application of chitin and chitosan derivatives in the pharmaceutical field. Curr. Pharm. Biotechnol., 303–309.

Katsoyannis, P.G., 1973. The scientific work of Leonidas Zervas. In: Katsoyannis, P.G. (Ed.), The Chemistry of Polypeptides. Springer, Boston, pp. 1–20.

Kawabe, K., 1934. Biochemical studies on carbohydrates. X. The fate of glucosamine in the animal body. J. Biochem. 20, 293–310.

Ke, C.L., Deng, F.S., Chuang, C.Y., Lin, C.H., 2021. Antimicrobial actions and applications of chitosan. Polymers 13, 904.

Kean, T., Thanou, M., 2010. Biodegradation, biodistribution and toxicity of chitosan. Adv. Drug Deliv. Rev. 62, 3–11.

Kean, T., Thanou, M., 2011. Chitin and chitosan: sources, production and medical applications. In: Williams, P.A. (Ed.), Renewable Resources for Functional Polymers and Biomaterials. Polysaccharides, Proteins and Polyesters. RSC Polymer Chemistry Series, RSC Publishing, pp. 292–318 (Chapter 10).

Keegan, G.M., Smart, J.D., Ingram, M.J., Barnes, L.M., Burnett, G.R., Rees, G.D., 2012. Chitosan microparticles for the controlled delivery of fluoride. J. Dent. 40, 229–240.

Kent, P.W., 1964. Chitin and mucosubstances. In: Florkin, M., Mason, H.S. (Eds.), Comparative Biochemistry. A Comprehensive Treatise. vol. VII. Academic Press, New York, pp. 93–136 (Chapter 2).

Kent, P.W., Whitehouse, M.W., 1955. Biochemistry of the Amino-Sugars. Butterworths Scientific Publications, London. 311 p.

Khajavian, M., Vatanpour, V., Castro-Muñoz, R., Boczkaj, G., 2022. Chitin and derivative chitosan-based structures—preparation strategies aided by deep eutectic solvents: a review. Carbohydr. Polym. 275, 118702.

Khan, A., Alamry, K.A., 2021. Recent advances of emerging green chitosan-based biomaterials with potential biomedical applications: a review. Carbohydr. Res. 506, 108368.

Khor, E., 2001. In: Khor, E. (Ed.), Chitin: Fulfilling a Biomaterials Promise, first ed. Elsevier Ltd. 148 p.

Khouvine, Y., 1932. Étude aux rayons X de la chitine d'*Aspergillus niger*, de *Psalliota campestris* et d'*Armillaria mellea*. C. R. Séances Hebd. Acad. Sci. 195, 396–397.

Kim, S.K. (Ed.), 2011. Chitin, Chitosan, Oligosaccharides and their Derivatives: Biological Activities and Applications. CRC Press, Taylor & Francis Group, LLC, Boca Raton, ISBN: 9781439816042. 666 p.

Kim, S.K. (Ed.), 2012. Chitosan and fish collagen as biomaterials for regenerative medecine—biocompatibility and allergy. In: Advances in Food and Nutrition Research Volume 65, Marine Medicinal Foods: Implications and Applications—Animals and Microbes. Elsevier.

Kim, S.K. (Ed.), 2014. Chitin and Chitosan Derivatives—Advances in Drug Discovery and Developments. CRC Press Taylor & Francis Group, LLC., Boca Raton, ISBN: 978-1-4665-6628-6. 511 p.

Kim, S.K. (Ed.), 2022. Chitooligosaccharides. Prevention and Control of Diseases. Springer, Cham, ISBN: 978-3-030-92806-3. 352 p.

Kim, S.K., Karadeniz, F., 2013. Chitosan and its derivatives for treatment of diabetic complications. In: Kim, S.K. (Ed.), Marine Pharmacognosy. Trends and Applications. CRC Press, Taylor & Francis Group, Boca Raton, pp. 191–200 (Chapter 16).

Kim, S.K., Choi, J.H., Balmaceda, E.A., Rha, C.K., 1999. Chitosan. In: Kühtreiber, W.M., Lanza, R.P., Chick, W.L. (Eds.), Cell Encapsulation Technology and Therapeutics. Part II. Encapsulation systems. Section 1, Birkhäuser, Boston, ISBN: 978-1-4612-1586-8, pp. 151–172 (Chapter 13).

Klobb, T., 1907. Braconnot H. Compte Rendu de la Séance Solennelle de Rentrée de l'Université de Nancy. 8 novembre 1906, Imprimerie de l'Est, Nancy, p. 36.

Kmiec, M., Pighinelle, L., Tedesco, M.F., Silva, M.M., Reis, V., 2017. Chitosan-properties and applications in dentistry. Adv. Tissue Eng. Regen. Med. 2, 00035.

Knecht, E., Hibbert, E., 1926. Some observations relating to chitin. J. Soc. Dye. Colour. 42, 343–345.

Koch, C., 1932. Der nachweis des chitins in tierischen skeletsubstanzen. Z. Morphol. Okol. Tiere 25, 730–756.

Koch-Weser, J., Schechter, P.J., 1978. Schmiedeberg in Strassburg 1872-1918: the making of modern pharmacology. Life Sci. 22, 1361–1371.

Komori, Y., 1926. Zur kenntnis der glukosaminverbindungen. J. Biochem. 6, 1–20.

Kong, D., Foley, S.R., Wilson, L.D., 2022. An overview of modified chitosan adsorbents for the removal of precious metals species from aqueous media. Molecules 27, 978.

Kos, L., 2016. Use of chitosan for textile wastewater decolourization. Fibres Textiles 24, 130–135.

Kotake, J., Sera, Y., 1913. Über ein neue glukosaminverbindungen, zugleich ein beitrag zur konstitutionsfrage des chitins. Z. Physiol. Chem. 88, 56–72.

Kou, S.G., Peters, L.M., Mucalo, M.R., 2021. Chitosan: a review of source and preparation methods. Int. J. Biol. Macromol. 169, 85–94.

Kou, S., Peters, L., Mucalo, M., 2022. Chitosan: a review of molecular structure, bioactivities and interactions with the human body and micro-organisms. Carbohydr. Polym. 282, 119132.

Krajewska, B., 1991. Chitin and its derivatives as supports for immobilization of enzymes. Acta Biotechnol. 11, 269–277.

Krajewska, B., 2005. Membrane-base processes performed with use of chitin/chitosan materials. Sep. Purif. Technol. 41, 305–312.

Krawkow, N.P., 1892. Über verschiedenartige chitine. Z. Biol. 29, 177–198.

Kreger, D.R., 1954. Observations on cell walls of yeats and some other fungi by X-ray diffraction and solubility tests. Biochim. Biophys. Acta 13, 1–9.

Krishnaswami, V., Kandansamy, R., Alagarsamy, S., Palanisamy, R., Natesan, S., 2018. Biological macromolecules for ophthalmic drug delivery to treat ocular diseases. Int. J. Biol. Macromol. 110, 7–19.

Krukenberg, 1886. Die angebliche löslichkeit des chitins. Z. Biol. 22, 480–488.

Kucharska, M., Sikora, M., Brzoza-Malczewska, K., Owczarek, M., 2019. Antimicrobial properties of chitin and chitosan. In: Van den Broek, L.A.M., Boeriu, C.G. (Eds.), Chitin and Chitosan: Properties and Applications. John Wiley & Sons, Ltd, pp. 169–187 (Chapter 7).

Kühnelt, W., 1928a. Studien über den mikrochemischen nachweis des chitins. Biol. Zentralbl. 48, 374–382.

Kühnelt, W., 1928b. Ein beitrag zur histochemie des insekten skelettes. Zool. Anz. 75, 111–115.

Kulagin, N.M., 1939. On the structure of chitin in insects. Proc. Lenin Acad. Agric. Sci. USSR Mosc. (7), 21–23.

Kumar, S., Madihally, S., 2022. Role of Chitosan and Chitosan-Based Nanomaterials in Plant Sciences. Elsevier Science, ISBN: 9780323853910. 500 p.

Kumar, A.B.V., Tharanathan, R.N., 2004. A comparative study on depolymerization of chitosan by proteolytic enzymes. Carbohydr. Polym. 58, 275–283.

Kumara, B.N., Shambhu, R., Prasad, K.S., 2021. Why chitosan could be apt candidate for glaucoma drug delivery—an overview. Int. J. Biol. Macromol. 176, 47–65.

Kumari, S., Kishor, R., 2020. Chitin and chitosan: origin, properties, and applications. In: Gopi, S., Thomas, S., Pius, A. (Eds.), Handbook of Chitin and Chitosan. Volume 1: Preparation and Properties. Elsevier, pp. 1–33 (Chapter 1).

Kunike, G., 1925. Nachweis und verbreitung organischer skeletsubstanzen bei tieren. Z. Vgl. Physiol. 2, 233–253.

Kunike, G., 1926a. Chitin und chitinseide. Kunstseide (Chemiefasern) 8, 182–188.

Kunike, G., 1926b. Chitin and chitin-silk. Chem. Zentralbl. 97, 2129–2130.

Kunike, G., 1926c. Chitin and chitin artificial silk. J. Soc. Dye. Colour. 42, 228–230.

Kurita, K., 1998. Chemistry and application of chitin and chitosan. Polym. Degrad. Stab. 59, 117–120.

Kurita, K., 2001. Controlled functionalization of the polysaccharide chitin. Prog. Polym. Sci. 26, 1921–1971.

Kurita, K., 2006. Chitin and chitosan: functional biopolymers from marine crustaceans. Mar. Biotechnol. 8, 203–226.

Kyzas, G.Z., Bikiaris, D.N., Mitropoulos, A.C., 2017. Chitosan adsorbents for dye removal: a review. Polym. Int. 66, 1800–1811.

Labrude, P., 1997. De quelques pharmaciens célèbres, aux noms sur des plaques de rues ou ignorées par elles, à Nancy et dans l'agglomération nancéienne. Bull. Acad. Soc. Lorraines Sci. 36, 171–181.

Labrude, P., Becq, C., 2003. Le pharmacien et chimiste Henri Braconnot. Rev. Hist. Pharm. 337, 61–78.

Lacroix, E., 1923. Texture chitineuse fondamentale de la coquille des foraminifères porcelanés. C. R. Hebd. Acad. Sci. 176, 1673–1674.

Lafon, M., 1941a. Sur la composition du tégument des crustacés. C. R. Séances Soc. Biol. 135, 1003–1007.

Lafon, M., 1941b. Sur l'élytre de quelques coléoptères. Contribution à l'étude chimique de tégument des insectes. Bull. Soc. Zool. Fr. 66, 173–182.

Lafon, M., 1943a. Recherches sur le cycle de la chitine chez quelques coléoptères. Arch. Zool. Exp. Gén. 83, 58–70.

Lafon, M., 1943b. Recherches biochimiques et physiologiques sur le squelette tégumentaire des arthropodes. Ann. Sci. Nat. Zool. Biol. Anim. 5, 113–146.

Lafon, M., 1943c. Sur la structure et la composition chimique du tégument de la Limule (*Xiphosura polyphemus* L.). Bull. Inst. Océanogr. (850). 11 p.

Lafon, M., 1948. Nouvelles recherches biochimiques et physiologiques sur le squelette tégumentaire des crustacés. Bull. Inst. Océanogr. Monaco 45, 1–28.

Lahiji, A., Sohrabi, A., Hungerford, D.S., Frondoza, C.G., 2000. Chitosan supports the expression of extracellular matrix proteins in human osteoblasts and chondrocytes. J. Biomed. Mater. Res. 51, 586–595.

Lang, X.Q., Wang, T., Sun, M.J., Chen, X.G., Liu, Y., 2020. Advances and applications of chitosan-based nanomaterials as oral delivery carriers: a review. Int. J. Biol. Macromol. 154, 433–445.

Laroque, B.R.L., 1860. Notice sur Jean-Louis Lassaigne. J. Vét. Midi 3, 491–500.

Larson, L.L., 1939. Manufacture of Paper. US Patent 2,184,307.

Larson, L.L., 1940. Manufacture of Paper. US Patent 2,216,845.

Lassaigne, J.L., 1843a. Mémoire sur un procédé simple pour constater la présence d'azote dans des quantités minimes de matière organique. C. R. Séances Hebd. Acad. Sci. 16, 387–391.

Lassaigne, J.L., 1843b. Sur le tissu tégumentaire des insectes des différents ordres. C. R. Séances Hebd. Acad. Sci. 16, 1087–1089.

Lassaigne, J.L., 1843c. Über die hautgewebe der insecten verschiedener ordnungen. J. Prakt. Chem. 29, 323–326.

Latreille, P.A., 1831. Cours D'entomologie ou L'histoire Naturelle des Crustacés, des Arachnides, des Myriapodes et des Insectes. Imprimerie de Crapelet, Paris. 669 p.

Leboucq, H., 1913. E. Gilson (1890). In: Notices Biographiques. vol. II. Université de Gand. Maison d'édition I. Vanderpoorten, pp. 575–577.

Ledderhose, G., 1876. Ueber salzsaures glycosamin. Ber. Dtsch. Chem. Ges. 9, 1200–1201.

Ledderhose, G., 1878. Ueber chitin und seine spaltungsprodukte. Z. Physiol. Chem. 2, 213–227.

Ledderhose, G., 1880a. Ueber Glykosamin. Dissertation der medicinischen Facultät der Kaiser Wilhelms. Universität Strassburg zur Erlangung der Doctorwürde. 24 p.

Ledderhose, G., 1880b. Ueber glykosamin. zeitschrift für physiologische. Chemie 4, 139–159.

Ledderhose, G., 1894. ueber Zerreisungen der Plantarfascie. Arch. Klin. Chir. 48, 853–856.

Levene, P.A., 1925. Hexosamines and Mucoproteins. Monographs on Biochemistry, Longmans, Green and Co., London.

Levene, P.A., La Forge, F.B., 1914. On chondroitin sulphuric acid: third paper. J. Biol. Chem. 18, 123–130.

Levene, P.A., La Forge, F.B., 1915. On chondroitin sulphuric acid: fourth paper. J. Biol. Chem. 20, 433–444.

Lezica, R.P., Quesada-Allué, L., 1990. Chitin. In: Dey, P.M. (Ed.), Methods in Plant Biochemistry. Volume 2: Carbohydrates. Academic Press Limited, London, ISBN: 0-12-461012-9, pp. 443–481 (Chapter 13).

Lian, Z., Pan, R., Wang, J., 2016. Microencapsulation of norfloxacin in chitosan/chitosan oligosaccharides and its application in shrimp culture. Int. J. Biol. Macromol. 92, 587–592.

Liao, F.H., Shieh, M.J., Chang, N.C., Chien, Y.W., 2007. Chitosan supplementation lowers serum lipids and maintains normal calcium, magnesium, and iron status in hyperlipidemic patients. Nutr. Res. 27, 146–151.

Liaqat, F., Eltem, R., 2018. Chitooligosaccharides and their biological activities: a comprehensive review. Carbohydr. Polym. 184, 243–259.

Lichtenthaler, F.W., 2002. Emil Fischer, his personality, his achievements, and his scientific progeny. Eur. J. Org. Chem., 4095–4122.

Lichtfouse, E., Morin-Crini, N., Fourmentin, M., Zemmouri, H., do Carmo Nascimento, I. O., Queiroz, L.M., MYM, T., Picos-Corrales, L.A., Pei, H., Wilson, L.D., Crini, G., 2019. Chitosan for direct bioflocculation of wastewater. Environ. Chem. Lett. 17, 1603–1621.

Lima, E.L., Muñoz, L.C., Harris, R.E., Caballaro, A.M.H., 2012. Potential applications of chitosan as a marine cosmeceutical. In: Kim, S.K. (Ed.), Marine Cosmeceuticals. Trends and Prospects. CRC Press, Taylor & Francis Group LLC, Boca Raton, pp. 319–335 (Chapter 24).

Lima, B.V., Oliveira, M.J., Barbosa, M.A., Gonçalves, R.M., Castro, F., 2021. Immunomodulatory potential of chitosan-based materials for cancer therapy: a systematic review of in vitro, in vivo and clinical studies. Biomater. Sci. 9, 3209.

Liming, Z. (Ed.), 2019. Oligosaccharides of Chitin and Chitosan: Bio-Manufacture and Applications. Springer Verlag, Singapore, ISBN: 978-981-13-9401-0.

Lin, F., Zhu, X., Luo, Y., Liu, M., 2019. Improvement of activated sludge dewatering properties using green conditioners: chitosan hydrochloride and lysozyme. RSC Adv. 9, 6936–6945.

Liu, C., Bai, R., 2014. Recent advances in chitosan and its derivatives as adsorbents for removal of pollutants from water and wastewater. Curr. Opin. Chem. Eng. 4, 62–70.

Lizundia, E., Nguyen, T.D., Winnick, R.J., MacLachlan, M.J., 2021. Biomimetic photonic materials derived from chitin and chitosan. J. Mater. Chem. C 9, 796.

Lobry de Bruyn, C.A., Franchimont, A.P.N., 1898. Das freie chitosamin. Berichte 31, 2476–2477.

Lobry de Bruyn, C.A., van Ekenstein, A.W., 1899. La chitosamine libre. Recl. Trav. Chim. Pays-Bas 18, 77–85.

Lobry de Bruyn, C.A., van Ekenstein, A.W., 1899. In: Masson et Cie (Ed.), Sur la Chitosamine. Bulletin de la Société Chimique. vol. XXI. Libraires de l'Académie de Médecine, Paris, pp. 748–749.

LogithKumar, R., KeshavNarayan, A., Dhivya, S., Chawla, A., Saravanan, S., Selvamurugan, N., 2016. A review of chitosan and its derivatives in bone tissue engineering. Carbohydr. Polym. 151, 172–188.

Looss, 1885. Neue lösungsmittel des chitins. Zool. Anz. 8, 333–334.

Lopes, C., Soares, J., Tavaria, F., Duarte, A., Correia, O., Sokhatska, O., Severo, M., Silva, D., Pintado, M., Delgado, L., Moreira, A., 2015. Chitosan coated textiles may improve atopic dermatitis severity by modulating skin staphylococcal profile: a randomized controlled trial. PLoS ONE 10, e0142844.

Lotmar, W., Picken, L.E.R., 1950. A new crystallographic modification of chitin and its distribution. Experientia 6, 58–59.

Löwy, E., 1909. Über kristallinisches chitosansulfat. Biochem. Z. 23, 47–60.

Lubs, H.A., 1937. Paper. US Patent 2,085,163.

Ludwig, H., 1861. Chitin. Arch. Pharm. 157, 76–77.

Ma, J.H., Faqir, Y., Tan, C.G., Khaliq, G., 2022. Terrestrial insects as a promising source of chitosan and recent developments in its application for various industries. Food Chem. 373, 131407.

Mahari, W.A.W., Waiho, K., Fazhan, H., Necibi, M.C., Hafsa, J., Mrid, R.B., Fal, S., El Arroussi, H., Peng, W.X., Tabatabei, M., Aghbashlo, M., Almomani, F., Lam, S.S., Sillanpää, M., 2022. Progress in valorisation of agriculture, aquaculture and shellfish biomass into biochemicals and biomaterials towards sustainable bioeconomy. Chemosphere 291, 133036.

Maldonado-Cabrera, B., Sánchez-Machado, D.I., López-Cervantes, J., Osuna-Chávez, R. F., Escárcega-Galaz, A.A., Robles-Zepeda, R.E., Sanches-Silva, A., 2021. Therapeutic

effects of chitosan in veterinary dermatology: a systematic review of the literature. Prev. Vet. Med. 190, 105325.
Maleki, G., Woltering, E.J., Mozafari, M.R., 2022. Applications of chitosan-based carrier as an encapsulating agent in food industry. Trends Food Sci. Technol. 120, 88–99.
Malerba, M., Cerana, R., 2019. Recent applications of chitin-and chitosan-based polymers in plants. Polymers 11, 839.
Malerba, M., Cerana, R., 2020. Chitin-and chitosan-based derivatives in plant protection against biotic and abiotic stresses and in recovery of contaminated soil and water. Polysaccharides 1, 21–30.
Maluin, F.N., Hussein, M.Z., 2020. Chitosan-based agronanochemicals as a sustainable alternative in crop protection. Molecules 25, 1611.
Manigandan, V., Karthik, R., Ramachandran, S., Rajapopal, S., 2018. Chitosan applications in food industry. In: Biopolymers for Food Design. Handbook of Food Bioengineering. Academic Press, pp. 469–491 (Chapter 15).
Marasco, M., 1938. Antistatic Photographic Film. US Patent 2,139,689.
Marasco, M., 1939. Photographic Film. US Patent 2,182,814.
Marchessault, R.H., Sarko, A., 1967. X-ray structure of polysaccharides. Adv. Carbohydr. Chem. 22, 421–482.
Mark, H., 1943. The investigation of high polymers with X-rays. In: Chemistry of Large Molecules. Interscience, New York, pp. 33–71.
Mark, H., 1952. In memoriam—Professor K. H. Meyer. J. Polym. Sci. 9, 193–195.
Marpu, S.B., Benton, E.N., 2018. Shining light on chitosan: a review on the usage of chitosan for photonics and nanomaterials research. Int. J. Mol. Sci. 19, 1795.
Massella, D., Giraud, S., Guan, J.P., Ferri, A., Salaun, F., 2019. Textiles for health: a review of textile fabrics treated with chitosan microcapsules. Environ. Chem. Lett. 17, 1787–1800.
Mathews, A.P., 1898. The life and work of Felix Hoppe-Seyler. Pop. Sci. Mon. 53, 542–552.
Mati-Baouche, N., Elchinger, P.H., de Baynast, H., Pierre, G., Delattre, C., Michaud, P., 2014. Chitosan as an adhesive. Eur. Polym. J. 60, 198–212.
Matica, A., Menghiu, G., Ostafe, V., 2017. Toxicity of chitosan based products. New Front. Chem. 26, 65–74.
Maxwell, R.W., 1939. Adhesive. US Patent 2,182,524.
McLintok, J., 1945. The insect cuticle—a review. In: Proceedings of the Entomological Society of Manitoba. vol. 1. University of Manitoba, pp. 16–29.
McQueen, D.M., Merill, W.J., 1936. Sizes. US Patent 2,047,217.
Meng, Q.G., Sun, Y., Cong, H.L., Hu, H., Xu, F.J., 2021. An overview of chitosan and its application in infectious diseases. Drug Deliv. Transl. Res. 11, 1340–1351.
Merker, E., 1929a. Die durchlässigkeit des chitins für ultravioletten licht. Verh. Dtsch. Zool. Ges. 33, 181–186.
Merker, E., 1929b. Die fluoreszenz in insektenauge, die fluoreszenz des chitins der insekten und seine durchlässigkeit für ultravioletten licht. Zool. Jahrb. 46, 483–574.
Meyer, H.H., 1922. Oswald Schmiedeberg. Naturwissenschaften 5, 105–107.
Meyer, K., 1938. The chemistry and biology of mucopolysaccharides and glycoproteins. Cold Spring Harb. Symp. Quant. Biol. 6, 91–102.
Meyer, K.H., 1942. Chitin. In: Natural and Synthetic High Polymers. Part F: Substances Related to or Associated with Cellulose, first ed. Interscience Publishers, Inc., New York, pp. 381–386 (Chapter VI).
Meyer, K.H., 1950. Chitin and other polysaccharides containing amino sugars. In: Natural and Synthetic High Polymers, second ed. Interscience Publishers, Inc., New York, pp. 448–455 (Chapter F, Part VI).
Meyer, K.H., Mark, H., 1928. Über den aufbau des chitins. Ber. Dtsch. Chem. Ges. A B Ser. 61, 1936–1939.

Meyer, K.H., Pankow, G.W., 1935. Sur la constitution et la structure de la chitine. Helv. Chim. Acta 18, 589–598.

Meyer, K.H., Wehrli, H., 1937. Comparaison chimique de la chitine et de la cellulose. Helv. Chim. Acta 20, 353–362.

Minagawa, T., Okamura, Y., Shigemasa, Y., Minami, S., Okamoto, Y., 2007. Effects of molecular weight and deacetylation degree of chitin/chitosan on wound healing. Carbohydr. Polym. 67, 640–644.

Minke, R., Blackwell, J., 1978. The structure of α-chitin. J. Mol. Biol. 120, 167–181.

Mirande, M., 1905. Sur la présence d'un corps réducteur dans le tégument chitineux des arthropodes. Arch. Anat. Microsc. Paris 7, 207–231.

Mohammadi, Z., Eini, M., Rastegari, A., Tehrani, M.R., 2021. Chitosan as a machine for biomolecule delivery: a review. Carbohydr. Polym. 256, 117414.

Mohammed, S.A., Shaaban, E.I.A., 2022. Efficacious nanomedicine track toward combating COVID-19. Rev, Nanotechnol, https://doi.org/10.1515/ntrev-2022-0036.

Mohan, K., Ganesan, A.R., Muralisankar, T., Jayakumar, R., Sathishkumar, P., Uthayakumar, V., Chandirasekar, R., Revathi, N., 2020. Recent insights into the extraction, characterization, and bioactivities of chitin and chitosan from insects. Trends Food Sci. Technol. 105, 17–42.

Morgan, W.T.J., Elson, L.A., 1934. A colourimetric method for the determination of N-acetylglucosamine and N-acetylchondrosamine. Biochem. J. 28, 988–995.

Morgulis, S., 1916. The chemical constitution of chitin. Science 44, 866–867.

Morgulis, S., 1917. An hydrolytic study of chitin. Am. J. Phys. 43, 328–342.

Morin-Crini, N., Lichtfouse, E., Torri, G., Crini, G., 2019. Applications of chitosan in food, pharmaceuticals, medicine, cosmetics, agriculture, textiles, pulp and paper, biotechnology, and environmental chemistry. Environ. Chem. Lett. 17, 1667–1692.

Mujtaba, M., Khawar, K.M., Camara, M.C., Carvalho, L.B., Fraceto, L.F., Morsi, R.E., Elsabee, M.Z., Kaya, M., Labibi, J., Ullah, H., Wang, D.P., 2020. Chitosan-based delivery systems for plants: a brief overview or recent advances and future directions. Int. J. Biol. Macromol. 154, 683–697.

Müller, F., 1901. Contributions to the knowledge of mucin and with that several related protein substances. Z. Biol. 24, 468–564.

Müller, A., 1928. Further X-ray investigation of long-chain compounds (n-hydrocarbons). Proc. R. Soc. Lond. A 120, 437–459.

Muñoz, L.C., Lima, E.L., Harris, R.E., Mengíbar, M.A.L., Contreras, N.A., Caballero, A. M.H., 2012. Chemical properties of chitosan as a marine cosmeceutical. In: Kim, S.K. (Ed.), Marine Cosmeceuticals. Trends and Prospects. CRC Press, Taylor & Francis Group LLC, Boca Raton, pp. 39–50 (Chapter 3).

Muñoz-Bonilla, A., Cerrada, M.L., Fernández-García, M. (Eds.), 2014. Antimicrobial activity of chitosan in food, agriculture and biomedicine. In: Polymeric Materials With Antimicrobial Activity. From Synthesis to Applications. RSC Polymer Chemistry Series, vol. 10. RSC Publishing. The Royal Society of Chemistry, Cambridge, pp. 22–53 (Chapter 2).

Muñoz-Núñez, C., Fernández-García, M., Muñoz-Bonilla, A., 2022. Chitin nanocrystals: Environmentally friendly materials for the development of bioactive films. Coatings 12, 144.

Musarurwa, H., Tavengwa, N.T., 2022. Advances in the application of chitosan-based metal organic frameworks as adsorbents for environmental remediation. Carbohydr. Polym. 283, 119153.

Muscholl, E., 1995. The second evolution of experimental pharmacology as a biological science: the pioneering work of Buchheim and Schmiedeberg. Br. J. Pharmacol. 116, 2155–2159.

Muscholl, E., 2001. Schmiedeberg Oswald. In: eLs. 1 p.

Muthu, M., Gopal, J., Chun, S., Devadoss, A.J.P., Hasan, N., Sivanesan, I., 2021. Crustacean waste-derived chitosan: antioxidant properties and future perspective. Antioxidants 10, 228.
Mutreja, R., Thakur, A., Goyal, A., 2020. Chitin, chitosan, marine to market. In: Handbook of Chitin and Chitosan. Volume 3: Chitin and Chitosan Based Polymer Materials for Various Applications, pp. 401–417 (Chapter 13).
Muzzarelli, R.A.A., 1973. Natural Chelating Polymers. Alginic Acid, Chitin and Chitosan. Pergamon Press, Oxford.
Muzzarelli, R.A.A., 1977. Chitin. Pergamon Press Ltd., Oxford, ISBN: 9780080203676.
Muzzarelli, R.A.A., 2010. Chitins and chitosans as immunoadjuvants and non-allergenic drug carriers. Mar. Drugs 8, 292–312.
Muzzarelli, R.A.A., Muzzarelli, C., 2005. Chitosan chemistry: relevance to the biomedical sciences. Adv. Polym. Sci. 186, 151–209.
Muzzarelli, R.A.A., Muzzarelli, C., 2009. Chitin and chitosan hydrogels. In: Handbook of Hydrocolloids, second ed. Woodhead Publishing Series in Food Science, Technology and Nutrition, pp. 849–888 (Chapter 31).
Muzzarelli, R.A.A., Pariser, E.R., 1978. Proceedings of the First International Conference on Chitin/Chitosan. Massachusetts Institute of Technology, MIT Sea Grant Program, Cambridge, USA.
Muzzarelli, R., Jeuniaux, C., Gooday, G.W. (Eds.), 1986. Chitin in Nature and Technology. Plenum Publishing Corporation, New York.
Muzzarelli, R.A.A., Boudrant, J., Meyer, D., Manno, N., DeMarchi, M., Paoletti, M.G., 2012. Current views on fungal chitin/chitosan, human chitinases, food preservation, glucans, pectins and inulin: a tribute to Henri Braconnot, precursor of the carbohydrate polymers science, on the chitin bicentennial. Carbohydr. Polym. 87, 995–1012.
Nabel, K., 1939. The membranes of the lower fungi especially of Rhizidiomyces bivellatus nov. sp. Arch. Mikrobiol. 10, 515–541.
Nakano, Y., Bin, Y., Bando, M., Nakashima, T., Okuno, T., Kurosu, H., Matsuo, M., 2007. Structure and mechanical properties of chitosan/poly(vinyl alcohol) blend films. Macromol. Symp. 258, 63–81.
Nasar, A., 2018. Chitosan-Based Adsorbents for Wastewater Treatment. Materials Research Forum LLC Publisher, Millersville, ISBN: 978-1-945291-74-6.
Nasri, H., Baradaran, A., Shirzad, H., Rafieian-Kopaei, M., 2014. New concepts in nutraceuticals as alternative for pharmaceuticals. Int. J. Prev. Med. 5, 1487–1499.
Navarro-Suarez, S., Flores-Palma, A., Flores-Ruiz, R., Gutiérrez-Pérez, J.L., Torres-Lagares, 2018. Nanobiomaterials in dentistry. In: Nanobiomaterials. Nanostructured Materials for Biomedical Applications. Elsevier, pp. 297–318.
Nayak, A.K., Hasnain, M.S., Beg, S., 2021. Chitosan in Biomedical Applications. Elsevier, ISBN: 9780128212448. 424 p.
Nechita, P., 2017. Applications of chitosan in wastewater treatment. In: Shalaby, E.A. (Ed.), Biological Activities and Application of Marine Polysaccharides. InTech, Rijeka, Croatia, pp. 209–228 (Chapter 10).
Needham, D.M., 1929. The chemical changes during the metamorphosis of insects. Biol. Rev. Biol. Proc. Camb. Philos. Soc. 4, 307–326.
Negi, H., Verma, P., Singh, R.K., 2021. A comprehensive review on the applications of functionalized chitosan in petroleum industry. Carbohydr. Polym. 266, 118125.
Neuberg, C., 1902. Über d-glucosamin and chitose. Berichte 35, 4009–4023.
Neuberg, C., Wolff, H., 1903. Ueber α- und β-2-amino-d-glucoheptonsäure. Berichte 36, 618–620.
Neuberger, A., Pitt Rivers, R., 1939. Preparation and configurative relationships of methylglucosaminides. J. Chem. Soc., 122–126.
Neville, A.C., 1967. Chitin orientation in cuticle and its control. Adv. Insect Physiol. 4, 213–286.

Neville, A.C., 1970. Cuticle ultrastructure in relation to the whole insect. In: Neville, A.C. (Ed.), Insect Ultrastructure. Blackwell, Oxford, pp. 17–39.

Neville, A.C., 1975. Biology of the Arthropod Cuticle. Springer-Verlag, Berlin, ISBN: 978-3-642-80912-5. 448 p.

Neville, A.C., Luke, B.M., 1969a. Molecular architecture of adult locust cuticle at the electron microscope level. Tissue Cell 1, 355–363.

Neville, A.C., Luke, B.M., 1969b. A two-system model for chitin-protein complexes in insect cuticles. Tissue Cell 1, 689–707.

Neville, A.C., Parry, D.A.D., Woodhead-Galloway, J., 1976. The chitin crystallite in arthropod cuticle. J. Cell Sci. 21, 73–82.

Nezakati, T., Seifalian, A., Tan, A., Seifalian, A.M., 2018. Conductive polymers: opportunities and challenges in biomedical applications. Chem. Rev. 118, 6766–6843.

Nicklès, J., 1856. Braconnot. Sa vie et ses Travaux. Extrait des Mémoires de l'Académie de Stanislas. Grimblot et Veuve Raybois, Imprimeurs-Libraires, Nancy, pp. 1–136.

Nicklès, J., 1856. Discours prononcé sur la tombe de M. Braconnot, le 15 janvier 1855. J. Pharm. Chem. 27, 219–224.

Nilsen-Nygaard, J., Strand, S.P., Vårum, K.M., Draget, K.I., Nordgård, C.T., 2015. Chitosan: gels and interfacial properties. Polymers 7, 552–579.

Niu, J., Lin, H.Z., Jiang, S.G., Chen, X., Wu, K.C., Liu, Y.J., Wang, S., Tian, L.X., 2013. Comparison of effect of chitin, chitosan, chitosan oligosaccharide and N-acetyl-D-glucosamine on growth performance, antioxidant defenses and oxidative stress status of Penaeus monodon. Aquaculture 372–375, 1–8.

No, H.K., Meyers, S.P., 1995. Preparation and characterization of chitin and chitosan—a review. J. Aquat. Food Prod. Technol. 4, 27–52.

Norman, A.G., Peterson, W.H., 1932. The chemistry of mould tissue. 2. The resistant cell wall material. Biochem. J. 26, 1946–1953.

Noyer-Weidner, M., Schaffner, W., 1995. Felix Hoppe-Seyler (1825-1895). A pioneer of biochemistry and molecular biology. Biol. Chem. Hoppe Seyler 376, 447–448.

Nwe, N., Furuike, T., Tamura, H., 2011a. Chitin and chitosan from terrestrial organisms. In: Kim, S.K. (Ed.), Chitin, Chitosan, Oligosaccharides and their Derivatives: Biological Activities and Applications. CRC Press, Taylor & Francis Group LLC, Boca Raton, pp. 3–10 (Chapter 1).

Nwe, N., Furuike, T., Tamura, H., 2011b. Chitosan from aquatic and terrestrial organisms and microorganisms. Production, properties and applications. In: Johnson, B.M., Berkel, Z.E. (Eds.), Biodegradable Materials. Nova Science Publishers Inc, pp. 29–50 (Chapter 2).

Nwe, N., Furuike, T., Tamura, H., 2013. Isolation and characterization of chitin and chitosan. In: Kim, S.K. (Ed.), Marine Biomaterials. Characterization, Isolation and Applications. CRC Press. Taylor & Francis Group LLC, Boca Raton, pp. 45–60 (Part I, Chapter 4).

Nwe, N., Furuike, T., Tamura, H., 2014. Isolation and characterization of chitin and chitosan from marine origin. Adv. Food Nutr. Res. 72, 1–15 (Chapter 1).

Oenological Codex, 2009. International Oenological Codex. Official Journal of the European Union, C 409, 5 December 2019. https://www.oiv.int/en/technical-standards-and-documents/oenological-products/international-oenological-codex.

Odier, A., 1823. Mémoire sur la composition chimique des parties cornées des insectes. In: Mémoires de la Société d'Histoire Naturelle de Paris. vol. 1. Baudouin Frères Libraires-É diteurs, Paris, pp. 29–42.

Offer, T.R., 1907. Ueber chitin. Biochem. Z. 7, 117–127.

Okuyama, K., Noguchi, K., Miyazawa, T., Yui, T., Ogawa, K., 1997. Molecular and crystal structure of hydrated chitosan. Macromolecules 30, 5849–5855.

Oladzadabbasabadi, N., Nafchi, A.M., Ariffin, F., Wijekoon, M.M.J.O., Al-Hassan, A.A., Dheyab, M.A., Ghasemlou, M., 2022. Recent advances in extraction, modification, and application of chitosan in packaging industry. Carbohydr. Polym. 277, 118876.

Orr, S.F.D., 1954. Infra-red spectroscopic studies of some polysaccharides. Biochim. Biophys. Acta 14, 173–181.

Orzali, L., Corsi, B., Forni, C., Riccioni, L., 2017. Chitosan in agriculture: a new challenge for managing plant disease. In: Shalaby, E. (Ed.), Biological Activities and Applications of Marine Polysaccharides. InTech Publishers, pp. 17–36 (Chapter 2).

Osman, Z., Arof, A.K., 2017. Chitosan and phthaloylated chitosan in electrochemical devices. In: Shalaby, E.A. (Ed.), Biological Activities and Application of Marine Polysaccharides. InTech, Rijeka, Croatia, pp. 17–36 (Chapter 14).

Oyatogun, G.M., Esan, T.A., Akpan, E.I., Adeosun, S.O., Popoola, A.P.I., Imasogie, B.I., Soboyejo, W.O., Afonja, A.A., Ibitoye, S.A., Abere, V.D., Oyatogun, A.O., Oluwasegun, K.M., Akinwole, I.E., Akinluwade, K.J., 2020. Chitin, chitosan, marine to market. In: Handbook of Chitin and Chitosan. Volume 3: Chitin and Chitosan Based Polymer Materials for Various Applications, pp. 341–381 (Chapter 11).

Özel, N., Elibol, M., 2021. A review on the potential uses of deep eutectic solvents in chitin and chitosan related processes. Carbohydr. Polym. 262, 117942.

Packard, A.S., 1886. On the nature and origin of the so-called spiral thread of tracheae. Am. Nat. 20, 438–442.

Packard, A.S., 1898. A Textbook of Entomology. MacMillan, New-York. 729 p.

Pakdel, P.R., Peighambardoust, S.J., 2018. Review on recent progress in chitosan-based hydrogels for wastewater treatment application. Carbohydr. Polym. 201, 264–279.

Pakizeh, M., Moradi, A., Ghassemi, T., 2021. Chemical extraction and modification of chitin and chitosan from shrimp shells. Eur. Polym. J. 159, 110709.

Pal, P., Pal, A., Nakashima, K., Yadav, B.K., 2021. Applications of chitosan in environmental remediation: a review. Chemosphere 266, 128934.

Paliwal, R., Paliwal, S.R., Sulakhiya, K., Kurmi, B.D., Kenwat, R., Mamgain, A., 2019. Chitosan-based nanocarriers for ophthalmic applications. In: Polysaccharide Carriers for Drug Delivery. vol. 4, pp. 79–104.

Patti, A.M., Katsiki, N., Nikolic, D., Al-Rasadi, K., Rizzo, M., 2015. Nutraceuticals in lipid-lowering treatment: a narrative review on the role of chitosan. Angiology 66, 416–421.

Payen, A., 1839. Mémoire relatif à la composition de la matière ligneuse. In: Comptes Rendus Hebdomadaires des Séances de l'Académie des Sciences. vol. 8. Bachelier, Imprimeur-Libraire, Paris, pp. 51–53.

Payen, A., 1840. Complément d'un mémoire sur la composition chimique du tissu propre des végétaux, et sur les différents états d'agrégation de tissu. In: Annales des Sciences Naturelles. vol. XIII. Crochard & Cie, Libraires-Éditeurs, Paris, pp. 305–310. Seconde Série.

Payen, A., 1842. Propriétés distinctives entre les membranes épidermiques et cellulaires des végétaux et les tissus tégumentaires des insectes et des crustacés. In: Mémoires sur les Développements des Végétaux. Extrait des Mémoires de l'Académie Royale des Sciences. Imprimerie Royale, Paris, pp. 370–375. Tome VIII des Savants Étrangers, septième mémoire.

Payen, A., 1843. Propriétés distinctives entre les membranes végétales et les enveloppes des insectes et des crustacés. C. R. Seances Acad. Sci. 17, 227–231.

Pearson, F.G., Marchessault, R.H., Liang, C.Y., 1960. Infrared spectra of crystalline polysaccharides. V. Chitin. J. Polym. Sci. 43, 101–116.

Peers, S., Montembault, A., Ladaviere, C., 2020. Chitosan hydrogels for sustained drug delivery. J. Control. Release 326, 150–163.

Peligot, E., 1858. Sur la composition de la peau des vers à soies. C. R. Acad. Sci. XLVII, 1034–1039.
Pellá, M.C.G., Lima-Tenorio, M.K., Tenorio-Neto, E.T., Guilherme, M.R., Muniz, E.C., Rubira, A.F., 2018. Chitosan-based hydrogels: from preparation to biomedical applications. Carbohydr. Polym. 196, 233–245.
Pelletier, E., Bonnet, C., Lemarchand, K., 2009. Biofouling growth in cold estuarine waters and evaluation of some chitosan and copper anti-fouling paints. Int. J. Mol. Sci. 10, 3209–3223.
Peniche, C., Argüelles-Monal, W., Goycoolea, F.M., 2008. Chitin and chitosan: major sources, properties and applications. In: Belgacem, M.N., Gandini, A. (Eds.), Monomers, Polymers and Composites from Renewable Resources. Elsevier, Amsterdam, pp. 517–542 (Chapter 25).
Percot, A., Viton, C., Domard, A., 2003. Optimization of chitin extraction from shrimp shells. Biomacromolecules 4, 12–18.
Peter, M.G., 1995. Applications and environmental aspects of chitin and chitosan. J. Macromol. Sci. 32, 629–640.
Peter, S., Lyczko, N., Gopakumar, D., Maria, H.J., Nzihou, A., Thomas, S., 2021. Chitin and chitosan based composites for energy and environmental applications: a review. Waste Biomass Valoriz. 12, 4777–4804.
Petrova, B., Cartwright, Z.M., Edwards, C.G., 2016. Effectiveness of chitosan preparations against *Brettanomyces bruxellensis* grown in culture media and red wines. J. Int. Sci. Vigne Vin 50, 49–56.
Phan, N.T.S., Le, K.K.A., Nguyen, T.V., Le, N.T.H., 2012. Chitosan as a renewable heterogeneous catalyst for the Knoevenagel reaction in ionic liquid as green solvent. Org. Chem., 928484.
Philibert, T., Lee, B.H., Fabien, N., 2017. Current status and new perspectives on chitin and chitosan as functional biopolymers. Appl. Biochem. Biotechnol. 181, 1314–1337.
Phillips, M., 1940. Anselme Payen, distinguished French chemist and pioneer investigator of the chemistry of lignin. J. Wash. Acad. Sci. 30, 65–71.
Phillips, J., 2017. Chitin: Properties, Applications and Research. Series: Polymer Science and Technology, NOVA Science Publishers.
Picken, L.E.R., 1940. The fine structure of biological systems. Biol. Rev. 15, 133–167.
Picken, L.E.R., 1960. The Organization of Cells. Clarendon Press, Oxford University, London. 629 p.
Picken, L.E.R., Lotmar, W., 1950. Oriented protein in chitinous structures. Nature 165, 599–600.
Plisko, E.A., Baranova, V.N., Nud'ga, L.A., 1974. Electrical Insulating Paper. USSR Patent 428.053.
Prabhu, A., Crapnell, R.D., Eersels, K., van Grinsven, B., Kunhiraman, A.K., Singla, P., McClements, J., Banks, C.E., Novakovic, K., Peeters, M., 2022. Reviewing the use of chitosan and polydopamine for electrochemical sensing. Curr. Opin. Electrochem. 32, 100885.
Prathibhani, H.M., Kumarihami, C., Kim, Y.H., Kwack, Y.B., Kim, J., Kim, J.G., 2022. Application of chitosan as edible coating to enhance storability and fruit quality of Kiwifruit: a review. Sci. Hortic. 292, 110647.
Prévost, M., D'Amat, R., 1956. Braconnot. Dictionnaire de Biographies Françaises. vol. 7 Letouzey et Ané, Paris, pp. 132–133.
Pringsheim, H., Krüger, D., 1932. Chitin. In: Klein, G. (Ed.), Handbuch der Pflanzenanalyse. Spezielle Analyse: Organishce stoffe II. vol. 3. Springer-Verlag GmBH, Wien, pp. 69–79. Part I.
Priyadarshi, R., Rhim, J.W., 2020. Chitosan-based biodegradable functional films for food packaging applications. Innovative Food Sci. Emerg. Technol. 62, 102346.

Proskuriakow, N.J., 1926. Über die Beteiligung des Chitins am Aufbau der Pilzzellwand. Biochem. Z. 167, 68–76.

Prudden, J.F., Migel, P., Hanson, P., Friedrich, L., Balassa, L., 1970. The discovery of a potent chemical wound-healing accelerator. Am. J. Surg. 119, 560–564.

Purchase, E.R., Braun, C.E., 1946. D-glucosamine hydrochloride. Org. Synth. 26, 36–37.

Qing, Y., Tamo, F. (Eds.), 2019. Targeting Chitin-Containing Organisms. Advances in Experimental Medicine and Biology, vol. 1142. Springer Singapore, ISBN: 978-981-13-7317-6.

Qu, B., Luo, Y.C., 2020. Chitosan-based hydrogel beads: preparations, modifications and applications in food and agriculture sectors. a review. Int. J. Biol. Macromol. 152, 437–448.

Queiroz, J., Fernandes, S.K.S.C., Azevedo, E.P., Barbosa, A.A., Fook, M.V.L., 2015. Chitosan: applicability in preventive dentistry. Dent. Mater. 31, e58–e59.

Rabea, E.I., Badawy, M.E.T., Stevens, C.V., Smagghe, G., Steurbault, W., 2003. Chitosan as antimicrobial agent: applications and mode of action. Biomacromolecules 4, 1457–1465.

Rahangdale, D., Kumar, A., 2018. Derivatized chitosan: fundamentals to applications. In: Biopolymer Grafting Applications. Elsevier, pp. 251–284 (Chapter 7).

Ramachandran, G.N., Ramakrishnan, C., 1962. The structure of chitin. Biochim. Biophys. Acta 63, 307–309.

Ramakrishnan, C., Prasad, N., 1972. Rigid-body refinement and conformation of α-chitin. Biochim. Biophys. Acta 261, 123–135.

Rammelberg, G., 1931. Beitrag zur kenntnis des chitins der pilze und krabben. Bot. Arch. 32, 1–37.

Rauh, F., Dornish, M., 2006. Chitosan. In: Guelcher, S.A., Hollinger, J.O. (Eds.), An Introduction to Biomaterials. CRC Press. Taylor & Francis Group, Boca Raton, pp. 249–260 (Chapter 14).

Ravi Kumar, M.N.V., 2000. A review of chitin and chitosan applications. React. Funct. Polym. 46, 1–27.

Ravi Kumar, M.N.V., Muzzarelli, R.A.A., Muzzarelli, C., Sashiwa, H., Domb, A.J., 2004. Chitosan chemistry and pharmaceutical perspectives. Chem. Rev. 104, 6017–6087.

Raza, Z.A., Khalil, S., Ayub, A., Banat, I.M., 2020. Recent developments in chitosan encapsulation of various active ingredients for multifunctional applications. Carbohydr. Res. 492, 108004.

Reese, E.T., 1963. Advances in Enzymic Hydrolysis of Cellulose and Related Materials. Pergamon Press, Oxford. 302 p.

Reese, T.A., Liang, H.E., Tager, A.M., Luster, A.D., Van Rooijen, N., Voehringer, D., Locksley, R.M., 2007. Chitin induces accumulation in tissue of innate immune cells associated with allergy. Nature 447, U92–U97.

Renault, F., Sancey, B., Badot, P.M., Crini, G., 2009. Chitosan for coagulation/flocculation processes—an eco-friendly approach. Eur. Polym. J. 45, 1337–1348.

Rezaei, F.S., Sharifianjazi, F., Esmaeilkhanian, A., Salehi, E., 2021. Chitosan films and scaffolds for regenerative medicine applications: a review. Carbohydr. Polym. 273, 118631.

Ribeiro, M.P., Espiga, A., Silva, D., Baptista, P., Henriques, J., Ferreira, C., Silva, J.C., Borges, J.P., Pires, E., Chaves, P., Correia, I.J., 2009. Development of a new chitosan hydrogel for wound dressing. Wound Repair Regen. 17, 817–824.

Richards, A.G., 1947a. The organization of arthropod cuticle. A modified interpretation. Science 105, 170–171.

Richards, A.G., 1947b. Studies on arthropod cuticle. 1. The distribution of chitin in lepidopterous scales, and its bearing on the interpretation of arthropod cuticle. Ann. Entomol. Soc. Am. 40, 227–240.

Richards, A.G., 1949. Studies on arthropod cuticle. 3. The chitin of *Limulus*. Science 109, 591–592.

Richards, A.G., 1951. The Integument of Arthropods. The Chemical Components and their Properties, the Anatomy and Development, and the Permeability. University of Minnesota Press, Minneapolis. 411 p.
Richards, A.G., 1952. Studies on arthropod cuticle. VII. Patent and masked carbohydrate in the epicuticle of insects. Science 115, 206–208.
Richards, A.G., 1958. The cuticle of arthropods. Ergeb. Biol. 20, 1–26.
Richards, A.G., Anderson, T.F., 1942. Electron microscope studies of insect cuticle, with a discussion of the application of electron optics to this problem. J. Morphol. 71, 135–184.
Richards, A.G., Cutkomp, L.K., 1946. Correlation between the possession of a chitinous cuticle and sensitivity to DDT. Biol. Bull. 90, 97–108.
Richards, A.G., Korda, F.H., 1948. Studies on arthropod cuticle. 2. Electron microscope studies of extracted cuticles. Biol. Bull. 94, 212–235.
Richards, A.G., Pipa, R.L., 1958. Studies on the molecular organization of insect cuticle. Smithson. Misc. Collect. 137, 247–262.
Rigby, G.W., 1936a. Carbohydrate Derivatives and Process of Making the Same. US Patent 2,033,787.
Rigby, G.W., 1936b. Substantially Undegraded Deacetylated Chitin and Process for Producing the Same. US Patent 2,040,879.
Rigby, G.W., 1936c. Process for the Preparation of Films and Filaments and Products Thereof. US Patent 2,040,880.
Rigby, G.W., 1936d. Emulsion. US Patent 2,047,225.
Rigby, G.W., 1936e. Chemical Process and Chemical Compounds Derived Therefrom. US Patent 2,047,226.
Rigby, G.W., 1937. Chemical Products and Process of Preparing the Same. US Patent 2,072,771.
Rinaudo, M., 2006. Chitin and chitosan: properties and applications. Prog. Polym. Sci. 31, 603–632.
Rinaudo, M., Goycoolea, F.M., 2019. Advances in Chitin/Chitosan Characterization and Applications. MDPI Basel, ISBN: 978-3-03897-803-9.
Rinaudo, M., Pérez, S., 2019. From Chitin to Chitosan. GlycoPedia, pp. 1–30.
Rippel, W.A., 1908. Chitin bei mikroorganismen. Eine richtigstellung. Biochem. Z. 290, 444.
Roberts, G.A.F., 1992. In: Roberts, G.A.F. (Ed.), Chitin Chemistry, first ed. Macmillan Press, London, Houndmills, UK, ISBN: 9781349115471.
Rocha, M.A.M., Coimbra, M.A., Nunes, C., 2017. Applications of chitosan and their derivatives in beverages: a critical review. Curr. Opin. Food Sci. 15, 61–69.
Roche, J., 1972. Eloge—Paul Karrer (1889-1971). Bull. Acad. Natl Med. 156 (4–5), 89–92.
Roelofsen, P.A., Hoette, I., 1951. Chitin in the cell wall of yeasts. Antonie Van Leeuwenhoek 17, 297–313.
Rosenheim, O., 1905. Chitin in the carapace of *Pterygotus osiliensis*, from the Silurian rocks of Oesel. Proc. R. Soc. London Ser. B 76, 398–400.
Rouget, C., 1859. Des substances amylacées dans les tissus des animaux, spécialement les articulés (chitine). C. R. Hebd. Seances Acad. Sci. 48, 792–795.
Roy, J.C., Salaün, F., Giraud, S., Ferri, A., 2017a. Solubility of chitin: solvents, solution behaviors and their related mechanisms. In: Solubility of Polysaccharides. IntechOpen, pp. 109–127 (Chapter 7).
Roy, J., Salaün, F., Giraud, S., Ferri, A., Guan, J., 2017b. Chitosan-based sustainable textile technology: process, mechanism, innovation, and safety. In: Shalaby, E.A. (Ed.), Biological Activities and Application of Marine Polysaccharides. InTech, Rijeka, Croatia, pp. 251–278 (Chapter 12).
Roy, B.K., Tahmid, I., Rashid, T.U., 2021. Chitosan-based materials for supercapacitor applications: a review. J. Mater. Chem. 9, 17592.

Rudall, K.M., 1955. The distribution of collagen and chitin. Symp. Soc. Exp. Biol. 9, 49–71.
Rudall, K.M., 1963. The chitin/protein complexes of insect cuticles. In: Advances in Insect Physiology. vol. 1. Academic Press, London, pp. 257–313.
Rudall, K.M., 1965. Skeletal structure in insects. In: Goodwin, T.W. (Ed.), Aspects of Insect Biochemistry. Academic Press, London, pp. 83–92.
Rudall, K.M., 1967. Conformation in chitin-protein complexes. In: Ramachadran, G.N. (Ed.), Conformation of Biopolymers. Papers Read at an International Symposium Held at the University of Madras, 18-21 January 1967. vol. 2. Academic Press, London, pp. 751–765.
Rudall, K.M., 1969. Chitin and its association with other molecules. J. Polym. Sci. Part C Polym. Symp. 28, 83–102.
Rudall, K.M., Kenchington, W., 1973. The chitin system. Biol. Rev. 49, 597–636.
Runham, N.W., 1961a. Investigations into the histochemistry of chitin. J. Histochem. Cytochem. 9, 87–92.
Runham, N.W., 1961b. The histochemistry of the radula of *Patella vulgata*. Q. J. Microsc. Sci. 102, 371–380.
Ruser, M., 1933. Beiträge zur kenntnis des chitins und der muskulatur der zecken (Ixodidae). Z. Morphol. Okol. Tiere 27, 199–261.
Sadiq, A.C., Olasupo, A., Ngah, W.S.W., Rahim, N.Y., Suah, F.B.M., 2021. A decade development in the application of chitosan-based materials for dye adsorption: a short review. Int. J. Biol. Macromol. 191, 1151–1163.
Safarzadeh, M., Sadeghi, S., Azizi, M., Rastegari-Pouyani, M., Pouriran, R., Hoseini, M.H. M., 2021. Chitin and chitosan as tools to combat COVID-19: a triple approach. Int. J. Biol. Macromol. 183, 235–244.
Şahan, G., Demir, A., 2014. Forms of chitosan biopolymer and their textile applications. In: Proceedings: XIIIth International Izmir Textile and Apparel Symposium April 2-5, 2014, pp. 233–236.
Sahariah, P., Másson, M., 2017. Antimicrobial chitosan and chitosan derivatives: a review of the structure-activity relationship. Biomacromolecules 18, 3846–3868.
Saheed, I.O., Oh, W.D., Suah, F.B.M., 2021. Chitosan modifications for adsorption of pollutants—a review. J. Hazard. Mater. 408, 124889.
Sahoo, D., Nayak, P.L., 2011. Chitosan: the most valuable derivative of chitin. In: Kalia, S., Avérous, L. (Eds.), Biopolymers: Biomedical and Environmental Applications. Scrivener Publishing LLC, New Jersey, pp. 129–169 (Chapter 6).
Salehi, E., Daraei, P., Shamsabadi, A.A., 2016. A review on chitosan-based adsorptive membranes. Carbohydr. Polym. 152, 419–432.
Samoila, P., Humelnicu, A.C., Maria, I., Cojocaru, C., Harabagiu, V., 2019. Chitin and chitosan for water purification. In: Van den Broek, L.A.M., Boeriu, C.G. (Eds.), Chitin and Chitosan: Properties and Applications. John Wiley & Sons, Ltd, pp. 429–476 (Chapter 17).
Samyn, P., Barhoum, A., Ohlund, T., Dufresne, A., 2018. Review: nanoparticles and nanostructured materials in papermaking. J. Mater. Sci. 53, 146–184.
Sánchez-Machado, D.I., López-Cervantes, J., Martínez-Ibarra, D.M., Escárcega-Galaz, A. A., Vega-Cázarez, C.A., 2022. The use of chitosan as a skin-regeneration agent in burns injuries: a review. E-Polymers 22, 75–86.
Santos, V.P., Marques, N.S.S., Maia, P.C.S.V., de Lima, M.A.B., de Oliveira, F.L., de Campos-Takaki, G.M., 2020. Saefood waste as attractive source of chitin and chitosan production and their applications. Int. J. Mol. Sci. 21, 4290.
Sarmento, B., das Neves, J., 2012. In: Sarmento, B., das Neves, J. (Eds.), Chitosan-Based Systems for Biopharmaceuticals. John Wiley & Sons, ISBN: 9781119964070. 600 p.
Sarode, S., Upadhyay, P., Khosa, M.A., Mak, T., Shakir, A., Song, S., Ullah, A., 2019. Overview of wastewater treatment methods with special focus on biopolymer chitin-chitosan. Int. J. Biol. Macromol. 121, 1086–1100.

Sashiwa, H., Harding, D. (Eds.), 2015. Advances in Marine Chitin and Chitosan. MDPI AD. 484 p.
Satitsri, S., Muanprasat, C., 2020. Chitin and chitosan derivatives as biomaterial resources for biological and biomedical applications. Molecules 25, 5961.
Savvaidis, I., 2022. Chitosan Applications in Food Systems. Elsevier Science, ISBN: 9780128216637. 240 p.
Schmidt, C., 1845. Zur Vergleichenden Physiologie der Wirbellosen Thiere. Eine Physiologisch-Chemische Untersuchung. Annalen der Chemie und Pharmacie, vol. LIV Vieweg u. Sohn, Braunschweig, pp. 298–311.
Schmidt, M., 1936. Makrochemische untersuchungen über das vorkommen von chitin bei mikroorganismen. Arch. Mikrobiol. 7, 241–260.
Schmidt, L., Waschkau, A., Ludwig, E., 1928a. Alkaliverbindungen von mehrwertigen alkoholen und kohlenhydraten. Monastsh. Chem. Verw. Teile Anderer Wiss. 49, 107–110.
Schmidt, L., Waschkau, A., Ludwig, E., 1928b. Das vorkommen von chitin und seine verwertung als systematisch-phylogenetisches merkmal im pflanzenreich. In: Sitzungsberichte der Kaiserlichen Akademie der Wissenschaften Mathematisch Naturwissenchaftliche Klasse Abt IIB. vol. 137, pp. 107–109.
Schmiedeberg, O., 1891. Ueber die chemische zusammensetzung des knorpels. Arch. Exp. Pathol. Pharmakol. 28, 355–404.
Schmiedeberg, O., 1920. Über chitin und chitinabkömmlinge des tier- und pflanzenreichs. Arch. Exp. Pathol. Pharmakol. 87, 74–85.
Scholl, E., 1908a. Die Reindarstellung des chitins aus *Boletus edulis*. Monatsh. Chem. Verw. Teile Anderer Wiss. 29, 1023–1036.
Scholl, E., 1908b. Die Reindarstellung des chitins aus Boletus edulis. In: Hölder, A. (Ed.), Sitzungsberichte der Thematisch Naturwissenchaftlichen Klasse. Akademie der Wissenschaftren. vol. 117. Abt IIB, Wien, pp. 547–560.
Schorigin, P., Hait, E., 1934. Über die nitrierung von chitin. Ber. Dtsch. Chem. Ges. A B Ser. 67, 1712–1714.
Schorigin, P., Hait, E., 1935. Über die acetylierung des chitins (Vorläufig. Mitteil.). Ber. Dtsch. Chem. Ges. A B Ser. 68, 971–973.
Schorigin, P., Makarowa-Semljanskaja, N.N., 1935a. Über die desaminierung von chitin und glucosamin. Ber. Dtsch. Chem. Ges. A B Ser. 68, 965–969.
Schorigin, P., Makarowa-Semljanskaja, N.N., 1935b. Über die methyläther des chitins (Vorläufig. Mitteil.). Ber. Dtsch. Chem. Ges. A B Ser. 68, 969–971.
Schulze, P., 1921. Ein neues verfahren zum bleichen und erweichen tierischer hartgebilde. In: Sitzungsberichte der Gesellschaft Naturforschender Freunde zu Berlin, pp. 135–139.
Schulze, P., 1922a. Über beziehungen zwischen pflanzlichen und tierischen skelettensubstanzen, über eine neue chitinreaktion une eine method zur bleichen und erweichen tierischer hartgebilde. Verh. Dtsch. Zool. Ges. 27, 71–73.
Schulze, P., 1922b. Über bezichungen zwischen pflanzlichen und tierischen skelettsubstanzen und über chitinreaktion. Biol. Zentralbl. 42, 388–394.
Schulze, P., 1926. Das chitin, sein aufbau, seine verbreitung, sein nachweis und seine behandlung bei der entomologischen preparation. Entomol. Mitt. 15, 420–423.
Schulze, P., Kunike, G., 1923. Zur mikrochemie tierischer skelettsubstanzen. Biol. Zentralbl. 43, 556–559.
Sebastian, J., Rouissi, T., Brar, S.K., 2020. Fungal chitosan: prospects and challenges. In: Handbook of Chitin and Chitosan, Volume 1: Preparation and Properties, pp. 419–452 (Chapter 14).
Seetharaman, K., Bertoft, E., 2012. Perspectives on the history of research on starch. Part I: on the linkages in starch. Starch/Stärke 64, 677–682.

Şenel, S., 2011. Applications of chitosan and its derivatives in veterinary medicine. In: Kim, S.K. (Ed.), Chitin, Chitosan, Oligosaccharides and their Derivatives: Biological Activities and Applications. CRC Press, Taylor & Francis Group LLC, Boca Raton, pp. 460–480 (Chapter 33).

Şenel, S., McClure, S.J., 2004. Potential applications of chitosan in veterinary medicine. Adv. Drug Deliv. Rev. 56, 1467–1480.

Senevirathne, M., Ahn, C.B., Kim, S.K., Je, J.Y., 2012. Cosmeceutical applications of chitosan and its derivatives. In: Kim, S.K. (Ed.), Marine Cosmeceuticals. Trends and Prospects. CRC Press, Taylor & Francis Group LLC, Boca Raton, pp. 169–179 (Chapter 13).

Serrano-Castañeda, P., Escobar-Chávez, J.J., Rodríguez-Cruz, I.M., Melgoza-Contreras, L. M., Martínez-Hernández, J.M., 2018. Microneedles as enhancer of drug absorption through the skin and applications in medicine and cosmetology. J. Pharm. Pharm. Sci. 21, 73–93.

Shahidi, F., Arachchi, J.K.V., Jeon, Y.J., 1999. Food applications of chitin and chitosans. Trends Food Sci. Technol. 10, 37–51.

Shahrajabian, M.H., Chaski, C., Polyzos, N., Tzortzakis, N., Petropoulos, S.A., 2021. Sustainable agriculture systems in vegetable production using chitin and chitosan as plant biostimulants. Biomolecules 11, 819.

Shamshina, J.L., 2019. Chitin in ionic liquids: historical insights into the polymer's dissolution and isolation. A review. Green Chem. 21, 3974–3993.

Shamshina, J.L., Berton, P., 2020. Uses of ionic liquids in chitin biorefinery: a systematic review. Front. Bioeng. Biotechnol. 8, 11.

Shamshina, J.L., Zavgorodnya, O., Rogers, R.D., 2019. Advances in processing chitin as a promising biomaterial from ionic liquids. In: Applications of Ionic Liquids in Biotechnology. Advances in Biochemical Engineering Biotechnology, vol. 168, pp. 177–198.

Shamshina, J.L., Kelly, A., Oldham, T., Rogers, R.D., 2020. Agricultural uses of chitin polymers. Environ. Chem. Lett. 18, 53–60.

Shariatinia, Z., 2019. Pharmaceutical applications of chitosan. Adv. Colloid Interf. Sci. 263, 131–194.

Sharif, R., Mujtaba, M., Rahman, M.U., Shalmani, A., Ahmad, H., Anwar, T., Deng, T.C., Wang, X.P., 2018. The multifunctional role of chitosan in horticultural crops: a review. Molecules 23, 872.

Sharifianjazi, F., Khaksar, S., Esmaeilkhanian, A., Bazli, L., Eskandarinezhad, S., Salahshour, P., Sadeghi, F., Rostamnia, S., Vahdat, S.M., 2022. Advancements in fabrication and application of chitosan composites in implants and dentistry: a review. Biomolecules 12, 155.

Sharon, N., 1980. Carbohydrates. Sci. Am. 243, 80–97.

Sharp, R.G., 2013. A review of the applications of chitin and its derivatives in agriculture to modify plant-microbial interactions and improve crop yields. Agronomy 3, 757–793.

Shen, X.P., Shamshina, J.L., Berton, P., Gurau, G., Rogers, D., 2016. Hydrogels based on cellulose and chitin: fabrication, properties, and applications. Green Chem. 18, 53–75.

Sheth, Y., Dharaskar, S., Khalid, M., Sonawane, S., 2021. An environment friendly approach for heavy metal removal from industrial wastewater using chitosan based biosorbent: a review. Sustainable Energy Technol. Assess. 43, 100951.

Shi, C., Sun, W., Sun, Y., Chen, L., Xu, Y., Tang, M., 2019. Synthesis, characterization, and sludge dewaterability evaluation of the chitosan-based flocculant CCPAD. Polymers 11, 95.

Sieber, V., Hofer, M., Brück, W.M., Garbe, D., Brück, T., Lynch, C.A., 2018. ChiBio: an integrated bio-refinery for processing chitin-rich bio-waste to specialty chemicals. In: Rampelotto, P.H., Trincone, A. (Eds.), Grand Challenges in Marine Biotechnology. Springer International Publishing AG, part of Springer Nature, pp. 555–578 (Chapter 14).

Sikorski, P., Hori, R., Wada, M., 2009. Revisit of α-chitin crystal structure using high resolution X-ray diffraction data. Biomacromolecules 10, 1100–1105.

Silva, S.S., Mano, J.F., Reis, R.L., 2017. Ionic liquids in the processing and chemical modification of chitin and chitosan for biomedical applications. Green Chem. 19, 1208–1220.

Simonin, F., 1856. Notice biographique sur M. Henri Braconnot. In: Compte Rendus des Travaux de la Société de Médecine de Nancy. vols. 1834–1855, pp. 51–79.

Simonin, F., 1870. Notice Biographique sur M. Henri Braconnot. Imprimerie de Sordoillet et Fils, Nancy, pp. 5–34.

Simonin, F., 2013. Notice sur Henri Braconnot, Chimiste. Hachette Livre-BNF, Bibliothèque Nationale de France, ISBN: 2012468578. 44 p.

Singh, R., Shitiz, K., Singh, A., 2017. Chitin and chitosan: biopolymers for wound management. Int. Wound J. 14, 1276–1289.

Singh, R., Upadhyay, S.K., Singh, M., Sharma, I., Sharma, P., Kamboj, P., Saini, A., Voraha, R., Sharma, A.K., Upadhyay, T.K., Khan, F., 2021a. Chitin, chitinases and chitin derivatives in biopharmaceutical, agricultural and environmental perspective. Biointerface Res. Appl. Chem. 11, 9985–10005.

Singh, A., Mittal, A., Benjakul, S., 2021b. Chitosan nanoparticles: preparation, food applications and health benefits. ScienceAsia 47, 1–10.

Singla, A.K., Chawla, M., 2001. Chitosan: some pharmaceutical and biological aspects. An update. J. Pharm. Pharmacol. 53, 1047–1067.

Sirajudheen, P., Poovathumkuzhi, N.C., Vigneshwaran, S., Chelaveettil, B.M., Meenakshi, S., 2021. Applications of chitin and chitosan based biomaterials for the adsorptive removal of textile dyes from water—a comprehensive review. Carbohydr. Polym. 273, 118604.

Skjåk-Braek, G., Anthonsen, T., Sandford, P.A., 1989. In: Skjåk-Braek, G., Anthonsen, T., Sandford, P.A. (Eds.), Chitin and Chitosan. Sources, Chemistry, Biochemistry, Physical Properties and Applications. Elsevier Applied Science, New York. 835 p.

Slagel, R., Sinkovitz, G., 1973a. Paper Products of Improved Dry Strength. US Patent 3.709.780.

Slagel, R., Sinkovitz, G., 1973b. Chitosan Graft Copolymers for Making Paper Products of Improved Dry Strength. US Patent 3.770.673.

Smedley, E., Rose, H.J., Rose, H.J., 1845. Encyclopædia_Metropolitana. Volume VIII. General View of the Animal Kingdom, and of the Circumstances Distinguishing it from the Vegetable Kingdom. William Clowes & Sons, London, pp. 215–217. Section I.

Sollas, I.B.J., 1907. On the identification of chitin by its physical constants. Proc. R. Soc. London Ser. B 79, 474–481.

Song, Z., Li, G.D., Guan, F.X., Liu, W.X., 2018. Application of chitin/chitosan and their derivatives in the papermaking industry. Polymers 10, 389.

Souza, V.G.L., Pires, J.R.A., Rodrigues, C., Coehloso, I.M., Fernando, A.L., 2020. Chitosan composites in packaging industry—current trends and future challenges. Polymers 12, 417.

Speck Jr., J.C., 1958. The Lobry de Bruyn-Alberda van Ekenstein transformation. In: Wolfrom, M.L. (Ed.), Advances in Carbohydrate Chemistry. vol. 13. Academic Press Inc, New York, pp. 63–104.

Spedding, H., 1964. Infrared spectroscopy and carbohydrate chemistry. Adv. Carbohydr. Chem. 19, 23–49.

Spiller, J., Tilden, W.A., McLeod, H., Mills, E.J., Carey Foster, G., 1905. Obituary notices. J. Chem. Soc. Trans. 87, 565–618.

Spoială, A., Ilie, C.I., Ficai, D., Ficai, A., Andronescu, E., 2021. Chitosan-based nanocomposite polymeric membranes for water purification—a review. Materials 14, 2091.

Stacey, M., 1943. Mucopolysaccharides and related substances. Chem. Ind. 62, 110–112.
Stacey, M., 1946. The chemistry of mucopolysaccharides and mucoproteins. Adv. Carbohydr. Chem. 2, 161–201.
Städeler, G., 1859. Untersuchungen über das fibroïn, spongin und chitin, nebst bemerkungen über den thierischen schleim. Justus Liebigs Ann. Chem. 111, 12–28.
Stephen, J., Lekshmi, M., Nayak, B.B., Kumar, S., 2021. Biotechnological approaches to valorization of fish biowastes and their potential applications. In: Pandey, P.K., Parhi, J. (Eds.), Advances in Fisheries Biotechnology. Springer, Singapore, pp. 495–521.
Steudel, H., 1902. Eine neue methode zum nachweis von glucosamin und ihre anwendung auf die spaltungsprodukte der mucine. Z. Physiol. Chem. 34, 353–384.
Sticher, O., Hamburger, M., 2008. Albert Hofmann (1906-2008)—an obiturary. Planta Med. 74, 791–793.
Stolte, K., 1907. Ueber das verhalten des glucosamins und seines nächsten umwandlungsproduktes im thierkörper. Beitr. Chem. Physiol. Pathol. II, 19–34.
Straus-Durckeim, H., 1828. Considérations Générales sur L'anatomie Comparé des Animaux Articulés. Levrault FG, Strasbourg. 463 p.
Sudha, P.N., Aisverya, S., Gomathi, T., Vijayalakshmi, K., Saranya, M., Sangeetha, K., Latha, S., Thomas, S., 2017. Applications of chitin/chitosan and its derivatives as adsorbents, coagulants and flocculants. In: Ahmed, S., Ikram, S. (Eds.), Chitosan - derivatives, composites and applications. Scrivener Publishing LLC, Wiley, pp. 453–487 (Chapter 17).
Suginta, W., Khunkaewla, P., Schulte, A., 2013. Electrochemical biosensor applications of polysaccharides chitin and chitosan. Chem. Rev. 113, 5458–5479.
Sundwik, E.E., 1881. Zur constitution des chitins. Z. Physiol. Chem. 5, 384–394.
Tabassum, N., Ahmed, S., Ali, M.A., 2021. Chitooligosaccharides and their structural-functional effect on hydrogels: a review. Carbohydr. Polym. 261, 117882.
Takata, R., 1929. The utilization of microorganisms for human food materials. 10. Carbohydrates of the mycelium of *Aspergillus oryzae*. J. Soc. Chem. Ind. Jpn. 32, 245–247.
Tangkijngamvong, N., Phaiyarin, P., Wanichwecharungruang, S., Kumtornrut, C., 2020. The anti-sebum property of chitosan particles. J. Cosmet. Dermatol. 19, 2135–2140.
Tatlow, D., Tatlow, C., Tatlow, S., Tatlow, S., 2020. A novel concept for treatment and vaccination against Covid-19 with an inhaled chitosan-coated DNA vaccine encoding a secreted spike protein portion. Clin. Exp. Pharmacol. Physiol. 47, 1874–1878.
Tauber, O.E., 1934. The distribution of chitin in an insect. J. Morphol. 56, 51–58.
Teng, D., 2012. From chitin to chitosan. In: Yao, K., Li, J., Yao, F., Yin, Y. (Eds.), Chitosan-Based Hydrogels: Functions and Applications. CRC Press. Taylor & Francis Group, Boca Raton, FL, pp. 1–38 (Chapter 1).
Terayama, H., 1954. Applications of the method of colloid titration to the study of bacteria. Arch. Biochem. Biophys. 50, 55–63.
Thakur, V.K., Voicu, S.I., 2016. Recent advances in cellulose and chitosan based membranes for water purification: a concise review. Carbohydr. Polym. 146, 148–165.
Théodoridès, J., 1962. Auguste Odier (1802-1870) et la découverte de la chitine (1821-1823). In: Atti del Symposium Internazionale di Storia delle Science (Firenze, Vinci, 8-10 ottobre 1960). Baccini & Chiappi, Firenze.
Thierfelder, H., 1926. Felix-Hoppe Seyler. Mohr. Enke F, Tübingen. 18 p.
Thomas, S., Pius, A., Gopi, S., 2020. Handbook of Chitin and Chitosan. Preparation and Properties. vol. 1 Elsevier, Amsterdam, ISBN: 978-0-128-179710.
Thor, C.J.B., 1939a. Chitin Compounds and Regenerated Chitin Products Such as Filaments, Films, Tubes or Straws. US Patent 2,168,374.
Thor, C.J.B., 1939b. Chitin Compounds and Regenerated Chitin Products Such as Filaments, Films, Tubes or Straws. US Patent 2,168,375.
Thor, C.J.B., Henderson, W.F., 1940. The preparation of alkali chitin. Am. Dyest. Rep. 29, 461–464.

Tiemann, F., 1884. Einiges über den abbau von salzsaurem glucosamin. Ber. Dtsch. Chem. Ges. 17, 241–251.
Tiemann, F., 1886. Ueber Glucosamin. Ber. Dtsch. Chem. Ges. 19, 49–53.
Tiemann, F., Landolt, R.H., 1886. Specifisches drehungsvermögen und krystallform des bromwasserstoffsauren glucosamins. Ber. Dtsch. Chem. Ges. 19, 155–157.
Tipson, R.S., 1957. Phoebus Aaron Theodor Levene, 1869-1940. Adv. Carbohydr. Chem. 12, 1–12.
Tonda-Turo, C., Ruini, F., Argentati, M., Di Girolamo, N., Robino, P., Nebbia, P., Ciardelli, G., 2016. Porous CS membranes with improved antimicrobial properties for the treatment of infected wound in veterinary applications. Mater. Sci. Eng. C 60, 416–426.
Torkaman, S., Rahmani, H., Ashori, A., Najafi, S.H.M., 2021. Modification of chitosan using amino acids for wound healing purposes: a review. Carbohydr. Polym. 258, 117675.
Tóth, G., Zechmeister, L., 1939. Chitin content of the mandible of the snail (*Helix pomatia*). Nature 144, 1049.
Tracey, M.V., 1955. Chitin. In: Paech, K., Tracey, M.V. (Eds.), Modern Methods of Plant Analysis. vol. II. Springer-Verlag, Berlin, pp. 264–274.
Tracey, M.V., 1957. Chitin. In: Reviews of Pure and Applied Chemistry (Australia). vol. 7, pp. 1–14.
Trim, A.R., 1941. Studies of the insect cuticle. I. Some general observations on certain arthropod cuticles with special reference to the characterization of the proteins. Biochem. J. 35, 1088–1098.
Triunfo, M., Tafi, E., Guarnieri, A., Scieuzo, C., Hahn, T., Zibek, S., Salvia, R., Falabella, P., 2021. Insect chitin-based nanomaterials for innovative cosmectics and cosmeceuticals. Cosmetics 8, 40.
Tsugita, T., 1990. Chitin/chitosan and their application. In: Voigt, M.N., Botta, J.R. (Eds.), Advances in Fisheries Technology and Biotechnology for Increased Profitability. Technomic Publisher Inc., Basel, pp. 287–298.
Tucker, S., 1945. A lost centenary: Lassaigne's test for nitrogen. The identification of nitrogen, sulfur, and halogens in organic compounds. J. Chem. Educ. 22, 212–214.
Ummu Habeeba, A.A., Reshmi, C.R., Sujith, A., 2007. Chitosan immobilized cotton fibres for antibacterial textile materials. Polym. Renewable Resour. 8, 61–70.
Underwood, C., van Eps, A.W., 2012. Nanomedicine and veterinary science: the reality and the practicality. Vet. J. 193, 12–23.
Valachová, K., Šoltés, L., 2021. Versatile use of chitosan and hyaluronan in medicine. Molecules 26, 1195.
Vakili, M., Rafatullah, M., Salamatinia, B., Abdullah, A.Z., Ibrahim, M.H., Tan, K.B., Gholami, Z., Amouzgar, P., 2014. Application of chitosan and its derivatives as adsorbents for dye removal from water and wastewater: a review. Carbohydr. Polym. 113, 115–130.
van den Broek, L.A.M., Boeriu, C.G., 2020. Chitin and Chitosan—Properties and Applications. Wiley Series in Renewable Resources, John Wiley & Sons Ltd., Hoboken, ISBN: 9781119450436.
van den Broek, L.A.M., Knoop, R.J.I., Kappen, F.H.J., Boeriu, C.G., 2015. Chitosan films and blends for packaging material. Carbohydr. Polym. 116, 237–242.
van der Wyk, A.J.A., 1952. Kurt H. Meyer 1883-1952. Helv. Chim. Acta 35, 1418–1422.
van Iterson Jr., G., Meyer, K.H., Lotmar, W., 1936. Über den feinbau des pflanzlichen chitins. Recl. Trav. Chim. Pays-Bas 55, 61–63.
Van Tran, V., Park, D., Lee, Y.C., 2018. Hydrogel applications for adsorption of contaminants in water and wastewater treatment. Environ. Sci. Pollut. Res. 25, 24569–24599.

van Wisselingh, C., 1898. Mikrochemische Untersuchungen über die Zellwände der Fungi. Jahrb. Wiss. Bot. 31, 618–687.
van Wisselingh, C., 1898. Chitin. In: Linsbauer's Handbuch der Pflanzenanatomie. vol. 2, pp. 170–192.
van Wisselingh, C., 1924. Die Zellmenbran. In: Handbuch der Pflanzenanatomie, allg. Teil, Bd. III/2. vol. III. Gebrüder Borntraeger, Leipzig, pp. 101–112.
Vårum, K.M., Smidsrød, O., 2004a. Structure-property relationship in chitosan. In: Dumitriu, S. (Ed.), Polysaccharides. Structural Diversity and Functional Versatility. Marcel Dekker, New York, pp. 625–641 (Chapter 26).
Vårum, K.M., Smidsrød, O., 2004b. Chitosans. In: Stephen, A.M., Phillips, G.O., Williams, P.A. (Eds.), Food Polysaccharides and their Applications, second ed. Marcel Dekker, New York, pp. 497–520 (Chapter 14).
Vauquelin, L.N., 1813. Expériences sur les champignons. In: Klostermann, J. (Ed.), Recueil de Mémoires Concernant la Chimie et les Arts Qui en Dépendent et Spécialement la Pharmacie. 31 Janvier 1813. Annales de Chimie, vol. 85. Librairie des Ecoles Impériales Polytechnique et des Ponts et Chaussées, Paris, pp. 5–25.
Verlee, A., Mincke, S., Stevens, C.V., 2017. Recent developments in antibacterial and antifungal chitosan and its derivatives. Carbohydr. Polym. 164, 268–283.
Verloop, A., Ferrell, C.D., 1977. Benzoylphenyl ureas—a new group of larvicides interfering with chitin deposition. In: Plimmer, J.R. (Ed.), Pesticide Chemistry in the 20th Century. ACS Symposium Series, vol. 37. American Chemical Society, Washington, pp. 237–270.
Verma, M.L., Kumar, S., Das, A., Randhawa, J.S., Chamundeeswari, M., 2020. Chitin and chitosan-based support materials for enzyme immobilization and biotechnological applications. Environ. Chem. Lett. 18, 315–323.
Vinay, T.N., Bhat, S., Choudhury, T.G., Paria, A., Jung, M.H., Kallappa, G.S., Jung, S.J., 2018. Recent advances in application of nanoparticles in fish vaccine delivery. Rev. Fish. Sci. Aquac. 26, 29–41.
Vinodh, R., Sasikumar, Y., Kim, H.J., Atchudan, R., Yi, M., 2021. Chitin and chitosan based biopolymer derived electrode materials for supercapacitor applications: a critical review. J. Ind. Eng. Chem. 104, 155–171.
von Fürth, O., Russo, M., 1906. Über kristallinische chitosan-verbindungen aus sepienschulpen. Beitr. Chem. Physiol. Pathol. 8, 163–190.
von Fürth, O., Scholl, E., 1907. In: Hofmeister, F. (Ed.), Ueber Nitrochitine. Zeitschrift für die Gesamte Biochemie. vol. 10. Beiträge zur Chemischen Physiologie und Pathologie, Braunschweig, pp. 188–198.
von Lippmann, E.O., 1912. Über ein vorkommen von chitin. Ber. Dtsch. Chem. Ges. 44, 3716–3717.
von Weimarn, P.P., 1926a. Universal method for converting fibroin, chitin, casein and similar substances into ropyplastic state and into state of colloidal solution by means of concentrated aqueous solution of readily soluble salts, capable of strong hydration. J. Text. Inst. 17T, 642–644.
von Weimarn, P.P., 1926b. Eine allgemein anwendbare methode, fibroin, chitin, kasein und ähnliche substanzen mit hilfe konzentrierter wässeriger lösungen leicht löslicher und stark hydratisierter salze in den plastischen zustand und in den zustand kolloider lösung überzuführe. Kolloid Z. 40, 120–122.
von Weimarn, P.P., 1927. Conversion of fibroin, chitin, casein and similar substances into the ropy-plastic state and colloid solution. Ind. Eng. Chem. 19, 109–110.
von Wettstein, F., 1921. Das vorkommen von chitin und seine verwertung als systematisch-phylogenetisches merkmal im pflanzenreich. In: Sitzungsberichte der Kaiserlichen Akademie der Wissenschaften Mathematisch Naturwissenchaftliche Klasse Abt. vol. 130, pp. 3–20.

Voncina, B., Fras, L.Z., Ristic, T., 2016. Active textile dressings for wound healing. In: Advances in Smart Medical Textile. A Volume in Woodhead Publishing Series in Textiles, Elsevier, pp. 73–92 (Chapter 4).

Vouk, V., 1915. Zur kenntis der mikrochemischen chitinreaktion. Ber. Dtsch. Bot. Ges. 33, 413–415.

Walton, A.G., Blackwell, J., 1973. Biopolymers. Academic Press, New York, p. 467.

Wang, K., 2012. Enzyme immobilization on chitosan-based supports. In: Yao, K., Li, J., Yao, F., Yin, Y. (Eds.), Chitosan-Based Hydrogels: Functions and Applications. CRC Press, Taylor & Francis Group, Boca Raton, pp. 339–406 (Chapter 8).

Wang, C., Li, J., Yao, F., 2012. Application of chitosan-based biomaterials in tissue engineering. In: Yao, K., Li, J., Yao, F., Yin, Y. (Eds.), Chitosan-Based Hydrogels: Functions and Applications. CRC Press. Taylor & Francis Group, Boca Raton, FL, pp. 407–468 (Chapter 9).

Wang, H.X., Qan, J., Ding, F.Y., 2018. Emerging chitosan-based films for food packaging applications. J. Agric. Food Chem. 66, 395–413.

Wang, H.X., Ding, F.Y., Ma, L., Zhang, Y.H., 2021. Edible films from chitosan-gelatin: physical properties and food packaging application. Food Biosci. 40, 100871.

Wang, J.K., Teng, C.C., Yan, L.F., 2022. Applications of deep eutectic solvents in the extraction, dissolution, and functional materials of chitin: research progress and prospects. Green Chem. 24, 552.

Wei, H., Gao, B.Q., Ren, J., Li, A.M., Yang, H., 2018. Coagulation/flocculation in dewatering of sludge: a review. Water Res. 143, 608–631.

Wertz, J.L., Perez, S., 2021. Chitin and Chitosans in the Bioeconomy. CRC Press, ISBN: 9781032128481. 172 p.

Wester, D.H., 1909. Studien über das chitin. Arch. Pharm. 247, 282–307.

Wester, D.H., 1910. Ueber die verbreitung und lokalisation des chitins im tierreiche. Zool. Jahrb. 28, 531–538.

Wettstein, A., 1972. Paul Karrer 1889-1971. Helv. Chim. Acta 55, 313–328.

Whistler, R.L., 1973. Chitin. In: Whistler, R.L., BeMiller, J.N. (Eds.), Industrial Gums: Polysaccharides and their Derivatives. Academic Press, Inc., San Diego, pp. 601–604 (Chapter 22).

Whistler, R.L., Smart, C.L., 1953. Chitin. In: Whistler, R.L., Smart, C.L. (Eds.), Polysaccharide Chemistry. Academic Press Inc., New York, pp. 395–405 (Chapter XXVIII).

White, C.F.A., 1944. Colour Photography. US Patent 2,363,764.

Wigglesworth, V.B., 1933. The physiology of the cuticle and ecdysis in *Rhodnius prolixus* (Tritomidae, Hemiptara); with special reference to the function of the oenocytes and of the dermal glands. Q. J. Microsc. Sci. 76, 269–318.

Wigglesworth, V.B., 1948. The insect cuticle. Biol. Rev. 23, 408–451.

Wigglesworth, V.B., 1957. The physiology of insect cuticle. Annu. Rev. Entomol. 2, 37–54.

Willson, K., Atala, A., Yoo, J.J., 2021. Bioprinting Au natural: the biologics of bioinks. Biomolecules 11, 1593.

Winterowd, J.G., Sandford, P.A., 1995. Chitin and chitosan. In: Stephen, A.M. (Ed.), Food Polysaccharides and Their Applications. Marcel Dekker, Inc., New York, pp. 441–461 (Chapter 13).

Winterstein, E., 1893. Zur kenntniss der pilzcellulose. Ber. Dtsch. Bot. Ges. 11, 441–445.

Winterstein, E., 1894. Ueber ein stickstoffhaltiges spaltungsprodukt der pilzcellulose. Ber. Dtsch. Bot. Ges. 27, 3113–3115.

Winterstein, E., 1894. Zur kenntniss der in den membrane der pilze enthaltenen bestandtheile. Hoppe Seylers Z. Physiol. Chem. 19, 521–562.

Winterstein, E., 1894. Berichtigung. Ber. Dtsch. Bot. Ges. 27, 3508–3509.

Winterstein, E., 1895. Zur kenntniss der in den membrane der pilze enthaltenen bestandtheile. Hoppe Seylers Z. Physiol. Chem. 19, 134–151.

Winterstein, E., 1895. Ueber pilzcellulose. Ber. Dtsch. Chem. Ges. 13, 65–70.

Winterstein, E., 1895. Sur un Produit de Décomposition azote de la Cellulose de Champignons. Bulletin de la Société Chimique de Paris, vol. XIV Masson et Cie, libraires de l'Académie de Médecine, Paris, pp. 502–503. Troisième série.

Winterstein, E., 1895. Sur les Produits de Dédoublement de la Cellulose de Champignons. Bulletin de la Société Chimique de Paris, vol. XIV Masson et Cie, libraires de l'Académie de Médecine, Paris, pp. 902–903. Troisième série.

Wirth, M., 2013. László Zechmeister. His Life and Pioneering Work in Chromatography. Springer Briefs in Molecular Science. History of Chemistry, Springer, Dordrecht, ISBN: 978-3-319-00641-3.

Wisniak, J., 2004. Anselme Payen. Educ. Quim. 16, 568–580.

Wisniak, J., 2007. Henri Braconnot. Rev. Cienc. Quím. 38, 345–355.

Wisniak, J., 2014. Jean-Louis Lassaigne. Rev. Cienc. Biol. 45, 119–130.

Wisniak, J., 2015. Charles Hatchett: the discoverer of niobium. Educ. Quim. 26, 346–355.

Witt, O.N., 1901. Ferdinand Tiemann. Ein Lebensbild. Ber. Dtsch. Chem. Ges., 4402–4455.

Wolfrom, M.L., 1958. US Patent 2,832,766 (Chem Abstracts 52, 20913).

Wolfrom, M.L., Han, T.M.S., 1959. The sulfonation of chitosan. J. Am. Chem. Soc. 81, 1764–1766.

Wolfrom, M.L., Shen, T.M., Summers, C.G., 1953. Sulfated nitrogenous polysaccharides and their anticoagulant activity. J. Am. Chem. Soc. 75, 1519–1591.

Wolfrom, M.L., Maher, G.G., Chaney, A., 1958. Chitosan nitrate. J. Org. Chem. 23, 1990–1991.

Wu, D.J., Zhu, L.X., Li, Y., Zhang, X.L., Xu, S.M., Yang, G.S., Delair, T., 2020. Chitosan-based colloidal polyelectrolyte complexes for drug delivery: a review. Carbohydr. Polym. 238, 11626.

Xie, S.M., Yuan, L.M., 2019. Recent development trends for chiral stationary phases based on chitosan derivatives, cyclofructan derivatives and chiral porous materials in high performance liquid chromatography. J. Sep. Sci. 42, 6–20.

Xing, K., Zhu, X., Peng, X., Qin, S., 2015. Chitosan antimicrobial and eliciting properties for pest control in agriculture: a review. Agron. Sustain. Dev. 35, 569–588. Springer Verlag/EDP Sciences/INRA.

Yang, S.Q., Jiang, Z.Q., Liu, J., Ma, S., 2019. Preparation of chitin oligosaccharides and its monomer. In: Zhao, L. (Ed.), Oligosaccharides of Chitin and Chitosan. Springer, pp. 55–81 (Chapter 4).

Yang, J., Shen, M.Y., Luo, Y., Wu, T., Chen, X.X., Wang, Y.X., Xie, J.H., 2021. Advanced applications of chitosan-based hydrogels: from biosensors to intelligent food packaging system. Trends Food Sci. Technol. 110, 822–832.

Yang, W., Tu, A., Ma, Y., Li, Z., Xu, J., Lin, M., Zhang, K., Jing, L., Fu, C., Jiao, Y., Huang, L., 2022. Chitosan and whey protein bio-inks for 3D and 4D printing applications with particular focus on food industry. Molecules 27, 173.

Yao, K., Li, J., Yao, F., Yin, Y. (Eds.), 2012. Chitosan-Based Hydrogels: Functions and Applications. CRC Press. Taylor & Francis Group, Boca Raton, FL, ISBN: 978-1-4398-2114-5. 511 p.

Yin, H., Du, Y., 2011. Mechanism and application of chitin/chitosan and their derivatives in plant protection. In: Kim, S.K. (Ed.), Chitin, Chitosan, Oligosaccharides and Their Derivatives: Biological Activities and Applications. CRC Press, Taylor & Francis Group LLC, Boca Raton, pp. 605–618 (Chapter 41).

Yin, Z., Chu, R., Zhu, L., Li, S., Mo, F., Hu, D., Liu, C., 2021. Application of chitosan-based flocculants to harvest microalgal biomass for biofuel production: a review. Renew. Sust. Energ. Rev. 145, 111159.

Ylitalo, R., Lehtinen, S., Wuolijoki, E., Ylitalo, P., Lehtimaki, T., 2002. Cholesterol-lowering properties and safety of chitosan. Arzneimittelforschung 52, 1–7.

Yong, S.K., Shrivastava, M., Srivastava, P., Kunhikrishnan, A., Bolan, N., 2015. Environmental applications of chitosan and its derivatives. In: Whitacre, D.M. (Ed.), Book Series: Reviews of Environmental Contamination and Toxicology. 233. Springer, pp. 1–43.

Yonge, C.M., 1932. On the nature and permeability of chitin. I. The chitin lining the foregut of decapod crustacea and the function of the tegumental glands. Proc. R. Soc. London Ser. B 111, 298–329.

Yonge, C.M., 1936. On the nature and permeability of chitin II—the permeability of the uncalcified chitin lining the foregut of Homarus. Proc. R. Soc. Lond. Ser. B Biol. Sci. 120, 15–41.

Yonge, C.M., 1938. Recent work on the digestion of cellulose and chitin by invertebrates. Sci. Prog. 32, 638–647.

Younes, I., Rinaudo, M., 2015. Chitin and chitosan preparation from marine sources. Structure, properties and applications. Mar. Drugs 13, 1133–1174.

Yu, C., Kecen, X., Xiaosai, Q., 2018. Grafting modification of chitosan. In: Biopolymer Grafting. Synthesis and Properties. Elsevier, pp. 298–364 (Chapter 7).

Yu, J., Wang, D., Geetha, N., Khawar, K.M., Jogaiah, S., Mujtaba, M., 2021. Current trends and challenges in the synthesis and applications of chitosan-based nanocomposites for plants: a review. Carbohydr. Polym. 261, 117904.

Yu, D., Feng, J., You, H., Zhou, S., Bai, Y., He, J., Cao, H., Che, Q., Guo, J., Su, Z., 2022. The microstructure, antibacterial and antitumor activities of chitosan oligosaccharides and derivatives. Mar. Drugs 20, 69.

Yuan, X.B., Zheng, J.P., Jiao, S.M., Cheng, G., Feng, C., Du, Y.G., Liu, H.T., 2019. A review on the preparation of chitosan oligosaccharides and application to human health, animal husbandry and agricultural production. Carbohydr. Polym. 220, 60–70.

Zaitschek, A., 1904. Versuche über die verdaulichkeit des chitins und den nahrwert der insekten. Arch. Gesammte Physiol. Menschen Tiere 104, 612–623.

Zaki, M.A., Salem, M.E.S., Gaber, M.M., Nour, A.M., 2015. Effect of chitosan supplemented diet on survival, growth, feed utilization, body composition, histology of sea bass (Dicentrarchus labrax). World J. Eng. Technol. 3, 38–47.

Zamboulis, A., Nanaki, S., Michailidou, G., Koumentakou, I., Lazaridou, M., Ainali, N.M., Xanthopoulou, E., Bikiaris, D.N., 2020. Chitosan and its derivatives for ocular delivery formulations: recent advances and developments. Polymers 12, 1519.

Zander, E., 1897. Vergleichende und kritische untersuchungen zum verständniss der iodreaktion des chitins. Pflügers Arch. Gesammte Physiol. Menschen Thiere 66, 545–573.

Zarrintaj, P., Saeb, M.R., 2018. Chitosan in dermatology. Clin. Dermatol. Open Access J. 3, 000154.

Zdanowicz, M., Wilpiszewska, K., Spychaj, T., 2018. Deep eutectic solvents for polysaccharides processing. A review. Carbohydr. Polym. 200, 361–380.

Zechmeister, L., Tóth, G., 1931. Zur kenntnis der hydrolyse von chitin mit salzsäure (I. Mitteil.). Ber. Dtsch. Chem. Ges. A B Ser. 64, 2028–2032.

Zechmeister, L., Tóth, G., 1932. Zur kenntnis der hydrolyse von chitin mit salzsäure (II. Mitteil.). Ber. Dtsch. Chem. Ges. A B Ser. 65, 161–162.

Zechmeister, L., Tóth, G., 1933. Ein beitrag zur desamidierung des glucosamins. Ber. Dtsch. Chem. Ges. A B Ser. 66, 522–525.

Zechmeister, L., Tóth, G., 1934. The comparison of herbal and animal chitin. Hoppe-Seylers Z. Physiol. Chem. 223, 53–56.

Zechmeister, L., Tóth, G., 1939a. Chitin und seine spaltprodukte. In: Fortschritte der Chemie Organischer Naturstoffe. vol. 2. Verlag von Julius Springer, Wien, pp. 212–247.

Zechmeister, L., Tóth, G., 1939b. Chromatographische zerlegung der chitinase. Naturwissenschaften 27, 367.

Zechmeister, L., Tóth, G., 1939c. Chromatographie der in der chitinreihe wirksamen enzyme des emulsins. Enzymologia 7, 165–169.

Zechmeister, L., Grassmann, W., Tóth, G., Bender, R., 1932. Über die verknüpfungsart der glucosamin-reste im chitin. Ber. Dtsch. Chem. Ges. A B Ser. 65, 1706–1708.

Zechmeister, L., Tóth, G., Balint, M., 1939a. Uber die chromatographische trennung einiger enzyme des emulsins. Enzymologia 5, 302–306.

Zechmeister, L., Tóth, G., Vajda, E., 1939b. Chromatographie der in der chitinreihe wirksamen enzyme der weinbergsgschneck (*Helix pomatia*). Enzymologia 7, 170–171.

Zemmouri, H., Mameri, N., Lounici, H., 2014. Chitosan use in chemical conditioning for dewatering municipal-activated sludge. Water Sci. Technol. 71, 810–816.

Zhang, L., Zeng, Y.X., Cheng, Z.J., 2016. Removal of heavy metal ions using chitosan and modified chitosan: a review. J. Mol. Liq. 214, 175–191.

Zhang, E.H., Xing, R.G., Liu, S., Qin, Y.K., Li, K.C., Li, P.C., 2019a. Advances in chitosan-based nanoparticles for oncotherapy. Carbohydr. Polym. 222, 115004.

Zhang, J., Hu, Q., Lu, J., Lin, S., 2019b. Study on the effect of chitosan conditioning on sludge dewatering. Water Sci. Technol. 79, 501–509.

Zhang, J.X., Jin, J.H., Wan, J.Q., Jiang, S.H., Wu, Y.Z., Wang, W.F., Gong, X., Wang, H. W., 2021. Quantum dots-based hydrogels for sensing applications. Chem. Eng. J. 408, 127351.

Zhang, J.H., Xie, S.M., Yuan, L.M., 2022. Recent progress in the development of chiral stationary phases for high-performance liquid chromatography. J. Sep. Sci. 45, 51–77.

Zhao, L., 2019. Oligosaccharides of Chitin and Chitosan. Bio-Manufacture and Applications. Springer Nature Singapore Pte Ltd, ISBN: 978-981-13-9401-0.

Zhao, D.Y., Yu, S., Sun, B.N., Gao, S., Guo, S.H., Zhao, K., 2018. Biomedical applications of chitosan and its derivative nanoparticles. Polymers 10, 462.

Zhuo, S.H., Wu, J.J., Zhao, L., Li, W.H., Zhao, Y.F., Li, Y.M., 2022. A chitosan-mediated inhalable nanovaccine against SARS-CoV-2. Nano Res.

Zikakis, J.P., 1984. In: Zikakis, J.P. (Ed.), Chitin, Chitosan, and Related Enzymes. Academic Press, Inc., Orlando, FL, ISBN: 9780323149976. 448 p.

Zoebelein, G., Hamman, I., Sirrenberg, W., 1980. BAY SIR 8514 a new chitin synthesis inhibitor. Z. Angew. Entomol. 89, 289–297.

Index

Note: Page numbers followed by *f* indicate figures *t* indicate tables and *s* indicate schemes.

FSC
www.fsc.org
MIX
Papier | Fördert
gute Waldnutzung
FSC® C083411

Kolibri 360 GmbH
Ferdinand-Jühlke-Straße 7
99095 Erfurt, Deutschland
produktsicherheit@kolibri360.de